Infosys Science Foundation Series

Infosys Science Foundation Series in Mathematical Sciences

The **Infosys Science Foundation Series in Mathematical Sciences,** a Scopus-indexed book series, focuses on high-quality content in the domain of pure and applied mathematics, biomathematics, financial mathematics, operations research and theoretical computer science. With this series, Springer hopes to provide readers with monographs, textbooks, edited volume, handbooks and professional books of the highest academic quality on current topics in relevant disciplines. Literature in this series will appeal to a wide audience of researchers, students, educators and professionals across the world.

Bang-Yen Chen · Majid Ali Choudhary ·
Mohammad Nazrul Islam Khan
Editors

Geometry of Submanifolds and Applications

Springer

Editors
Bang-Yen Chen
Department of Mathematics
Michigan State University
East Lansing, MI, USA

Majid Ali Choudhary
Department of Mathematics
Maulana Azad National Urdu University
Hyderabad, Telangana, India

Mohammad Nazrul Islam Khan
Department of Computer Engineering
Qassim University, College of Computer
Buraydah, Saudi Arabia

ISSN 2363-6149 ISSN 2363-6157 (electronic)
Infosys Science Foundation Series
ISSN 2364-4036 ISSN 2364-4044 (electronic)
Infosys Science Foundation Series in Mathematical Sciences
ISBN 978-981-99-9749-7 ISBN 978-981-99-9750-3 (eBook)
https://doi.org/10.1007/978-981-99-9750-3

Mathematics Subject Classification: 53C50, 53C55, 53C25, 53C15, 53C40, 34B10, 34B15, 11C20, 53B25, 53C42, 53C05, 58A30, 83D05, 83C05, 53D10

This Springer imprint is published by the registered company Springer Nature Singapore Pte Ltd.
The registered company address is: 152 Beach Road, #21-01/04 Gateway East, Singapore 189721, Singapore

Paper in this product is recyclable.

Preface

Beginning with the theory of curves and surfaces in ordinary Euclidean 3-space, the theory of submanifolds as a subarea of differential geometry dates back as far as differential geometry itself. The theory has evolved in the last century into a very important area of mathematics with many applications, and making use of a variety of techniques. Since the late 1950s, the theory has a rapid development which attracted more and more researchers. Consequently, a lot of interesting results have been achieved (see, e.g., [7] for an extensive survey).

In 1956, John F. Nash proved the following famous embedding theorem.

Theorem 1. [17] Every Riemannian manifold can be isometrically embedded in a Euclidean space with sufficiently high codimension.

Nash's embedding theorem is aimed for in the hope that if each Riemannian manifold could always be isometrically embedded as an Euclidean submanifold, then it could yield the opportunity to use help from extrinsic geometry. But this hope was not materialized for many years, as pointing out by M. Gromov in his 1985 survey article [15]. There were several reasons why it is so difficult to apply Nash's theorem. One reason is that it requires in general very large codimension for a Riemannian manifold to admit an isometric embedding in an Euclidean space, On the other hand, submanifolds of higher codimension are difficult to understand. The second reason is that at that time there did not exist general optimal relationships between known intrinsic invariants and main extrinsic invariants for an arbitrary Euclidean submanifold. These difficulties lead to the following fundamental problem in submanifold theory (see, e.g., [4, 5, 6, 8]).

Problem 1. Find simple optimal relationship between extrinsic invariants and intrinsic invariants of a submanifold and find their applications.

In order to give some answers to this fundamental problem, one of the editors of this book introduced in the early 1990s new types of Riemannian invariants, i.e., δ-invariants (also known as Chen invariants). He was also able to provide solutions to Problem 1 by establishing optimal inequalities between his δ-invariants and the squared mean curvature and gave many applications (see, e.g., [7, 8, 9]). A special case of his inequalities is known today as the Chen–Ricci inequality. The main purpose of the first chapter of this book is to provide a comprehensive survey on recent developments on the Chen–Ricci inequality.

Differential geometry has many nice applications to Einstein's general relativity theory. The second chapter discusses perfect fluid spacetimes fulling $f(R; T)$-gravity when Ricci and gradient Ricci solitons, Yamabe and gradient Yamabe solitons, η-Ricci and gradient η-Ricci solitons are its metrics.

A $2n$-dimensional manifold is called symplectic if it admits a non-degenerate closed 2-form Ω. An n-dimensional submanifold M of a symplectic $2n$-manifold $(M;\Omega)$ is called Lagrangian if $\Omega|_{\mathrm{TM}} = 0$. Lagrangian submanifolds appear naturally in mathematical physics. Chapter 3 provides a nice survey on Lagrangian submanifolds of the nearly Kähler 6-sphere.

A hypersurface of a Riemannian or Lorentzian space form is called a Pythagorean hypersurface [2] if its first, second, and third fundamental forms satisfy a Pythagorean-like formula: $\mathrm{I}^2 + \mathrm{II}^2 = \mathrm{III}^2$. In Chap. 4, Pythagorean hypersurfaces in real space forms were extended to Pythagorean submanifolds. The authors of this chapter provide examples and show that Pythagorean submanifolds are isoparametric where the principal curvatures are given in terms of the Golden ratio. They also classify Pythagorean hypersurfaces in real space forms.

The notion of f-structure was introduced by K. Yano in 1961, which generalized both almost-complex and almost-contact structures. Chapter 5 is devoted to the study of the class of manifolds which admit an f-structure with two-dimensional parallelizable kernel.

A submersion is a differentiable map between two manifolds whose differential is everywhere surjective. Riemannian submersions are submersions equipped with compatible Riemannian metrics. Chapter 6 focuses on Kähler submersions which are special case of almost-Hermitian submersions. Among others, the authors of this chapter prove, under certain conditions, that the fibers and the base manifold of such submersions are almost-Yamabe solitons.

For a surface M in a Euclidean 4-space, P. Wintgen [20] proves that the Gauss curvature G and normal curvature K^D of M satisfy $G + |K^D| \leq \|H\|^2$; where $\|H\|^2$ is the squared mean curvature. Chapter 7 studies generalized Wintgen inequality for a contact pseudo-slant submanifold of (ϵ)-para Sasakian space form. The authors describe those submanifolds satisfying the equality case of the inequality. They also derive an inequality for Ricci solitons to discover some connections between intrinsic and extrinsic invariants.

S. Decu, S. Haesen, and L. Verstraelen [13] introduced δ-Casorati curvatures in 2017. Today, there are many interesting results on δ-Casorati curvatures obtained by various authors [10]. Chapter 8 presents two inequalities involving δ-Casorati curvatures of quaternion bi-slant submanifolds in quaternion space forms. The authors of this chapter also characterize those submanifolds satisfying the equality cases of the inequalities.

Chapter 9 provides another application of differential geometry to general relativity. The authors of this chapter study gravity and dark matter in the framework of metric-affine geometry.

Chapter 10 studies and analyzes the properties of metallic structure on manifolds using the metallic ratio which is a generalization of the Golden proportion. The author of this chapter discusses statistical metallic manifolds and statistical submersions.

Chapter 11 studies tangent bundle of three-dimensional quasi-Sasakian manifolds endowed with a quarter-symmetric non-metric ξ-connection. The authors of this chapter prove several special curvature properties for such manifolds with the connection on its tangent bundle.

In Chap. 12, the author introduces and studies a Darboux mate of a given spherical Legendre curve in the Euclidean unit sphere S^2. The authors also investigate a triple sequence of curvatures provided by the higher-order derivatives of the last Frenet equation for the frontal of the Legendre curve.

In the last chapter, the authors investigate Riemannian 3-manifolds which admit conformal η-Ricci–Yamabe solitons or gradient conformal η-Ricci–Yamabe solitons. Several results in this respect are presented.

Given the enormous amount of work on the theory of submanifolds and their applications that has been published over the past few decades, the editors decided it would be reasonable to invite a number of experts to contribute their work in the form of chapters to highlight the current advances in differential geometry with a focus on the geometry of submanifolds. Several geometers responded to our call. The editors wish to thank each and every contribution.

The editors hope that readers will find this book useful as a resource for conducting creative and fruitful research as well as an approachable introduction to the geometry of submanifolds and its applications.

East Lansing, USA — Bang-Yen Chen
Hyderabad, India — Majid Ali Choudhary
Buraydah, Saudi Arabia — Mohammad Nazrul Islam Khan

References

1. M.E. Aydin, A. Mihai, I. Mihai, Some inequalities on submanifolds in statistical manifolds of constant curvature. Filomat 29(3), 465–476 (2015)
2. M.E. Aydin, A. Mihai, I. Mihai, Pythagorean isoparametric hypersurfaces in Riemannian and Lorentzian space forms. Axioms, 11, 59 (2022)
3. S. Amari, *Differential-Geometrical Methods in Statistics, Lecture Notes in Statistics*, vol. 28. (Springer-Verlag, New York, 1985)
4. B.-Y. Chen, Some pinching and classification theorems for minimal submanifolds. Arch. Math. 60(6), 568–578 (1993)
5. B.-Y. Chen, Mean curvature and shape operator of isometric immersions in real-space-forms. Glasgow Math. J. 38 (1996), 87–97.
6. B.-Y. Chen, Relations between Ricci curvature and shape operator for sub-manifolds with arbitrary codimensions, Glasgow Math. J. 41(1), 33–44 (1999)
7. B.-Y. Chen, *Riemannian Submanifolds. Handbook of Differential Geometry*, vol. I, (North-Holland, Amsterdam, 2000), pp. 187–418
8. B.-Y. Chen, Some new obstructions to minimal and Lagrangian isometric immersions. Japan. J. Math. 26(1), 105–127 (2000)

9. B.-Y. Chen, Pseudo-Riemannian geometry, δ-invariants and applications; World Scientific, (Hackensack, NJ, 2011)
10. B.-Y. Chen, Recent developments in δ-Casorati curvature invariants. Turkish J. Math. 45(1), 1–46 (2021)
11. B.-Y. Chen, S. Deshmukh, Ricci solitons and concurrent vector fields, Balkan J. Geom. Appl. 20(1) 14–25 (2015)
12. B.-Y. Chen, S. Deshmukh, Classification of Ricci solitons on Euclidean hypersurfaces, Int. J. Math. 25(11), 1450104, 22, (2014)
13. S. Decu, S. Haesen, L. Verstraelen, Optimal inequalities involving Casorati curvatures. Bull. Transil. Univ. Braşov. Ser. III. Math. Inform. Phys. 14(49), 85–93 (2007)
14. H. Furuhata, Hypersurfaces in statistical manifolds, Differential Geom. Appl. 27, 420–429 (2009)
15. M. Gromov, Isometric immersions of Riemannian manifolds. In: The mathematical heritage of Élie Cartan (Lyon, 1984), Astérisque 1985, Numéro Hors Série, pp. 129–133.
16. A. Mihai, C. Ozgur, Chen inequalities for submanifolds of real space forms with a semi-symmetric metric connection, Taiwanese J. Math. 14(4), 1465–1477 (2010)
17. J. F. Nash, The imbedding problem for Riemannian manifolds, Ann. Math. 63, 20–63 (1956)
18. K. Uohashi, A. Ohara, T. Fujii, 1-conformally at statistical submanifolds, Osaka J. Math. 37(2) 501–507 (2000)
19. P. W. Vos, Fundamental equations for statistical submanifolds with applications to the Bartlett correction, Ann. Inst. Statist. Math. 41(3), 429–450 (1989)
20. P. Wintgen, Sur l'inégalité de Chen-Willmore, C. R. Acad. Sci. Paris 288, 993–995 (1979)

Contents

Recent Developments on Chen–Ricci Inequalities in Differential Geometry

Bang-Yen Chen and Adara M. Blaga

1 Introduction

Nash [148] proved in 1956 his famous embedding theorem.

Theorem 1.1 *Every Riemannian n-manifold can be isometrically embedded in a Euclidean m-space with dimension* $m = \frac{n}{2}(n+1)(3n+11)$.

This embedding theorem aimed for in the hope that if each Riemannian manifold could always be regarded as a Euclidean submanifold, then it could yield the opportunity to use help from extrinsic geometry. However, this hope was not materialized for many years, as pointed out in Gromov's 1985 survey article [84].

There were several reasons why it is so difficult to apply Nash's theorem. One reason is that it requires in general very large codimension for a Riemannian manifold to admit an isometric embedding in Euclidean spaces. On the other hand, submanifolds of higher codimension are difficult to understand. Another reason is that at that time there did not exist general optimal relationships between known intrinsic invariants and main extrinsic invariants for arbitrary Euclidean submanifolds (see [53]). This leads to the following fundamental problem in submanifold theory (see [38, 42]).

Problem 1.1 *Find simple relationship between main extrinsic invariants and main intrinsic invariants of a submanifold.*

In order to give some answers to this fundamental problem, the first author introduced in the early 1990s new types of Riemannian invariants, known as δ-invariants

B.-Y. Chen (✉)
Department of Mathematics, Michigan State University, East Lansing, Michigan 48824-1027, USA
e-mail: chenb@msu.edu

A. M. Blaga
Department of Mathematics, West University of Timişoara, Timişoara 300223, Romania

B.-Y. Chen et al. (eds.), *Geometry of Submanifolds and Applications*, Infosys Science Foundation Series, https://doi.org/10.1007/978-981-99-9750-3_1

or Chen invariants, which are very different in nature from the "classical" Ricci and scalar curvatures (see, e.g., [37, 41, 43]). Using his invariants, Chen was able to give solutions to Problem 1.1 by establishing sharp inequalities between his δ-invariants and the squared mean curvature. Using these inequalities, he introduced and investigated the notion of ideal immersions.

Later in 1996, the first author established in [38] a sharp inequality involving Ricci curvature Ric and squared mean curvature $\|H\|^2$ for any submanifold M in a real space form $\widetilde{M}^m(c)$ of constant sectional curvature c, namely,

$$\mathrm{Ric}(X) \leq (n-1)c + \frac{n^2}{4}\|H\|^2, \quad n = \dim M,$$

for any unit vector X tangent to M. Since then, this inequality has drawn the attention of many geometers around the world. Consequently, many inequalities of similar type were proved by many geometers for different kinds of submanifolds in various ambient manifolds. The main purpose of this article is thus to present a comprehensive survey on recent developments in this inequality. It is both authors' intention that this survey article will provide a useful reference for graduate students as well as researchers working on this interesting research topic in differential geometry.

2 Preliminaries

Throughout this article, by a manifold we mean a connected differentiable manifold. Manifolds, maps, functions, vector fields, etc. are assumed to be of class C^∞. Let $\mathfrak{X}(M)$ denote the space of all vector fields on a manifold M, $\mathcal{F}(M)$ the space of functions, and T^1M the unit tangent bundle of M. For the basic knowledge on Riemannian manifolds and submanifolds, we refer to the books [54, 55, 154, 167].

2.1 *Basic on Riemannian Manifolds*

Let (M, g) be an n-dimensional Riemannian manifold. We denote by ∇ the Levi–Civita connection of (M, g). Then the Riemann curvature tensor R is the $(0, 3)$-tensor field defined by

$$R(X, Y)Z = \nabla_X \nabla_Y Z - \nabla_Y \nabla_X Z - \nabla_{[X,Y]} Z$$

for $X, Y, Z \in \mathfrak{X}(M)$, where $[\ ,\ ]$ is the Lie bracket. For orthonormal vectors X, Y of a 2-plane $\pi \subset T_pM$, the sectional curvature of the plane spanned by X, Y is

$$K(\pi) = g(R(X, Y)Y, X).$$

The *Ricci tensor* at $p \in M$, denoted by Ric_p, is defined by

$$\mathrm{Ric}_p(Y, Z) = \mathrm{Trace}\{X \mapsto R(X, Y)Z\}$$

equivalently,

$$\mathrm{Ric}(X, Y) = \sum_{\ell=1}^{n} \langle R(E_i, X)Y, E_i \rangle,$$

where $\{E_1, \ldots, E_n\}$ is an orthonormal basis of T_pM. A Riemannian manifold is called an *Einstein manifold* if $\mathrm{Ric} = cg$ for some constant c.

For $X \in T^1M$, the *Ricci curvature* Ric at X is defined by $\mathrm{Ric}(X) = \mathrm{Ric}(X, X)$. The *scalar curvature* τ of M is given by $\tau = \sum_{i<j} K(E_i \wedge E_j)$. The Ricci and scalar curvatures are independent of the choice of the orthonormal basis $\{E_1, \ldots, E_n\}$.

The Ricci curvature was extended to k-Ricci curvature as follows [42]: For a unit vector X in a k-plane section L^k, we choose an orthonormal basis $\{E_1, \ldots, E_k\}$ of L^k such that $E_1 = X$. Then the *k-Ricci curvature* Ric_{L^k} of L^k at X is defined by

$$\mathrm{Ric}_{L^k}(X) = \sum_{j=2}^{k} K_{1j}, \quad K_{ij} = K(E_i \wedge E_j). \tag{2.1}$$

The *scalar curvature* $\tau(L^k)$ of L^k is $\tau(L^k) = \sum_{1\le i<j\le k} K_{ij}$. And the *k-Ricci curvature invariant* θ_k was defined in [42] to be

$$\theta_k(p) = \left(\frac{1}{k-1}\right) \inf_{L^k, X} \mathrm{Ric}_{L^k}(X), \quad p \in M, \tag{2.2}$$

where L^k runs over all k-plane sections $\subset T_pM$ and X runs over unit vectors in L^k.

2.2 Basic on Riemannian Submanifolds

Let M be a submanifold of a Riemannian manifold $\widetilde{M}$. Denote by ∇ and $\widetilde{\nabla}$ the Levi–Civita connections of M and $\widetilde{M}$, respectively. The formulas of Gauss and Weingarten are given, respectively, by

$$\widetilde{\nabla}_X Y = \nabla_X Y + h(X, Y), \quad \widetilde{\nabla}_X \xi = -A_\xi X + D_X \xi$$

for $X, Y \in \mathfrak{X}(M)$ and ξ a normal vector field of M, where h, A, and D are the *second fundamental form*, the *shape operator*, and the *normal connection*.

For each normal vector ξ of M at $p \in M$, the shape operator A_ξ is a symmetric endomorphism of the tangent space T_pM. The shape operator and the second fundamental form are related simply by $\langle h(X, Y), \xi \rangle = \langle A_\xi X, Y \rangle$.

The *mean curvature vector* of M in $\widetilde{M}$ is given by

$$H = \frac{1}{n} \operatorname{Trace} h, \quad n = \dim M.$$

A submanifold is said to be *minimal* (respectively, *totally geodesic*) if its mean curvature vector (respectively, its second fundamental form) vanishes identically.

Let R and $\widetilde{R}$ denote the Riemann curvature tensors of M and $\widetilde{M}$, respectively. If $\widetilde{M}$ is of constant curvature c, then the three fundamental equations of Gauss, Codazzi, and Ricci are given, respectively, by

$$\begin{aligned}\langle R(X, Y)Z, W\rangle = \ & \langle A_{h(Y,Z)}X, W\rangle - \langle A_{h(X,Z)}Y, W\rangle \\ & + c\,(\langle X, W\rangle\langle Y, Z\rangle - \langle X, Z\rangle\langle Y, W\rangle),\end{aligned}$$

$$(\overline{\nabla}_X h)(Y, Z) = (\overline{\nabla}_Y h)(X, Z),$$

$$\langle R^{\perp}(X, Y)\xi, \eta\rangle = \langle [A_\xi, A_\eta]X, Y\rangle$$

for $X, Y, Z, W \in \mathfrak{X}(M)$ and ξ, η normal vector fields of M, where $R^{\perp}$ denotes the normal curvature tensor associated with D and $\overline{\nabla} h$ is given by

$$(\overline{\nabla}_X h)(Y, Z) = D_X h(Y, Z) - h(\nabla_X Y, Z) - h(Y, \nabla_X Z).$$

2.3 Submanifolds of Kähler Manifolds

There exist three important classes of submanifolds M in an almost Hermitian manifold $(\widetilde{M}, J, g)$; namely, complex, totally real, and slant submanifolds based on the action of the almost complex structure J of $\widetilde{M}$ acting on the tangent bundle TM defined as follows: A submanifold M of $\widetilde{M}$ is called *complex* if TM is invariant under the action of J, i.e., $J(T_pM) = T_pM$ for any $p \in M$; M is called *totally real* if J maps each T_pM into the corresponding normal space, i.e., $J(T_pM) \subseteq T_p^{\perp}M$ for any $p \in M$. A totally real submanifold M of $\widetilde{M}$ is called *Lagrangian* if $\dim_{\mathbb{R}} M = \dim_{\mathbb{C}} \widetilde{M}$. We refer to [46] for general results on Lagrangian submanifolds.

For a vector $0 \neq X \in T_pM$, the angle $\theta(X)$ between JX and T_pM is called the *Wirtinger angle* of X. A submanifold M of $\widetilde{M}$ is called *slant* or (θ*-slant*) [35, 36] if $\theta(X)$ is independent of the choice of $X \in T_pM$ and of $p \in M$. In this case, the angle θ is called the *slant angle*. Clearly, complex or totally real submanifolds are slant submanifolds with slant angle $\theta = 0$ or $\theta = \frac{\pi}{2}$, respectively. A slant submanifold is called *proper* if it is neither complex nor totally real. From J-action points of view, slant submanifolds are the most natural submanifolds of almost Hermitian manifolds.

For any vector X tangent to a submanifold M of an almost Hermitian manifold, we put

$$JX = PX + FX,$$

where PX and FX denote the tangential component and normal component of JX, respectively. A proper slant submanifold of a Kähler manifold $\widetilde{M}$ is called *Kählerian slant* [36] if the endomorphism P is parallel with respect to the Levi–Civita connection ∇, i.e., $\nabla P = 0$.

2.4 *H*-Umbilical Submanifolds

The notion of H-umbilical submanifold was introduced in [39, 40] as follows: An n-dimensional Lagrangian submanifold M of an almost Hermitian manifold is called H*-umbilical* if its second fundamental form h takes the form:

$$h(E_1, E_1) = \lambda J E_1, \ h(E_2, E_2) = \cdots = h(E_n, E_n) = \mu J E_1,$$
$$h(E_1, E_j) = \mu J E_j, \ h(E_j, E_k) = 0, \ 2 \le j \ne k \le n,$$

for some suitable functions λ and μ with respect to an orthonormal local frame $\{E_1, \ldots, E_n\}$. The ratio $\lambda : \mu$ is called the *ratio* of the H-umbilical submanifold M.

3 Chen–Ricci Inequalities for Submanifolds in Riemannian Manifolds

The *relative null subspace* $\mathcal{N}_p$ of a submanifold M in a Riemannian manifold at a point $p \in M$ is defined by (see [112])

$$\mathcal{N}_p = \{X \in T_pM : h(X, Y) = 0 \ \forall Y \in T_pM\},$$

where h is the second fundamental form of M.

In 1996, the first author established a pointwise sharp relationship between Ricci curvature and the squared mean curvature for submanifolds in a real space form.

Theorem 3.1 ([38]) *If M is an n-dimensional submanifold of a real space form $R^m(c)$, then*

(1) *For any $X \in T_p^1 M$, we have:*

$$\mathrm{Ric}(X) \le \frac{n^2}{4}\|H\|^2 + (n-1)c. \tag{3.1}$$

(2) *If* $H(p) = 0$, *then a unit tangent vector* $X \in T_p^1 M$ *satisfies the equality case of* (3.1) *if and only if* X *lies in the relative null subspace* $\mathcal{N}_p$.
(3) *The equality case of* (3.1) *holds identically for all unit vectors in* $T_p^1 M$ *if and only if either* p *is a totally geodesic point, or* $n = 2$ *and* p *is a totally umbilical point.*

To prove answers to Problem 1, the first author introduced his δ-invariants, $\delta(n_1, \ldots, n_k)$, and proved the following optimal inequality (see also [37, 41, 44, 52]).

Theorem 3.2 ([48]) *Let* $\phi : M \to \widetilde{M}$ *be an isometric immersion between Riemannian manifolds. Then we have:*

$$\delta(n_1, \ldots, n_k) \leq c(n_1, \ldots, n_k)\|H\|^2 + b(n_1, \ldots, n_k) \max \widetilde{K},$$

where $\max \widetilde{K}_p$ *denotes the maximum sectional curvature of* $\widetilde{M}$ *restricted to 2-plane sections of the tangent space* $T_p M$, *and*

$$c(n_1, \ldots, n_k) = \frac{n^2(n + k - 1 - \sum_{j=1}^k n_j)}{2(n + k - \sum_{j=1}^k n_j)},$$

$$b(n_1, \ldots, n_k) = \frac{1}{2}n(n - 1) - \frac{1}{2}\sum_{j=1}^{k} n_j(n_j - 1).$$

If we denote the maximum of the Ricci curvature of M by $\widehat{\mathrm{Ric}}$, then, due to

$$\delta(n - 1) = \widehat{\mathrm{Ric}}, \quad b(n - 1) = n - 1, \quad c(n - 1) = \frac{n^2}{4},$$

we obtain immediately the following result from Theorem 3.2.

Proposition 3.3 *Let* M *be an* n*-dimensional submanifold of a Riemannian manifold* $\widetilde{M}$. *Then, for any* $X \in T^1 M$, *we have:*

$$\mathrm{Ric}(X) \leq \frac{n^2}{4}\|H\|^2 + (n - 1) \max \widetilde{K}. \tag{3.2}$$

For submanifolds of complex space forms, (3.2) implies the following.

Corollary 3.4 *If* M *is an* n*-dimensional submanifold of a complex space form* $\widetilde{M}(c)$, *then, for any* $X \in T^1 M$, *we have:*

$$\widehat{\mathrm{Ric}} \leq \frac{n^2}{4}\|H\|^2 + (n - 1)c \quad \text{for} \quad c \geq 0, \tag{3.3}$$

$$\widehat{\mathrm{Ric}} \leq \frac{n^2}{4}\|H\|^2 + \frac{n - 1}{4}c \quad \text{for} \quad c < 0. \tag{3.4}$$

Remark 3.5 Inequality (3.4) was also obtained by Sasahara [168].

In 1999, B. D. Suceavă proved the following.

Theorem 3.6 ([186]) *Let M be an n-dimensional submanifold of a Riemannian manifold $\widetilde{M}$. Then the scalar curvature τ of M satisfies*

$$\tau \leq \frac{n(n-1)}{2}\|H\|^2 + \sum_{1\leq i<j\leq n} \widetilde{K}(e_i, e_j), \tag{3.5}$$

where $\widetilde{K}(e_i, e_j)$ denotes the sectional curvature of $\widetilde{M}$ restricted to the plane section spanned by $\{e_i, e_j\}$ for an orthonormal basis $\{e_1, \ldots, e_n\}$.

Applying the method used in Theorem 3.6, Tripathi refined Proposition 3.3 and Theorem 3.1 in 2005 to the following.

Theorem 3.7 ([101, 192]) *Let M be a submanifold of a Riemannian manifold $\widetilde{M}$. Then*

(1) *For any $X \in T_p^1 M$, we have:*

$$\mathrm{Ric}(X) \leq \frac{n^2}{4}\|H\|^2 + (n-1)\,\widetilde{\mathrm{Ric}}(X), \quad n = \dim M, \tag{3.6}$$

where $\widetilde{\mathrm{Ric}}_{T_pM}(X)$ is the n-Ricci curvature of $\widetilde{M}$ restricted to T_pM (see (2.1)).

(2) *If $H(p) = 0$, then the equality case of (3.6) is satisfied for $X \in T_p^1 M$ if and only if $X \in \mathcal{N}_p$.*

(3) *The equality case of (3.6) holds for all $X \in T_p^1 M$ if and only if either p is a totally geodesic point, or $n = 2$ and p is a totally umbilical point.*

Remark 3.8 There exists no direct relationship between the Ricci curvature $\widetilde{\mathrm{Ric}}(X)$ and the n-Ricci curvature $\widetilde{\mathrm{Ric}}_{TM}(X)$ of $\widetilde{M}$ in general.

In [98], Hineva obtained the lower estimate of the Ricci curvature of a submanifold M of an arbitrary Riemannian manifold $\widetilde{M}$:

$$\begin{aligned}\mathrm{Ric}(X) \geq\ & \widetilde{\mathrm{Ric}}(X) \\ & + \frac{n-1}{n}\left[2n\|H\|^2 - S - \frac{n-2}{\sqrt{n-1}}\sqrt{n\|H\|^2(S - n\|H\|^2)}\right],\end{aligned} \tag{3.7}$$

where S is the squared norm of the second fundamental form $\|h\|^2$.

For a non-totally geodesic submanifold M, the equality in the lower estimate (3.7) is achieved only when M is quasi-umbilical with flat normal bundle, and for codimensions p with $1 \leq p \leq n-3$, the equality in (3.7) holds only when M is conformally flat with a flat normal bundle.

In [42], the first author extended Theorem 3.1 to the following.

Theorem 3.9 ([42]) *Let M be an n-dimensional submanifold of a real space form $R^m(c)$. Then, for any integer k, $2 \le k \le n$, we have:*

$$\theta_k \le \frac{n^2(k-1)}{4(n-1)} \|H\|^2 + (k-1)c, \tag{3.8}$$

where θ_k is the k-Ricci curvature invariant defined by (2.2).

The equality case of (3.8) *holds identically at a point $p \in M$ if and only if either p is a totally geodesic point, or $n = k = 2$ and p is a totally umbilical point.*

In [42], the first author also established the following sharp relationship between the shape operator A_H and θ_k for any submanifold in real space form $R^m(c)$:

$$\begin{cases} A_H > \frac{n-1}{n}(\theta_k(p) - c)I & \text{if } \theta_k(p) \ne c; \\ A_H \ge 0 & \text{if } \theta_k(p) = c. \end{cases}$$

4 Chen–Ricci Inequalities for Submanifolds in Manifolds of Quasi-constant Curvature

Riemannian manifolds of quasi-constant curvature were introduced by the first author and Yano in [62] as follows: A Riemannian m-manifold $(\widetilde{M}, g)$ is said to be of *quasi-constant curvature* if there exist a unit vector field ξ, called the *generator*, and functions κ, μ such that the curvature tensor $\widetilde{R}$ of $\widetilde{M}$ satisfies

$$\begin{aligned} \widetilde{R}(X, Y, Z, W) = & \kappa[g(Y, Z)g(X, W) - g(X, Z)g(Y, W)] \\ & + \mu[g(X, W)\zeta(Y)\zeta(Z) - g(X, Z)\zeta(Y)\zeta(W) \\ & + g(Y, Z)\zeta(X)\zeta(W) - g(Y, W)\zeta(X)\zeta(Z)], \end{aligned}$$

where ζ is the 1-form dual to a unit vector field ξ. Denote such manifold by $\widetilde{M}^m_{\kappa,\mu}(\zeta)$.

Özgür [156] studied the Chen–Ricci inequality for submanifolds in $\widetilde{M}^m_{\kappa,\mu}(\zeta)$. He proved the following.

Theorem 4.1 *Let M be an n-dimensional ($n \ge 3$) submanifold in $\widetilde{M}^m_{\kappa,\mu}(\zeta)$. Then, for any integer k, $2 \le k \le n$, we have:*

$$\|H\|^2 \ge \theta_k - \kappa - \frac{2\mu}{n} \|\zeta^T\|^2,$$

where ζ^T denotes the tangential component of ζ.

As a generalization of Theorem 3.9, Zhang et al. [217] proved the following.

Theorem 4.2 *Let M be an n-dimensional ($n \geq 3$) in $\widetilde{M}^m_{\kappa,\mu}(\zeta)$ with a semi-symmetric metric connection. Then, for any integer k, $2 \leq k \leq n$, we have:*

$$\|H\|^2 \geq \theta_k - \kappa - \frac{2\mu}{n}\|\zeta^T\|^2 + \frac{2}{n}\lambda,$$

where λ is the trace of the $(0, 2)$*-tensor field s.*

In [217], Zhang et al. also established several extensions of Theorem 3.1.

In 2009, De and Gazi [67] defined a Riemannian manifold of *nearly quasi-constant curvature* as a Riemannian manifold whose curvature tensor satisfies

$$\begin{aligned}\widetilde{R}(X, Y, Z, W) = &\kappa[g(Y, Z)g(X, W) - g(X, Z)g(Y, W)]\\ &- \mu[g(X, W)B(Y, Z) - g(X, Z)B(Y, W)\\ &+ g(Y, Z)B(X, W) - g(Y, W)B(X, Z)],\end{aligned}$$

where κ, μ are functions and B is a non-zero symmetric tensor of type $(0, 2)$.

In [215], Zhang proved the following result.

Theorem 4.3 *Let M be an n-dimensional ($n \geq 3$) in a Riemannian manifold of nearly quasi-constant curvature. Then, for any $X \in T^1M$, we have:*

$$\mathrm{Ric}(X) \leq (n-1)\kappa + \mu[(n-2)B(X, X) + \lambda] + \frac{n^2}{4}\|H\|^2, \tag{4.1}$$

where λ is the trace of B.

The equality case of (4.1) *holds identically for all unit vectors in T^1_pM if and only if either p is a totally geodesic point, or $n = 2$ and p is a totally umbilical point.*

For further results in this direction, see also [89, 216].

5 Chen–Ricci Inequalities for Real Space Forms with Non-symmetric Connections

Hayden [93] introduced the notion of a semi-symmetric metric connection on a Riemannian manifold. Since then, non-symmetric metric connections have been studied by many geometers. Let (M, g) be a Riemannian manifold. A linear connection $\widetilde{\nabla}$ on M is called a *semi-symmetric connection* if the torsion tensor $\widetilde{T}$ of $\widetilde{\nabla}$ given by $\widetilde{T}(X, Y) = \widetilde{\nabla}_X Y - \widetilde{\nabla}_Y X - [X, Y]$ satisfies

$$\widetilde{T}(X, Y) = \omega(Y)X - \omega(X)Y, \tag{5.1}$$

for a 1-form ω. Let $\widetilde{P}$ be the vector field dual to ω, i.e., $\omega(X) = g(X, \widetilde{P})$. A semi-symmetric connection $\widetilde{\nabla}$ is called a *semi-symmetric metric connection* if $\widetilde{\nabla}g = 0$

holds identically. Associated with ∇ and $\widetilde{P}$, there exists a unique semi-symmetric metric connection $\widetilde{\nabla}$ satisfying

$$\widetilde{\nabla}_X Y = \nabla_X Y + \omega(Y)X - g(X, Y)\widetilde{P}.$$

We put

$$\alpha(X, Y) = (\mathring{\widetilde{\nabla}}_X \omega)Y - \omega(X)\omega(Y) + \frac{1}{2}\omega(\widetilde{P})g(X, Y),$$

where $\mathring{\widetilde{\nabla}}$ is the Levi–Civita connection of $(\widetilde{M}, g)$.

Yano [209] proved that a Riemannian manifold is conformally flat if and only if it admits a semi-symmetric metric connection whose curvature tensor vanishes identically.

Now, let us consider a Riemannian manifold $(\widetilde{M}, g)$ endowed with a semi-symmetric non-metric connection $\widetilde{\nabla}$ and we denote the Levi–Civita connection of $(\widetilde{M}, g)$ by $\mathring{\widetilde{\nabla}}$.

If $(\widetilde{M}, g)$ is a real space form $R^m(c)$, then the curvature tensor $\mathring{\widetilde{R}}$ with respect to $\mathring{\widetilde{\nabla}}$ satisfies

$$\mathring{\widetilde{R}}(X, Y, Z, W) = c[g(X, W)g(Y, Z) - g(X, Z)g(Y, W)].$$

Let M be a submanifold of the real space form $R^m(c)$. Then the curvature tensor $\widetilde{R}$ with respect to the semi-symmetric non-metric connection $\widetilde{\nabla}$ can be written as

$$\widetilde{R}(X, Y, Z, W) = \mathring{\widetilde{R}}(X, Y, Z, W) + s(X, Z)g(Y, W) - s(Y, Z)g(X, W),$$

where s is a $(0, 2)$-tensor field defined by

$$s(X, Y) = (\mathring{\widetilde{\nabla}}_X \omega)Y - \omega(X)\omega(Y). \tag{5.2}$$

The next result was proved by Özgür and Mihai [158] and also by Zhang and Zhang [214].

Theorem 5.1 *Let M be an n-dimensional ($n \geq 3$) submanifold of a real space form $R^m(c)$ equipped with a semi-symmetric non-metric connection. Then*

(1) *For any $X \in T_p^1 M$, we have:*

$$\|H\|^2 \geq \frac{4}{n^2}\big[\mathrm{Ric}(X) - (n-1)(c - \Omega)\big], \tag{5.3}$$

where $\Omega = s(X, X) + g(\widetilde{P}^{\perp}, h(X, X))$.

(2) *If $H(p) = 0$, then a unit tangent vector $X \in T_p^1 M$ satisfies the equality case of (5.3) if and only if X lies in the relative null subspace $\mathcal{N}_p$.*

(3) *The equality case of* (5.3) *holds identically for all unit vectors in* $T_p^1 M$ *if and only if either* p *is a totally geodesic point, or* $n = 2$ *and* p *is a totally umbilical point.*

Remark 5.2 A further result in this direction was also obtained by Wang [206].

In [135], Mihai and Özgür also proved the following.

Theorem 5.3 *Let* M *be an* n*-dimensional* ($n \geq 3$) *submanifold of a real space form* $R^m(c)$ *with a semi-symmetric non-metric connection such that the vector field* $\widetilde{P}$ *is tangent to* M. *Then, for any integer* k, $2 \leq k \leq n$, *we have:*

$$\theta_k \leq \|H\|^2 + c - \frac{2}{n}\lambda,$$

where λ *is the trace of the* $(0, 2)$*-tensor field* s.

A linear connection $\widetilde{\nabla}$ on a real space form is called a *Ricci quarter-symmetric connection* if its torsion tensor $\widetilde{T}$ satisfies

$$\widetilde{T}(X, Y) = \omega(Y)LX - \omega(X)LY,$$

for a 1-form ω, where L is the Ricci operator given by $g(LX, Y) = \mathrm{Ric}(X, Y)$.

Poyraz and Yoldaş [166] studied the Ricci curvature of submanifolds of real space forms endowed with a Ricci quarter-symmetric metric connection. They extended Theorem 3.9 to the following.

Theorem 5.4 *Let* M *be an* n*-dimensional* ($n \geq 3$) *submanifold of a real space form* $R^m(c)$ *endowed with a Ricci quarter-symmetric metric connection. Then, for any integer* k, $2 \leq k \leq n$, *we have:*

$$\theta_k \leq \|H\|^2 + c\left[2 - \frac{4(n-1)m}{n}\right].$$

6 Chen–Ricci Inequalities for Totally Real and Lagrangian Submanifolds in Complex Space Forms

The curvature tensor of a complex space form $\widetilde{M}(c)$ of constant holomorphic sectional curvature c satisfies

$$\begin{aligned} R(X, Y)Z = \frac{c}{4}[&g(Y, Z)X - g(X, Z)Y + g(JY, Z)JX \\ &- g(JX, Z)JY + 2g(X, JY)JZ]. \end{aligned}$$

6.1 Totally Real Submanifolds of Complex Space Forms

For totally real submanifolds of a complex space form $\widetilde{M}(c)$, the first author proved the following.

Theorem 6.1 ([43]) *Let M be an n-dimensional totally real submanifold of a complex space form $\widetilde{M}(c)$. Then we have:*

$$\widehat{\mathrm{Ric}} \leq \frac{n^2}{4}\|H\|^2 + (n-1)c. \tag{6.1}$$

The equality case of (6.1) *holds identically if and only if either M is totally geodesic Lagrangian submanifold, or $n = 2$ and M is totally umbilical.*

In particular, for a Lagrangian submanifold M in $\widetilde{M}(c)$, we obtain the following.

Theorem 6.2 ([43]) *If M is an n-dimensional Lagrangian submanifold of a complex space form $\widetilde{M}(c)$, then we have:*

$$\widehat{\mathrm{Ric}} \leq \frac{n^2}{4}\|H\|^2 + (n-1)c. \tag{6.2}$$

If the equality case of (6.2) *holds identically, then M is minimal.*

6.2 Lagrangian Submanifolds of Complex Space Forms

Using an optimization technique, T. Oprea improved inequality (6.2) to the following.

Theorem 6.3 ([153]) *Let M be an n-dimensional Lagrangian submanifold of a complex space form $\widetilde{M}(c)$. Then we have:*

$$\widehat{\mathrm{Ric}} \leq \frac{n-1}{4}\big(n\|H\|^2 + c\big). \tag{6.3}$$

In [68], S. Deng provided an algebraic proof of the inequality (6.2). Furthermore, Deng also characterized Lagrangian submanifolds satisfying (6.3). In fact, he proved the following.

Theorem 6.4 ([68]) *Let M be an n-dimensional Lagrangian submanifold of a complex space form $\widetilde{M}(c)$. Then we have:*

$$\widehat{\mathrm{Ric}} = \frac{n-1}{4}\big(n\|H\|^2 + c\big). \tag{6.4}$$

The equality case of (6.4) *holds at $p \in M$ if and only if either p is a totally geodesic point, or $n = 2$ and p is a H-umbilical point with ratio 3.*

Remark 6.5 Note that the Whitney 2-sphere in $\mathbb{C}^2$ satisfies the equality case of (6.4) identically (see [39] for details).

Remark 6.6 Using curvature-like tensor as did in [58], Tripathi provided in [195] alternate proofs for Theorems 6.3 and 6.4.

The next optimal inequality involving the δ-invariant was proved by the first author together with Dillen et al. in [57].

Theorem 6.7 *Let M be an n-dimensional Lagrangian submanifold of a complex space form $\widetilde{M}(4c)$. Then, for any k-tuple $(n_1, \ldots, n_k)$, we have:*

(1) *If $\sum_{i=1}^k n_i < n$, then*

$$\delta(n_1, \ldots, n_k) \leq \frac{n^2\left[n - \sum_{i=1}^k n_i + 3k - 1 - 6\sum_{i=1}^k (2+n_i)^{-1}\right]}{2\left[n - \sum_{i=1}^k n_i + 3k + 2 - 6\sum_{i=1}^k (2+n_i)^{-1}\right]}\|H\|^2 + \frac{1}{2}\left[n(n-1) - \sum_{i=1}^k n_i(n_i - 1)\right]c\,. \tag{6.5}$$

(2) *If $\sum_{i=1}^k n_i = n$, then*

$$\delta(n_1, \ldots, n_k) \leq \frac{n^2\left[k - 1 - 2\sum_{i=2}^k (2+n_i)^{-1}\right]}{2\left[k - 2\sum_{i=2}^k (2+n_i)^{-1}\right]}\|H\|^2 + \frac{1}{2}\left[n(n-1) - \sum_{i=1}^k n_i(n_i - 1)\right]c\,. \tag{6.6}$$

Remark 6.8 It is easy to see that, for $\delta(n_1, \ldots, n_k) = \delta(n-1)$, the inequality (6.5) is exactly the inequality (6.3) in Theorem 6.3.

Remark 6.9 The equality cases of (6.5) and (6.6) were characterized in [57].

6.3 Submanifolds of Complex Space Forms with Non-symmetric Metric Connections

In this subsection, we follow the notations in Sect. 5.

Mihai [131] investigated the Chen–Ricci inequality for Lagrangian submanifolds in complex space forms endowed with a semi-symmetric metric connection. She obtained the following result.

Theorem 6.10 *Let M be an n-dimensional $(n \geq 2)$ Lagrangian submanifold of a complex space form $\widetilde{M}(4c)$ endowed with a semi-symmetric metric connection. Then, for any $X \in T^1M$, we have:*

$$\mathrm{Ric}(X) + (n-2)\alpha(X, X) + \mathrm{Trace}(\alpha) \leq (n-1)\left(c + \frac{n}{4}\|H\|^2\right). \tag{6.7}$$

The equality case of (6.7) *holds identically if and only if either M is totally geodesic, or $n = 2$ and M is an H-umbilical Lagrangian surface with ratio 3.*

A. Mihai and C. Özgür studied k-Ricci curvature invariant of a submanifold M in a complex space form endowed with a semi-symmetric metric connection such that the vector field $\widetilde{P}$ is tangent to M. They extended Theorem 3.9 to the following.

Theorem 6.11 ([134]) *Let M be an n-dimensional $(n \geq 3)$ Lagrangian submanifold of a complex space form $\widetilde{M}(4c)$ endowed with a semi-symmetric metric connection such that the vector field $\widetilde{P}$ is tangent to M. Then, for any integer k, $2 \leq k \leq n$, and any point $p \in M$, we have:*

$$\|H\|^2(p) \geq \theta_k(p) + \frac{2}{n}\lambda - c - \frac{3c}{n(n-1)}\|P\|^2,$$

where $\|P\|^2 = \sum_{i,j=1}^{n} g(JE_i, E_j)^2$ for an orthonormal basis $\{E_1, \ldots, E_n\}$ of T_pM and λ is the trace of the $(0, 2)$*-tensor field s.*

Remark 6.12 Ghişoiu [83] studied the Chen–Ricci inequality for submanifolds in a generalized complex space form with a semi-symmetric metric connection and obtained the corresponding inequality for such submanifolds.

Remark 6.13 In [6], Al-Khaldi et al., studied the Chen–Ricci inequality for submanifolds of a generalized complex space form with a quarter-symmetric metric connection. They extended Chen–Ricci inequality for such submanifolds.

7 Chen–Ricci Inequalities for Slant Submanifolds

7.1 Slant Submanifolds of Complex Space Forms

The Chen–Ricci inequality for slant submanifolds in complex space forms was proved by Matsumoto et al. [125].

Theorem 7.1 *Let M be an n-dimensional θ-slant submanifold of a complex space form $\widetilde{M}(4c)$. Then*

(1) *For any* $X \in T_p^1 M$, *we have:*

$$\mathrm{Ric}(X) \le \frac{n^2}{4}\|H\|^2 + (n-1)c + 3c\cos^2\theta. \tag{7.1}$$

(2) *If* $H(p) = 0$, *then a unit tangent vector* $X \in T_p^1 M$ *satisfies the equality case of* (7.1) *if and only if* X *lies in the relative null subspace* $\mathcal{N}_p$.

(3) *The equality case of* (7.1) *holds identically for all unit vectors in* $T_p^1 M$ *if and only if either* p *is a totally geodesic point, or* $n = 2$ *and* p *is a totally umbilical point.*

Mihai and Radulescu [136] and Tripathi [195] obtained independently the following result for Kählerian slant submanifolds.

Theorem 7.2 *If* M *is an* n*-dimensional Kählerian* θ*-slant submanifold in a complex space form* $\widetilde{M}(4c)$, *then, for any* $X \in T^1 M$, *we have:*

$$\mathrm{Ric}(X) \le (n-1)\left(c + \frac{n}{4}\|H\|^2\right) + 3c\cos^2\theta. \tag{7.2}$$

The equality case of (7.2) *holds identically if and only if either* $c = 0$ *and* M *is totally geodesic, or* $n = 2$, $c < 0$, *and* M *is a slant* H*-umbilical surface with ratio 3.*

7.2 Slant Submanifolds of Generalized Complex Space Forms

An almost Hermitian manifold $(\widetilde{M}, J, g)$ is called a *RK-manifold* if its curvature tensor $\widetilde{R}$ satisfies

$$\widetilde{R}(JX, JY, JZ, JW) = \widetilde{R}(X, Y, Z, W).$$

Let us put

$$\lambda(X, Y) = \widetilde{R}(X, Y, JX, JY) - \widetilde{R}(X, Y, X, Y).$$

An almost Hermitian manifold $\widetilde{M}$ is called of *pointwise constant type* if at any point $p \in \widetilde{M}$ and any vector $X \in T_p\widetilde{M}$, we have: $\lambda(X, Y) = \lambda(X, Z)$, where Y and Z are vectors in $T_p^1\widetilde{M}$ which are orthogonal to X and JX. And $\widetilde{M}$ is called *of constant type* if $\lambda(X, Y)$ is a constant function for any unit vectors fields X and Y satisfying $g(X, Y) = g(JX, Y) = 0$.

A *generalized complex space form* is a RK-manifold of constant holomorphic sectional curvature and of constant type. Every complex space form is a generalized complex space form, but the converse is not true. The simplest example is the nearly Kähler 6-sphere S^6.

Let $\widetilde{M}(c, \alpha)$ denote a generalized complex space form of constant holomorphic sectional curvature c and of constant type α. Then its curvature tensor $\widetilde{R}$ satisfies

$$\widetilde{R}(X,Y)Z = \frac{c+3\alpha}{4}[g(Y,Z)X - g(X,Z)Y] + \frac{c-\alpha}{4}\big[g(X,JZ)JY - g(Y,JZ)JX + 2g(X,JY)JZ\big].$$

It is easy to see that $\widetilde{M}(c,\alpha)$ is a real space form if $c=\alpha$ and it is a complex space form of constant holomorphic sectional curvature c if $\alpha = 0$.

For submanifolds of a generalized complex space form $\widetilde{M}(c,\alpha)$ we have the next result from [130].

Theorem 7.3 *Let M be an n-dimensional submanifold of a generalized complex space form $\widetilde{M}(c,\alpha)$. Then*

(1) *For any $X \in T_p^1 M$, we have:*

$$\mathrm{Ric}(X) \le \frac{1}{4}\left[n^2\|H\|^2 + (n-1)(c+3\alpha) + \frac{3}{2}(c-\alpha)\|PX\|^2\right]. \tag{7.3}$$

(2) *If $H(p) = 0$, then a unit tangent vector $X \in T_p^1 M$ satisfies the equality case of (7.3) if and only if X lies in the relative null subspace $\mathcal{N}_p$.*
(3) *The equality case of (7.3) holds identically for all unit vectors in $T_p^1 M$ if and only if either p is a totally geodesic point, or $n = 2$ and p is a totally umbilical point.*

For a θ-slant submanifold in a generalized complex space form, Song and Liu [183] and Hong and Tripathi [100] obtained the following.

Theorem 7.4 *Let M be an n-dimensional θ-slant submanifold of a generalized complex space form $\widetilde{M}(c,\alpha)$. Then, for any $X \in T^1M$, we have:*

$$\mathrm{Ric}(X) \le \frac{1}{4}\left[n^2\|H\|^2 + (n-1)(c+3\alpha) + \frac{3}{2}(c-\alpha)\cos^2\theta\right]. \tag{7.4}$$

If $H \neq 0$, then the equality case of (7.4) holds identically for all vectors in $T_p^1 M$ if and only if either p is a totally geodesic point, or $n = 2$ and p is a totally umbilical point.

If $\theta = \frac{\pi}{2}$, then Theorem 7.4 reduces to the following.

Corollary 7.5 ([100, 183]) *If M is an n-dimensional totally real submanifold in a generalized complex space form $\widetilde{M}(c,\alpha)$, then, for any $X \in T^1M$, we have:*

$$\mathrm{Ric}(X) \le \frac{1}{4}[n^2\|H\|^2 + (n-1)(c+3\alpha)]. \tag{7.5}$$

If $H \neq 0$, then the equality case of (7.5) holds identically for all unit tangent vectors at p if and only if either p is a totally geodesic point, or $n = 2$ and p is a totally umbilical point.

A. Mihai also obtained in [130] the following.

Theorem 7.6 *Let M be an n-dimensional submanifold of a generalized complex space form $\widetilde{M}(c, \alpha)$. Then, for any integer k, $2 \leq k \leq n$ and any $p \in M$, we have:*

$$\|H\|^2(p) \geq \theta_k(p) - \frac{1}{4}(c + 3\alpha) - \frac{3(c - \alpha)}{4n(n-1)}\|P\|^2.$$

8 Chen–Ricci Inequalities for Some Special Submanifolds in Kähler and Nearly Kähler Manifolds

Besides complex, totally real, and slant submanifolds, there exist another important class of submanifolds, called CR-submanifolds, introduced by Bejancu [30]. A submanifold M of an almost Hermitian manifold $\widetilde{M}$ is a *CR-submanifold* if there exist a complex distribution $\mathcal{D}$ (i.e., $J\mathcal{D} = \mathcal{D}$) and a totally real distribution $\mathcal{D}^{\perp}$ (i.e., $J\mathcal{D}^{\perp} \subset T^{\perp}M$) such that

$$TM = \mathcal{D} \oplus \mathcal{D}^{\perp}.$$

A CR-submanifold is called *proper* if it is neither a totally real nor a complex submanifold. Clearly, every real hypersurface of an almost Hermitian manifold $\widetilde{M}$ is a CR-submanifold with $\text{rank}(\mathcal{D}^{\perp}) = 1$. A real hypersurface M of a complex space form $\widetilde{M}(c)$ is called a *Hopf hypersurface* if its shape operator satisfies $A_{\xi}J\xi = \kappa J\xi$ for some function κ, where ξ is a unit normal vector field of M. The tangent vector field $J\xi$ is called the *Hopf vector field*. A Hopf hypersurface M is called *special* if $J\xi$ is an eigenvector of A_{ξ} with eigenvalue 0, i.e., $A_{\xi}(J\xi) = 0$.

8.1 Real Hypersurfaces of Complex Space Forms

For real hypersurfaces, F. R. Al-Solamy, the first author, and Deshmukh [7] proved the following.

Theorem 8.1 *Let M be a real hypersurface of a complex space form $\widetilde{M}^m(c)$. Then we have:*

$$\text{Ric}(J\xi, J\xi) \leq \frac{(2m-1)^2}{2}\|H\|^2 + (m-1)c, \tag{8.1}$$

where ξ is a unit normal vector field of M in $\widetilde{M}^m(c)$.

The equality case of (8.1) *holds identically if and only if M is a minimal special Hopf hypersurface.*

Theorem 8.1 implies the following.

Corollary 8.2 ([7]) *If M is a real hypersurface of a complex space form $\widetilde{M}^m(c)$ and M satisfies the equality case of* (8.1) *identically, then the holomorphic distribution $\mathcal{D}$ of M is non-integrable, unless $c = 0$ and M is totally geodesic.*

By applying Theorem 8.1, we have the following.

Theorem 8.3 ([7]) *If M is a real hypersurface of $\widetilde{M}^2(c)$, then we have:*

$$\mathrm{Ric}(J\xi, J\xi) \leq \frac{1}{2}(9\|H\|^2 + c). \tag{8.2}$$

The equality case of (8.2) *holds identically if and only if $c = 0$ and N is totally geodesic.*

A real hypersurface M of an almost Hermitian manifold $\widetilde{M}$ with $\dim_{\mathbb{C}} \widetilde{M} = n$ is called *totally real n-ruled* if for each $p \in M$ there exists an n-dimensional totally real, totally geodesic submanifold N_p^n of M through p.

For $c = 0$, Theorem 8.3 was improved in [7] to the following.

Theorem 8.4 *If M is a real hypersurface of $\mathbb{C}^3$, then we have:*

$$\mathrm{Ric}(J\xi, J\xi) \leq \frac{25}{2}\|H\|^2. \tag{8.3}$$

If the equality case of (8.3) *holds identically, then M is a totally real 3-ruled minimal submanifold of $\mathbb{C}^3$.*

For real hypersurfaces in $CP^3(4)$, we have the next proposition.

Proposition 8.5 ([7]) *If M is a real hypersurface of $CP^3(4)$, then we have:*

$$\mathrm{Ric}(J\xi, J\xi) \leq \frac{25}{2}\|H\|^2 + 4. \tag{8.4}$$

The equality case of (8.4) *holds identically if and only if locally there exists an orthonormal frame $\{E_1, E_2, E_3 = JE_1, E_4 = JE_2, E_5\}$ on M such that*

$$\begin{aligned} &\sigma(E_1, E_1) = \lambda\xi, \;\; \sigma(E_2, E_2) = -\lambda\xi, \;\; \sigma(E_3, E_3) = -\frac{1}{\lambda}\xi, \\ &\sigma(E_4, E_4) = \frac{1}{\lambda}\xi, \;\; \text{and } \sigma(E_a, E_b) = 0 \text{ otherwise}, \end{aligned} \tag{8.5}$$

where λ is a nowhere zero function.

8.2 Real Hypersurfaces of Complex Hyperbolic Spaces

For real hypersurfaces of $CH^m(-4)$, the first author proved the following.

Theorem 8.6 ([47]) *Let M be a real hypersurface of the complex hyperbolic space $CH^m(-4)$ of constant holomorphic sectional curvature -4 with $m \geq 2$. Then the maximal Ricci curvature of M satisfies*

$$\widehat{\mathrm{Ric}} \leq \frac{(2m-1)^2}{4}\|H\|^2 - 2(m-1). \tag{8.6}$$

The equality case of (8.6) *holds identically if and only if M is a Hopf hypersurface with constant mean curvature given by $2\kappa/(2m-1)$, where κ is the principal curvature associated with the Hopf vector field $J\xi$, i.e., $A_\xi J\xi = \kappa J\xi$.*

Moreover, if M has constant principal curvatures, then M satisfies the equality case of (8.6) *identically if and only if M is either an open portion of a horosphere in $CH^2(-4)$, $m \geq 3$ and M is the tubular hypersurface over a totally geodesic complex hypersurface $CH^{m-1}(-4)$ in $CH^m(-4)$ with radius $r = \tanh^{-1}(1/\sqrt{2m-3})$.*

If $m = 2$, the assumption of constant principal curvatures given in Theorem 1 holds automatically. In fact, we have the following result.

Corollary 8.7 ([47]) *If M is a real hypersurface of the complex hyperbolic space $CH^2(-4)$, then we have:*

$$\widehat{\mathrm{Ric}} \leq \frac{9}{4}\|H\|^2 - 2. \tag{8.7}$$

The equality case of (8.7) *holds identically if and only if M is an open portion of the horosphere in $CH^2(-4)$.*

From [7], we also have the following.

Proposition 8.8 *If M is a real hypersurface of $CH^3(-4)$, then we have:*

$$\mathrm{Ric}(J\xi, J\xi) \leq \frac{25}{2}\|H\|^2 - 4. \tag{8.8}$$

The equality case of (8.8) *holds identically if and only if locally there exists an orthonormal frame $\{E_1, E_2, E_3 = JE_1, E_4 = JE_2, E_5\}$ on M such that the second fundamental form h of M satisfies* (8.5).

8.3 CR-Submanifolds

For 3-dimensional submanifolds M of the nearly Kähler unit six-sphere $S^6(1)$, Theorem 3.1 implies that the maximum Ricci curvature of M satisfies

$$\widehat{\mathrm{Ric}} \leq \frac{9}{4}\|H\|^2 + 2. \tag{8.9}$$

For a 3-dimensional CR-submanifold M of the nearly Kähler unit six-sphere $S^6(1)$, T. Sasahara proved the next result in [168].

Theorem 8.9 *Let M be a 3-dimensional CR-submanifold of the nearly Kähler unit six-sphere $S^6(1)$. If M satisfies the equality case of* (8.9)*, then M is minimal and quasi-Einstein.*

Sasahara [168] also studied proper CR-submanifolds in complex hyperbolic space with $\mathrm{rank}(\mathcal{D}^\perp) = 1$. He had established the explicit representation for such submanifolds which are normal in an anti-de Sitter space-time via the Hopf fibration.

8.4 Submanifolds of Bochner–Kähler Manifolds

In 1949, S. Bochner introduced a Kählerian analogue of conformal curvature tensor for a Kähler manifold. A Kähler manifold with vanishing Bochner curvature tensor is simply called a *Bochner–Kähler manifold* by the first author in [34].

The Chen–Ricci inequality for submanifolds of Bochner–Kähler manifolds was proved by Lone et al. [121]. In particular, they proved the improved Chen–Ricci inequality for Kählerian slant submanifolds in Bochner–Kähler manifolds. They also characterized the submanifolds which satisfy the equality case.

9 Submanifolds of Locally Conformal Kähler Manifolds

A Hermitian manifold $(\widetilde{M}, J, g)$ is called a *locally conformal Kähler manifold* [72] if each point $p \in \widetilde{M}$ has an open neighborhood U with function $\sigma : U \to \mathbb{R}$ such that $\tilde{g} = e^{-\sigma} g|_U$ is a Kähler metric. If we choose $U = \widetilde{M}$, then $(\widetilde{M}, J, g)$ is called a *globally conformal Kähler manifold*. The *fundamental 2-form* ω is defined as $\omega(X, Y) = g(JX, Y)$ for $X, Y \in \mathfrak{X}(\widetilde{M})$.

The next proposition provides a necessary and sufficient condition for a Hermitian manifold to be locally conformal Kähler (see [72]).

Proposition 9.1 *A Hermitian manifold $(\widetilde{M}, J, g)$ is a locally conformal Kähler manifold if and only if there exists a global 1-form α satisfying*

$$(\widetilde{\nabla}_Z\omega)(Y, X) = \alpha(X)g(Z, Y) - \alpha(Y)g(Z, X) \\ - \alpha(JX)\omega(Z, Y) + \alpha(JY)\omega(Z, X)$$

for any vector fields $X, Y, Z \in \mathfrak{X}(\widetilde{M})$.

The 1-form α is called the *Lee form*. For a locally conformal Kähler manifold $(\widetilde{M}, J, g)$, one defines a symmetric (0, 2)-tensor $\widetilde{Q}$ by

$$\widetilde{Q}(Y, X) = -(\widetilde{\nabla}_Y\alpha)(X) - \alpha(Y)\alpha(X) + \frac{1}{2}\|\alpha\|^2 g(Y, X), \tag{9.1}$$

where $\|\alpha\|$ denotes the norm of the Lee form with respect to g.

Carriazo et al. [32] proved the following inequality for totally real submanifolds in a locally conformal Kähler space form (see also [130]).

Theorem 9.2 *Let M be an n-dimensional totally real submanifold of a locally conformal Kähler space form $\widetilde{M}(c)$ of constant holomorphic sectional curvature c. Then*

(1) *For any $X \in T_p^1M$, we have:*

$$\|H\|^2 \geq \frac{4}{n^2}\big[\mathrm{Ric}(X) - (n-1)(c + 6\sigma)\big], \tag{9.2}$$

where $\sigma = \frac{1}{n}\sum_{i=1}^{n} \widetilde{Q}(E_i, E_i)$ and $\{E_1, \ldots, E_n\}$ is an orthonormal frame of T_pM.

(2) *If $H(p) = 0$, then a unit tangent vector $X \in T_p^1M$ satisfies the equality case of* (9.2) *if and only if $X \in \mathcal{N}_p$.*

(3) *The equality case of* (9.2) *holds identically for all unit vectors in T_p^1M if and only if either p is a totally geodesic point, or $n = 2$ and p is a totally umbilical point.*

For the k-Ricci curvature invariant, A. Mihai proved the following.

Theorem 9.3 ([130]) *Let M be an n-dimensional totally real submanifold of a locally conformal Kähler space form $\widetilde{M}(c)$. Then, for any integer k, $2 \leq k \leq n$ and any $p \in M$, we have:*

$$\|H\|^2(p) \geq \theta_k(p) - \frac{c + 6\tilde{\rho}}{4},$$

where $\tilde{\rho}$ denotes the trace of the symmetric (0, 2)*-tensor $\widetilde{Q}$ defined by* (9.1).

Hong et al. [99] investigated Chen–Ricci inequalities for general, complex, and totally real submanifolds in a locally conformal Kähler space form. In the same article, they also characterized the submanifolds satisfying the equality case of the corresponding inequalities.

10 Quaternionic Kähler Manifolds

An *almost quaternionic Hermitian manifold* $(\overline{M}, \Sigma, \overline{g})$ is a Riemannian manifold endowed with a subbundle $\Sigma \subset \mathrm{End}(T\overline{M})$ with local basis $\{J_1, J_2, J_3\}$ satisfying

$$\overline{g}(J_\alpha X, J_\alpha Y) = \overline{g}(X, Y),\ J_\alpha^2 = -I,\ J_\alpha J_{\alpha+1} = -J_{\alpha+1} J_\alpha = J_{\alpha+2},\ \alpha = 1, 2, 3,$$

for vectors $X, Y \in T\overline{M}$. If the bundle Σ is parallel with respect to the Levi–Civita connection of $\overline{g}$, then $(\overline{M}, \Sigma, \overline{g})$ is said to be a *quaternionic Kähler manifold*. Clearly, such a manifold is of dimension $4m$ with $m \geq 1$.

For a quaternionic Kähler manifold $(\overline{M}, \Sigma, \overline{g})$ and a vector $0 \neq X \in T\overline{M}$, the 4-plane $Q(X)$ spanned by $\{X, J_1X, J_2X, J_3X\}$ is called a *quaternionic 4-plane*. Any 2-plane in $Q(X)$ is called a *quaternionic plane*. The sectional curvature of a quaternionic plane is called a *quaternionic sectional curvature*. A quaternionic Kähler manifold is called a *quaternionic space form* if it has constant quaternionic sectional curvature.

We denote by $\overline{M}(c)$ a quaternionic space form of constant quaternionic sectional curvature c. The curvature tensor $\overline{R}$ of $\overline{M}(c)$ satisfies

$$\begin{aligned}\overline{R}(X, Y)Z = \frac{c}{4}\{&\overline{g}(Z, Y)X - \overline{g}(X, Z)Y \\ &+ \sum_{\alpha=1}^{3}[\overline{g}(Z, J_\alpha Y)J_\alpha X - \overline{g}(Z, J_\alpha X)J_\alpha Y + 2\overline{g}(X, J_\alpha Y)J_\alpha Z]\}.\end{aligned}$$

A submanifold M of a quaternionic Kähler manifold $\overline{M}$ is called *slant* if, for each vector $0 \neq X \in T_pM$, the angle $\theta(X)$ between $J_\alpha X$ and T_pM, $\alpha = 1, 2, 3$, is a global constant. A slant submanifold of a quaternionic Kähler manifold is called *proper* (or proper θ-slant) if $\theta \neq 0, \frac{\pi}{2}$. A submanifold M of a quaternionic Kähler manifold $(\overline{M}, \Sigma, \overline{g})$ is called a *quaternionic CR-submanifold* [29] if there exist two orthogonal complementary distributions $\mathcal{D}$ and $\mathcal{D}^\perp$ on M such that $\mathcal{D}$ is invariant under the quaternionic structure and $\mathcal{D}^\perp$ is totally real.

For any vector X tangent to a submanifold M of a quaternionic Kähler manifold, we put

$$J_\alpha X = P_\alpha X + F_\alpha X, \quad P_\alpha X \in T_pM, \quad F_\alpha X \in T_p^\perp M.$$

The squared norm of P_α is $\|P_\alpha\|^2 = \sum_{i,j=1}^{n} g(P_\alpha E_i, E_j)^2$, where $\{E_1, \ldots, E_n\}$ is an orthonormal basis of T_pM.

11 Chen–Ricci Inequalities for Submanifolds in Quaternionic Space Forms

11.1 Submanifolds in Quaternionic Space Forms

For quaternionic CR-submanifolds of a quaternionic space form $\overline{M}(c)$, Mihai et al. proved the following.

Theorem 11.1 ([140]) *Let M be an n-dimensional quaternionic CR-submanifold of a quaternionic space form $\overline{M}(c)$. Then*

(1) *For any $X \in \mathcal{D}_p^{\perp}$, we have:*

$$\mathrm{Ric}(X) \leq \frac{n^2}{4}\|H\|^2 + \frac{n-1}{4}c. \tag{11.1}$$

(2) *For any $X \in \mathcal{D}_p$, we have:*

$$\mathrm{Ric}(X) \leq \frac{n^2}{4}\|H\|^2 + \frac{n+8}{4}c. \tag{11.2}$$

(3) *If $H(p) = 0$, then a unit tangent vector $X \in T_p^1 M$ satisfies the equality case of* (11.1) *(respectively,* (11.2)*) if and only if $X \in \mathcal{D}_p^{\perp} \cap \mathcal{N}_p$ (respectively, $X \in \mathcal{D}_p \cap \mathcal{N}_p$), where $\mathcal{N}_p$ is the relative null space of M at $p \in M$.*

Remark 11.2 If M is an n-dimensional totally real submanifold of a quaternionic space form $\overline{M}(c)$, then inequality (11.1) was due to [117] (see also [213]).

For θ-slant submanifolds, G. E. Vîlcu proved the following.

Theorem 11.3 ([203]) *Let M be an n-dimensional θ-slant submanifold of a quaternionic space form $\overline{M}(c)$. Then*

(1) *For any $X \in T_p^1 M$, we have:*

$$\mathrm{Ric}(X) \leq \frac{n^2}{4}\|H\|^2 + \frac{n-1}{4}c + \frac{3}{8}c\cos^2\theta. \tag{11.3}$$

(2) *If $H(p) = 0$, then a unit tangent vector $X \in T_p^1 M$ satisfies the equality case of* (11.3) *if and only if X belongs to the relative null space of M at p.*
(3) *The equality case of* (11.3) *holds identically for all unit tangent vectors at p if and only either p is a totally geodesic point, or $n = 2$ and p is a totally umbilical point.*

Remark 11.4 Inequality (11.3) is a better estimate than the inequality (3.1) given in [175, Theorem 3.1].

For Lagrangian submanifolds, S. Deng proved the following.

Theorem 11.5 ([69]) *Let M be an n-dimensional Lagrangian submanifold of a quaternionic space form $\overline{M}(c)$. Then, for any $X \in T_p^1 M$, we have:*

$$\mathrm{Ric}(X) \leq \frac{n-1}{4}\left(\|H\|^2 + c\right). \tag{11.4}$$

The equality case of (11.4) *holds for any $X \in T_p^1 M$ if and only if either p is a totally geodesic point, or $n = 2$ and p is an H-umbilical point with ratio 3.*

11.2 k-Ricci Inequalities for Submanifolds in Quaternionic Space Forms

For the k-Ricci curvature invariant, we have the next result proved by G. E. Vîlcu.

Theorem 11.6 ([202]) *Let M be an n-dimensional θ-slant submanifold of a quaternionic space form $\overline{M}(c)$. Then, for any $p \in M$ and any integer k, $2 \leq k \leq n$, we have:*

$$\|H\|^2 \geq \theta_k - \frac{c}{4}\left(1 + \frac{9}{n-1}\cos^2\theta\right).$$

11.3 Submanifolds of Quaternionic Space Forms with Semi-symmetric Connection

In [71], Decu-Marinescu investigated the Chen–Ricci inequality for quaternionic CR-submanifolds in quaternionic space forms endowed with semi-symmetric non-metric connection. She proved in [70, 71] the following.

Theorem 11.7 *Let M be an n-dimensional quaternionic CR-submanifold of a quaternionic space form $\overline{M}(c)$ with semi-symmetric non-metric connection. Then*

(1) *For any $X \in \mathcal{D}_p^{\perp}$, we have:*

$$\mathrm{Ric}(X) \leq \frac{n^2}{4}\|H\|^2 + \frac{n-1}{4}c - (2n-3)\lambda + (n-2)s(X,X) - \frac{1}{2}n^2\omega(H), \tag{11.5}$$

where ω is defined by (5.1), *s is defined by* (5.2), *and λ is the trace of s.*

(2) *For any $X \in \mathcal{D}_p$, we have:*

$$\mathrm{Ric}(X) \le \frac{n^2}{4}\|H\|^2 + \frac{n+8}{4}c - (2n-3)\lambda + (n-2)s(X,X) - \frac{1}{2}n^2\omega(H). \tag{11.6}$$

(3) *If $H(p) = 0$, then a unit tangent vector $X \in T_p^1 M$ satisfies the equality case of (11.5) (or (11.6)) if and only if $X \in \mathcal{D}_p^{\perp} \cap \mathcal{N}_p$ (or $X \in \mathcal{D}_p \cap \mathcal{N}_p$), where $\mathcal{N}_p$ is the relative null space of M at $p \in M$.*

In [120], Lone studied the Chen–Ricci inequality for arbitrary submanifolds in quaternionic space forms endowed with semi-symmetric metric connection. Among others, he obtained the following.

Theorem 11.8 ([120]) *Let M be an n-dimensional submanifold of a quaternionic space form $\overline{M}(c)$ equipped with a semi-symmetric metric connection. Then*

(1) *For any $X \in T_p^1 M$, we have:*

$$\begin{aligned}\mathrm{Ric}(X) \le\ & \frac{n^2}{4}\|H\|^2 + \frac{n-1}{4}c - \lambda + (n-2)s(X,X) \\ & - \frac{3}{4}c\sum_{j=2}^{n}\sum_{r=1}^{3} g(J_r X, E_j)^2,\end{aligned} \tag{11.7}$$

where $\{X, E_2, \ldots, E_n\}$ is an orthonormal basis of $T_p M$.

(2) *If $H(p) = 0$, then a unit tangent vector X at p satisfies the equality case of (11.7) if and only if $X \in \mathcal{N}_p$.*

12 Almost Contact Metric, Sasakian and Kenmotsu Manifolds

An odd-dimensional Riemannian manifold $(\overline{M}, g)$, with the Levi–Civita connection ∇, is called an *almost contact metric manifold* [31] if there exist a $(1, 1)$-tensor field ϕ, a vector field ξ (the structure vector field), and a 1-form η on $\overline{M}$ such that

$$\begin{aligned}&\eta(\xi) = 1,\ \ \phi^2(X) = -X + \eta(X)\xi,\ \ \phi\xi = 0,\ \ \eta \circ \phi = 0,\ \ \eta(X) = g(X, \xi),\\ &g(\phi X, \phi Y) = g(X, Y) - \eta(X)\eta(Y),\end{aligned}$$

for any $X, Y \in \mathfrak{X}(\overline{M})$. A submanifold M of an almost contact metric manifold $\overline{M}$ is called an *anti-invariant submanifold* if $\phi(T_p M) \subset T_p^{\perp} M$ for every $p \in M$.

An almost contact metric manifold is called *contact* if $g(X, \phi Y) = d\eta(X, Y)$. An almost contact metric manifold $\overline{M}$ is said to be *normal* if the tensor field $N_\phi = [\phi, \phi] + 2d\eta \otimes \xi$ vanishes identically, where $[\phi, \phi]$ is the Nijenhuis tensor of ϕ. A normal contact metric manifold is said to be a *Sasakian manifold*. An almost contact metric manifold is Sasakian if and only if the Levi–Civita connection ∇ satisfies $(\nabla_X \phi)Y = g(X, Y) - \eta(Y)X$.

An almost contact metric manifold is called *almost cosymplectic* if $d\eta = 0$ and $d\phi = 0$. An almost cosymplectic manifold is called *cosymplectic* if it satisfies $\nabla\phi = 0$ and $\nabla\xi = 0$.

An almost contact metric manifold $\overline{M}$ is called a *Kenmotsu manifold* if it satisfies $(\nabla_X \phi)Y = g(\phi X, Y)\xi - \eta(Y)\phi X$.

A Sasakian manifold is called *Sasakian space form* if it has constant ϕ-sectional curvature. The Riemann curvature tensor of a Sasakian space form $\overline{M}(c)$ of constant ϕ-sectional curvature c is given by

$$\begin{aligned}\overline{R}(X, Y)Z &= \frac{c+3}{4}\big[g(Y, Z)X - g(X, Z)Y\big] \\ &+ \frac{c-1}{4}\big[g(X, \phi Z)\phi Y - g(Y, \phi Z)\phi X + 2g(X, \phi Y)\phi Z\big] \\ &+ \frac{c-1}{4}\big[\eta(X)\eta(Z)Y - \eta(Y)\eta(Z)X + g(X, Z)\eta(Y)\xi - g(Y, Z)\eta(X)\xi\big].\end{aligned}$$

Sasakian space forms $\overline{M}(c)$ can be modeled based on $c > -3$, $c = -3$, or $c < -3$. $\mathbb{R}^{2m+1}$ admits a Sasakian structure with constant ϕ-sectional curvature -3, while S^{2m+1} is of constant ϕ-sectional curvature 1. We denote the above Sasakian space forms by $\mathbb{R}^{2m+1}(-3)$ and $S^{2m+1}(1)$, respectively.

Alegre et al. [9] defined generalized Sasakian space forms as follows: An odd-dimensional manifold $\overline{M}$ endowed with an almost contact metric structure (ϕ, ξ, η, g) is called a *generalized Sasakian space form* if there exist three functions f_1, f_2, f_3 on $\overline{M}$ such that the curvature tensor of $\overline{M}$ satisfies

$$\begin{aligned}\overline{R}(X, Y)Z &= f_1\big[g(Y, Z)X - g(X, Z)Y\big] \\ &+ f_2\big[g(X, \phi Z)\phi Y - g(Y, \phi Z)\phi X + 2g(X, \phi Y)\phi Z\big] \\ &+ f_3\big[\eta(X)\eta(Z)Y - \eta(Y)\eta(Z)X + g(X, Z)\eta(Y)\xi - g(Y, Z)\eta(X)\xi\big].\end{aligned}$$

Let us denote such a manifold simply by $\overline{M}(f_1, f_2, f_3)$.

A generalized Sasakian space form $\overline{M}(f_1, f_2, f_3)$ is a Sasakian space form if $f_1 = \frac{c+3}{4}$ and $f_2 = f_3 = \frac{c-1}{4}$ for a constant c.

An n-dimensional submanifold M of an almost contact metric manifold $\overline{M}^{2m+1}$ is called *C-totally real* if the structure vector field ξ is normal to M. It is well known that one has $\phi(T_pM) \subset T_p^{\perp}M$ for C-totally real submanifolds. A C-totally real submanifold is called *Legendrian* if $n = m$ (see [31]). Therefore, every Legendrian submanifold is C-totally real with the smallest possible codimension.

For any vector X tangent to a submanifold M of an almost contact metric manifold, we put

$$\phi X = PX + FX,$$

where PX and FX denote the tangential component and normal component of ϕX, respectively.

13 Chen–Ricci Inequalities for Submanifolds in Sasakian Space Forms

In this section, we present some results on Chen–Ricci inequalities for submanifolds in Sasakian space forms and in generalized Sasakian space forms.

13.1 Submanifolds of Sasakian Space Forms

For C-totally real submanifolds of a Sasakian space form, we have the following.

Theorem 13.1 ([116, 128, 138]) *Let M be an n-dimensional C-totally real submanifold of a Sasakian space form $\overline{M}(c)$. Then, for any $X \in T_p^1 M$, we have:*

$$\mathrm{Ric}(X) \leq \frac{1}{4}[n^2\|H\|^2 + (n-1)(c+3)]. \tag{13.1}$$

The equality case of (13.1) *holds identically if and only if either M is totally geodesic, or $n = 2$ and M is totally umbilical.*

Theorem 13.2 ([138]) *Let $\overline{M}(c)$ be a Sasakian space form and M an n-dimensional submanifold of $\overline{M}(c)$ tangent to ξ. Then, for any $X \in T_p^1 M$ orthogonal to ξ, we have:*

$$\mathrm{Ric}(X) \leq \frac{1}{4}\left[n^2\|H\|^2 + (n-1)(c+3) + \frac{1}{2}(3\|PX\|^2 - 2)(c-1)\right].$$

For contact slant submanifolds in Sasakian space forms, one has the following.

Corollary 13.3 ([138]) *If $\overline{M}(c)$ is a Sasakian space form and M an n-dimensional contact θ-slant submanifold of $\overline{M}(c)$, then, for any $X \in T_p^1 M$ orthogonal to ξ, we have:*

$$\mathrm{Ric}(X) \leq \frac{1}{4}\big[n^2\|H\|^2 + 3(\cos^2\theta - 2)(c-1) + (n-1)(c+3)\big].$$

I. Mihai and I. N. Rădulescu improved (13.1) for Legendrian submanifolds to the following.

Theorem 13.4 ([144]) *Let M be an n-dimensional Legendrian submanifold in a Sasakian space form $\overline{M}(c)$ of constant ϕ-sectional curvature c. Then, for any $X \in T^1 M$, we have:*

$$\mathrm{Ric}(X) \leq \frac{n-1}{4}[n\|H\|^2 + c + 3]. \tag{13.2}$$

The equality case of (13.2) holds identically if and only if either M is totally geodesic, or M is a H-umbilical Legendrian surface of ratio 3.

A proper contact θ-slant submanifold is called a *special contact θ-slant submanifold* [141] if $(\nabla_X P)Y = \cos^2\theta[g(X,Y)\xi - \eta(Y)X]$, for any $X, Y \in \mathfrak{X}(M)$. For special contact θ-slant submanifolds in a Sasakian space form $\overline{M}(c)$, Mihai and Ghisoiu [141] obtained the following.

Theorem 13.5 *Let M be an $(n+1)$-dimensional special contact θ-slant submanifold in a Sasakian space form $\overline{M}^{2n+1}(c)$. Then we have:*

$$\widehat{\mathrm{Ric}} \leq \frac{1}{4}\left[(n+1)^2\|H\|^2 + n(c+3) + \frac{1}{2}(3\cos^2\theta - 2)(c-1)\right]. \tag{13.3}$$

If the equality case of (13.3) holds identically, then M is a minimal submanifold.

Mihai and Radulescu [137] improved Theorem 13.5 to the following.

Theorem 13.6 *Let M be an $(n+1)$-dimensional special contact θ-slant submanifold in a Sasakian space form $\overline{M}^{2n+1}(c)$. Then, for any $X \in T^1M$ orthogonal to ξ, we have:*

$$\begin{aligned} \mathrm{Ric}(X) \leq \frac{1}{4}\Big[&\frac{(n-1)(n+1)^2}{n}\|H\|^2 + n(c+3) \\ &+ (3\cos^2\theta - 1)(c-1) - 4\sin^2\theta\Big]. \end{aligned} \tag{13.4}$$

The equality case of (13.4) holds for any $X \in T^1M$ orthogonal to ξ if and only if either M is a totally contact geodesic submanifold, i.e.,

$$h(X,Y) = \eta(X)h(Y,\xi) + \eta(Y)h(X,\xi)$$

for any $X, Y \in \mathfrak{X}(M)$, or $n = 2$ and M is 3-dimensional H-umbilical contact slant submanifold with ratio 3.

Remark 13.7 Tripathi et al. [198] derived the Chen–Ricci inequality for a submanifold M of a Sasakian space form $\overline{M}(c)$ for which the structure vector field ξ is tangent to M. They also discussed the equality case. Some applications of these results are given for slant, invariant, anti-invariant, and CR-submanifolds of $\overline{M}(c)$. Cioroboiu [64] also derived the Chen–Ricci inequality for a submanifold M of a Sasakian space form $\overline{M}(c)$ such that ξ is tangent to M and he applied his result to slant, semi-slant, and bi-slant submanifolds in Sasakian space forms. For further results in this respect, see also [100, 194, 199].

Remark 13.8 Tripathi and Ziabari [196, 223] defined conformal Sasakian space forms and derived Chen–Ricci inequalities for several kinds of slant submanifolds of a conformal Sasakian space form tangent to the structure vector ξ.

Remark 13.9 Malek and Balgeshir [122] studied slant submanifold of an almost contact metric 3-structure manifold. They provided some examples and characterized these submanifolds. Moreover, they defined and studied Sasakian slant submanifolds of an almost contact 3-structure manifold. They established a Chen–Ricci type inequality for Sasakian slant submanifolds.

Remark 13.10 Pitiş [162] defined *-Legendrian submanifold in a Sasakian space form. He derived upper and lower bounds for the Ricci curvature for such submanifolds.

13.2 *Submanifolds of Sasakian Space Forms with Additional Connection*

There are several articles which studied the Chen–Ricci inequality for submanifolds of Sasakian space forms equipped with an additional connection.

Mihai and Özgür [134] proved an inequality for k-Ricci curvature invariant θ_k for submanifolds M in a Sasakian space form with a semi-symmetric metric connection such that the vector field $\widetilde{P}$ dual to ω is tangent to M (see Sect. 5 for the definitions of α, ω, and $\widetilde{P}$). Also, Mihai [131] proved the following.

Theorem 13.11 *Let M be an n-dimensional ($n \geq 2$) Legendrian submanifold in a Sasakian space form $\overline{M}(c)$ with a semi-symmetric metric connection such that $\widetilde{P}$ is tangent to M. Then, for any $X \in T_p^1 M$, we have:*

$$\mathrm{Ric}(X) + (n-2)\alpha(X, X) + \lambda \leq \frac{n-1}{4}\left(c + 3 + n\|H\|^2\right), \tag{13.5}$$

where λ is the trace of α.

The equality case of (13.5) *holds identically if and only if either M is totally geodesic, or $n = 2$ and M is a H-umbilical Legendrian surface with ratio 3.*

Remark 13.12 Jin and Lee [103] defined the Tanaka–Webster connection for submanifolds in Sasakian manifolds by the naturally extended affine connection of Tanno's Tanaka–Webster connection. Involving the pseudo-Ricci curvature and the squared pseudo-mean curvature, they proved an inequality for submanifolds in a Sasakian space form tangent to the structure vector field in terms of the Tanaka–Webster connection. Also, Lee et al. [114] established an improved Chen–Ricci inequality for anti-invariant submanifolds between the pseudo-Ricci curvature and the square of pseudo-mean curvature with respect to the Tanaka–Webster connection in Sasakian space forms. They also characterized the equality case of the inequality.

13.3 Submanifolds of Generalized Sasakian Space Forms

As far as we know, there are ten papers that discussed the Chen–Ricci inequality for submanifolds in generalized Sasakian space forms. Here, we merely provide the list of those ten papers as follows:

Abedi et al. [1], Al-Dayel and Khan [5], Alegre et al. [8], Al-Khaldi et al. [6], Hong and Tripathi [101], Khan and Ozel [105], Li et al. [115], Malek and Nejadakbary [123], Nejadakbary Mehraban et al. [149], and Shukla and Tiwari [174]. For the detailed results, please refer to these papers.

14 Chen–Ricci Inequalities for Submanifolds in Cosymplectic Manifolds

A *cosymplectic space form* $\overline{M}(f_1, f_2, f_3)(c)$ of constant ϕ-sectional curvature c is a generalized Sasakian space form with $f_1 = f_2 = f_3 = \frac{c}{4}$.

14.1 Submanifolds of Cosymplectic Space Forms

D. W. Yoon studied θ-slant submanifolds in a cosymplectic space form. He proved the following.

Theorem 14.1 ([212]) *Let M be an n-dimensional θ-slant submanifold tangent to ξ in a cosymplectic space form $\overline{M}(c)$. Then*

(1) *For any $X \in T_p^1 M$ orthogonal to ξ, we have:*

$$\mathrm{Ric} \leq \frac{1}{4}\left[n^2\|H\|^2 + (n-1)c + \frac{1}{2}(3\cos^2\theta - 2)c\right]. \tag{14.1}$$

(2) *If $H(p) = 0$, then a vector $X \in T_p^1 M$ orthogonal to ξ at p satisfies the equality case of* (14.1) *if and only if X lies in the relative null subspace $\mathcal{N}_p$.*
(3) *The equality case of* (14.1) *holds identically for all $X \in T_p^1 M$ orthogonal to ξ at p if and only if p is a totally geodesic point.*

A submanifold M tangent to ξ in $\overline{M}(c)$ is called *bi-slant* if there exist two orthogonal distributions $\mathcal{D}_1$ and $\mathcal{D}_2$ on M such that TM is the orthogonal direct sum:

$$TM = \mathcal{D}_1 \oplus \mathcal{D}_2 \oplus \mathrm{Span}\{\xi\}$$

and $\mathcal{D}_i$ is slant distribution with slant angle θ_i $(i = 1, 2)$.

Yoon [212] and Liu and Zhou [119] proved the following.

Theorem 14.2 ([119, 212]) *Let M be an n-dimensional bi-slant submanifold satisfying $g(X, \phi Y) = 0$ for any $X \in \mathcal{D}_1$ and any $Y \in \mathcal{D}_2$, tangent to ξ, in a cosymplectic space form $\overline{M}(c)$. Then*

(1) *For any $X \in T_p^1 M$ orthogonal to ξ, we have:*

$$\mathrm{Ric} \leq \frac{1}{4}\left[n^2\|H\|^2 + (n-1)c + \frac{1}{2}(3\cos^2\theta_1 - 2)c\right], \ \text{if}\, X \in \mathcal{D}_1; \tag{14.2}$$

$$\mathrm{Ric} \leq \frac{1}{4}\left[n^2\|H\|^2 + (n-1)c + \frac{1}{2}(3\cos^2\theta_2 - 2)c\right], \ \text{if}\, X \in \mathcal{D}_2. \tag{14.3}$$

(2) *If $H(p) = 0$, then a vector $X \in T_p^1 M$ orthogonal to ξ at p satisfies the equality case of* (14.2) *or* (14.3) *if and only if $X \in \mathcal{N}_p$.*
(3) *The equality case of* (14.2) *or* (14.3) *hold identically for all $X \in T_p^1 M$ orthogonal to ξ at p if and only if p is a totally geodesic point.*

Remark 14.3 As consequences of Theorem 14.2, Yoon [212] and Liu and Zhou [119] stated the corresponding inequalities for semi-slant, CR, invariant, and anti-invariant submanifolds.

Remark 14.4 For semi-invariant submanifolds in cosymplectic space forms, see also [205] by Wang and Liu.

Remark 14.5 Ali et al. [11] derived the Chen–Ricci inequality for C-totally real submanifolds in cosymplectic space forms.

14.2 Submanifolds of Locally Conformal Almost Cosymplectic Manifolds

A conformal change of an almost contact metric structure on an almost contact metric manifold $\overline{M}$ is given by

$$\phi^* = \phi, \ \ \xi^* = e^{-\rho}\xi, \ \ \eta^* = e^{\rho}\eta, \ \ g^* = e^{2\rho}g,$$

for a function ρ. An almost contact metric manifold $\overline{M}$ is called a *locally conformal almost cosymplectic manifold* if each point of $\overline{M}$ has a neighborhood $\mathcal{U}$ such that $(\mathcal{U}, \phi^*, \xi^*, \eta^*, g^*)$ is almost cosymplectic for some function ρ on $\mathcal{U}$. Equivalently, $\overline{M}$ is a locally conformal almost cosymplectic manifold if there exists a 1-form ω such that $d\Phi = 2\omega \wedge \Phi$, $d\eta = \omega \wedge \eta$, and $d\omega = 0$, where $\Phi(X, Y) = g(X, \phi Y)$ (see [151]).

A locally conformal almost cosymplectic manifold $\overline{M}$ of dimension ≥ 5 is of pointwise constant ϕ-sectional curvature if and only if its curvature tensor $\overline{R}$ takes the following form:

$$
\begin{aligned}
\overline{R}(X,Y)Z &= \frac{c-3f^2}{4}\,[g(Y,Z)X - g(X,Z)Y] \\
&+ \frac{c+f^2}{4}\,[2g(X,\phi Y)\phi Z + g(X,\phi Z)\phi Y - g(Y,\phi Z)\phi X] \\
&+ \left(\frac{c+f^2}{4} + f'\right)[\eta(X)\eta(Z)Y - \eta(Y)\eta(Z)X + g(X,Z)\eta(Y)\xi - g(Y,Z)\eta(X)\xi],
\end{aligned}
$$

where c is the pointwise ϕ-sectional curvature of $\overline{M}$ and f is the function such that $\omega = f\eta$ and $f' = \xi f$.

In [193], Tripathi derived the following inequalities for submanifolds in locally conformal almost cosymplectic manifolds.

Theorem 14.6 ([193]) *Let M be an n-dimensional submanifold in a locally conformal almost cosymplectic manifold $\overline{M}(c)$ of pointwise constant ϕ-sectional curvature c tangential to the structure vector field ξ. Then*

(1) *For any $X \in T_p^1 M$, we have:*

$$
\begin{aligned}
\mathrm{Ric}(X) \leq \frac{1}{4}\Big[& n^2\|H\|^2 + (n-1)(c-3f^2) \\
& + \left(3\|PX\|^2 - (n-2)\eta(X)^2 - 1\right)\left(c+f^2\right)\Big] \\
& - \left[1 + (n-2)\eta(X)^2\right] f'.
\end{aligned}
\tag{14.4}
$$

(2) *For $H(p) = 0$, a vector $X \in T_p^1 M$ satisfies the equality case of* (14.4) *if and only if X lies in the relative null space $\mathcal{N}_p$ at p.*

(3) *The equality case of* (14.4) *holds for all $X \in T_p^1 M$, if and only if either p is a totally geodesic point, or $n = 2$ and p is a totally umbilical point.*

Theorem 14.6 implies the following.

Theorem 14.7 ([193]) *Let M be an n-dimensional submanifold in a locally conformal almost cosymplectic manifold $\overline{M}(c)$ of pointwise constant ϕ-sectional curvature c such that $\xi \in TM$. Then*

(1) *If M is θ-slant, then for any $X \in T_p^1 M$, we have:*

$$
\begin{aligned}
\mathrm{Ric}(X) \leq \frac{1}{4}\Big\{ & n^2\|H\|^2 + (n-1)\left(c-3f^2\right) \\
& + \left(3\cos^2\theta - \left(n + 3\cos^2\theta - 2\right)\eta(X)^2 - 1\right)\left(c+f^2\right)\Big\} \\
& - \left[1 + (n-2)\eta(X)^2\right] f'.
\end{aligned}
\tag{14.5}
$$

(2) *If M is invariant, then for any $X \in T_p^1 M$, we have:*

$$\mathrm{Ric}\,(X) \leq \frac{1}{4}\Big\{n^2\|H\|^2 + (n-1)(c-3f^2) + (2-(n+1)\eta(X)^2)(c+f^2)\Big\} - \left[1+(n-2)\eta(X)^2\right] f'. \tag{14.6}$$

(3) *If M is anti-invariant, then for any $X \in T_p^1 M$, we have:*

$$\mathrm{Ric}(X) \leq \frac{1}{4}\Big\{n^2\|H\|^2 + (n-1)(c-3f^2) - \left(1+(n-2)\eta(X)^2\right)\left(c+f^2+4f'\right)\Big\}. \tag{14.7}$$

(4) *If $H(p) = 0$, a unit vector $X \in T_p M$ satisfies the equality case of* (14.5), (14.6), *and* (14.7) *if and only if $X \in N_p$.*
(5) *The equality case of* (14.5), (14.6), *and* (14.7) *holds identically for all vectors $X \in T_p^1 M$ if and only if either p is a totally geodesic point, or $n = 2$ and p is a totally umbilical point.*

For k-Ricci curvature invariant, Tripathi proved the following.

Theorem 14.8 ([193]) *Let M be an n-dimensional submanifold in a locally conformal almost cosymplectic manifold $\overline{M}(c)$ of pointwise constant ϕ-sectional curvature c such that $\xi \in TM$. Then, for an integer k, $2 \leq k \leq n$ and a point $p \in M$, we have:*

$$\|H\|^2 \geq \theta_k(p) - \frac{c-3f^2}{4} - \frac{3\,\|P\|^2\,(c+f^2)}{4n(n-1)} + \frac{2}{n}\left(\frac{c+f^2}{4} + f'\right).$$

Remark 14.9 Results similar to Theorems 14.6, 14.7, and 14.8 were also established by Arslan et al. [20] and Kim et al. [111].

For bi-slant submanifolds in locally conformal almost cosymplectic manifolds, D. A. Yoon obtained the following.

Theorem 14.10 ([211]) *Let M be an n-dimensional bi-slant submanifold satisfying $g(X, \phi Y) = 0$ for any $X \in \mathcal{D}_1$ and any $Y \in \mathcal{D}_2$, tangent to ξ, in a locally conformal almost cosymplectic space form $\overline{M}(c)$. Then*

(1) *For any $X \in T_p^1 M$ orthogonal to ξ, if $X \in \mathcal{D}_1$ we have:*

$$\mathrm{Ric}(X) \leq \frac{1}{4}\Big[n^2\|H\|^2 + (n-1)(c-3f^2) + \frac{3}{2}(c+f^2)\cos^2\theta_1 - 4\left(\frac{c+f^2}{4} + f'\right)\Big]. \tag{14.8}$$

(2) *For any* $X \in T_p^1 M$ *orthogonal to* ξ, *if* $X \in \mathcal{D}_2$ *we have:*

$$\begin{aligned}\mathrm{Ric}(X) \leq \frac{1}{4}\Big[&n^2\|H\|^2 + (n-1)(c-3f^2)\\ &+ \frac{3}{2}(c+f^2)\cos^2\theta_2 - 4\left(\frac{c+f^2}{4} + f'\right)\Big].\end{aligned} \tag{14.9}$$

(3) *If* $H(p) = 0$, *then a vector* $X \in T_p^1 M$ *orthogonal to* ξ *at* p *satisfies the equality case of* (14.8) *or* (14.9) *if and only if* $X \in \mathcal{N}_p$.

(4) *The equality case of* (14.8) *or* (14.9) *holds identically for all* $X \in T_p^1 M$ *orthogonal to* ξ *at* p *if and only if* p *is a totally geodesic point.*

Remark 14.11 As immediate consequences, Yoon [211] stated the corresponding inequalities for semi-slant, CR, invariant, and anti-invariant submanifolds in a locally conformal almost cosymplectic space form.

Remark 14.12 For submanifolds of a locally conformal almost cosymplectic manifold of pointwise constant ϕ-sectional curvature endowed with a semi-symmetric metric connection, see Özgür and Murathan's articles [159, 160].

15 Chen–Ricci Inequalities for Submanifolds in Kenmotsu Manifolds

A *Kenmotsu space form* $\overline{M}(c)$ of constant ϕ-sectional curvature c is a generalized Sasakian space form with $f_1 = \frac{c-3}{4}$ and $f_2 = f_3 = \frac{c+1}{4}$.

Arslan et al. [19] studied the Chen–Ricci inequality for submanifolds in a Kenmotsu space form and proved the following.

Theorem 15.1 *Let* M *be an* n*-dimensional submanifold in a Kenmotsu space form* $\overline{M}(c)$ *tangent to* ξ. *Then*

(1) *For any* $X \in T^1 M$ *orthogonal to* ξ, *we have:*

$$\mathrm{Ric}(X) \leq \frac{1}{4}\Big[n^2\|H\|^2 + (n-1)(c-3) + \frac{1}{2}(3\|PX\|^2 - 2)(c+1)\Big]. \tag{15.1}$$

(2) *If* $H(p) = 0$, *then a vector* $X \in T_p^1 M$ *orthogonal to* ξ *at* p *satisfies the equality case of* (15.1) *if and only if* $X \in \mathcal{N}_p$.

(3) *The equality case of* (15.1) *holds identically for all* $X \in T_p^1 M$ *orthogonal to* ξ *at* p *if and only if* p *is a totally geodesic point.*

For k-Ricci curvature invariant, we also have the following result.

Theorem 15.2 ([19]) *Let M be an n-dimensional submanifold in a Kenmotsu space form $\overline{M}(c)$ tangent to ξ. Then, for an integer k, $2 \leq k \leq n$ and a point $p \in M$, we have:*

$$\|H\|^2(p) \geq \theta_k(p) - \frac{c-3}{4} - \frac{(3\|P\|^2 - 2n + 2)(c+1)}{4n(n-1)}. \tag{15.2}$$

In [19], inequalities (15.1) and (15.2) were specialized to slant, invariant, and anti-invariant submanifolds as well.

Costache and Zamfir [66] improved Theorem 15.1 to the following.

Theorem 15.3 ([66]) *Let M be an $(n+1)$-dimensional special contact θ-slant submanifold in a Kenmotsu space form $\overline{M}^{2n+1}(c)$ tangent to ξ. Then, for any $X \in T^1M$ orthogonal to ξ, we have:*

$$\mathrm{Ric}(X) \leq \frac{(n+1)^2(n-1)}{4n}\|H\|^2 - 1 + \frac{(n-1)(c-3)}{4} + \frac{3(c+1)}{4}\cos^2\theta. \tag{15.3}$$

The equality case of (15.3) *holds for any $p \in M$ and any vector T_p^1M orthogonal to ξ if and only if either M is a totally geodesic submanifold, or $n = 2$ and M is a H-umbilical submanifold with ratio 3.*

Remark 15.4 Mustafa et al. [147] derived the Chen–Ricci inequality for warped product semi-slant submanifolds in Kenmotsu space forms. Khan et al. [106] derived the inequality for bi-warped product submanifolds in Kenmotsu space forms.

16 Chen–Ricci Inequalities for Submanifolds in (κ, μ)-Contact Space Forms

A contact metric manifold $(\overline{M}, \phi, \xi, \eta, g)$ is called a generalized (κ, μ)-space if its curvature tensor satisfies

$$\overline{R}(X, Y)\xi = \kappa[\eta(Y)X - \eta(X)Y] + \mu[\eta(Y)hX - \eta(X)hY]$$

for some functions κ, μ, where $h = \frac{1}{2}\mathcal{L}_\xi\phi$ and $\mathcal{L}$ is the Lie derivative. If both κ, μ are constant, then $\overline{M}$ is called a (κ, μ)-space. A (κ, μ)-space is called a (κ, μ)-*contact space form* if it has constant ϕ-sectional curvature.

16.1 Submanifolds in (κ, μ)-Contact Space Forms

If $\kappa = 1$, then a (κ, μ)-contact space form is a Sasakian space form. For $\kappa < 1$, K. Arslan et al. proved the following.

Theorem 16.1 ([17]) *Let M be an n-dimensional submanifold normal to ξ in a (κ,μ)-contact space form $\overline{M}^{2m+1}(c)$. Then*

(1) *For any $X \in T_p^1 M$, we have:*

$$\operatorname{Ric}(X) \leq \frac{1}{4}\left[n^2\|H\|^2 - \|hX\|^2 + 2(n-1)(1-\kappa) + 3(\kappa+1)\|PX\|^2\right]. \tag{16.1}$$

(2) *If $H(p) = 0$, then a vector $X \in T_p^1 M$ satisfies the equality case of* (16.1) *if and only if $X \in \mathcal{N}_p$.*
(3) *The equality case of* (16.1) *holds all $X \in T_p^1 M$ if and only if either p is a totally geodesic point, or $n = 2$ and p is a totally umbilical point.*

In [17], Arslan et al. also specialized for invariant and anti-invariant submanifolds.

Remark 16.2 Tripathi and Kim [197] and Tripathi [194] derived the Chen–Ricci inequality for C-totally real submanifold in a (κ,μ)-contact space form. Also, Tripathi [194] proved that invariant submanifolds of non-Sasakian (κ,μ)-spaces and non-Sasakian (κ,μ)-space forms always satisfied the Chen–Ricci inequality. Tripathi and Kim [197] also derived the corresponding inequality for k-Ricci curvature invariant for C-totally real submanifold in a (κ,μ)-contact space form.

Remark 16.3 Shukla and Tiwari [175] derived the Chen–Ricci inequality for submanifold tangent to ξ of a (κ,μ)-contact space form. In [175], they also derived the corresponding inequalities for bi-slant submanifolds tangent to ξ in a (κ,μ)-contact space form.

16.2 Submanifolds in Generalized (κ,μ)-Contact Space Forms

Carriazo et al. [33] defined a *generalized* (κ,μ)*-space form* as an almost contact metric manifold whose curvature tensor $\overline{R}$ satisfies

$$\begin{aligned}\overline{R}(X,Y)Z = &f_1R_1(X,Y)Z + f_2R_2(X,Y)Z + f_3R_3(X,Y)Z\\ &+ f_4R_4(X,Y)Z + f_5R_5(X,Y)Z + f_6R_6(X,Y)Z,\end{aligned}$$

where $f_1,\ldots,f_6$ are functions and $R_1,\ldots,R_6$ are tensor fields defined by

$$R_1(X, Y)Z = g(Y, Z)X - g(X, Z)Y,$$
$$R_2(X, Y)Z = g(X, \phi Z)\phi Y - g(Y, \phi Z)\phi X + 2g(X, \phi Y)\phi Z,$$
$$R_3(X, Y)Z = \eta(X)\eta(Z)Y - \eta(Y)\eta(Z)X + g(X, Z)\eta(Y)\xi - g(Y, Z)\eta(X)\xi,$$
$$R_4(X, Y)Z = g(Y, Z)hX - g(X, Z)hY + g(hY, Z)X - g(hX, Z)Y,$$
$$R_5(X, Y)Z = g(hY, Z)hX - g(hX, Z)hY + g(\phi hX, Z)\phi hY - g(\phi hY, Z)\phi hX,$$
$$R_6(X, Y)Z = \eta(X)\eta(Z)hY - \eta(Y)\eta(Z)hX + g(hX, Z)\eta(Y)\xi - g(hY, Z)\eta(X)\xi.$$

For the Chen–Ricci inequality of submanifolds in a generalized (κ, μ)-contact space form, we refer to [75, 76] by Faghfouri and Ghaffarzadeh.

16.3 Submanifolds in (κ, μ)-Contact Space Forms with Semi-symmetric Connections

For the Chen–Ricci inequality of submanifolds in generalized (κ, μ)-contact space forms endowed with semi-symmetric connections, see [2, 3, 207].

17 Chen–Ricci Inequalities for Submanifolds in S-Space Forms

A tensor field f of type $(1, 1)$ of rank $2m$ on a Riemannian manifold (M, g) is called a *f-structure* if it satisfies $f^3 + f = 0$ [208]. A Riemannian $(2m + s)$-manifold (M, g) equipped with an f-structure is called an *metric f-manifold* if there exist s global vector fields $\xi_1, \ldots, \xi_s$ on M such that

$$f\xi_\alpha = 0, \quad \eta_\alpha \circ f = 0, \quad f^2 = -I + \sum_{\alpha=1}^{s} \eta_\alpha \otimes \xi_\alpha,$$
$$g(X, Y) = g(fX, fY) + \sum_{\alpha=1}^{s} \eta_\alpha(X)\eta_\alpha(Y)$$

for $X, Y \in \mathfrak{X}(M)$, where $\eta_1, \ldots, \eta_s$ are the dual 1-forms of $\xi_1, \ldots, \xi_s$. Let F be the fundamental 2-form on M defined by $F(X, Y) = g(X, fY)$. Then a metric f-manifold is called *normal* if the Nijenhuis tensor $[f, f]$ of f satisfies

$$[f, f] = -2\sum_{\alpha=1}^{s} \xi_\alpha \otimes d\eta_\alpha.$$

A normal metric f-manifold is called an *S-manifold* if $F = d\eta_\alpha$ for $\alpha \in \{1, \ldots, s\}$.

A plane section on a metric f-manifold M is called an *f-section* if it is spanned by a unit vector X orthogonal to all structure vector fields $\xi_1, \dots, \xi_s$ and fX. The sectional curvature of an f-section is called an *f-sectional curvature*. An S-manifold is called an *S-space form* if it has constant f-sectional curvature c, denoted by $\widehat{M}(c)$. A submanifold M of a S-manifold $\widehat{M}$ is called an *integral submanifold* if $\eta_\alpha(X) = 0$ holds for all $\alpha \in \{1, \dots, s\}$ and all $X \in TM$.

The next theorem was proved by Kim et al. [109].

Theorem 17.1 *Let M be an n-dimensional integral submanifold of a S-space form $\widehat{M}(c)$. Then*

(1) *For any $X \in T_p^1 M$, we have:*

$$\mathrm{Ric}(X) \leq \frac{1}{4}\left[n^2\|H\|^2 + (n-1)(c+3s)\right]. \tag{17.1}$$

(2) *If $H(p) = 0$, then a vector $X \in T_p^1 M$ satisfies the equality case of* (17.1) *if and only if $X \in \mathcal{N}_p$.*
(3) *The equality case of* (17.1) *holds for all $X \in T_p^1 M$ if and only if either p is a totally geodesic point, or $n = 2$ and p is a totally umbilical point.*

Remark 17.2 In [109], Kim et al. specialized their result for totally real and C-totally real submanifolds. For submanifolds tangent to the structure vector fields in S-space forms in this respect, see also [110].

Remark 17.3 Fernández and Hans-Uber [78] established the Chen–Ricci inequality for slant submanifolds in S-space forms. Several applications of their inequality were also given. For further results for submanifolds in S-space forms or generalized S-space forms, see [23, 74, 78, 127].

18 Statistical Manifolds and Submanifolds

A statistical structure on a smooth manifold M is a pair (g, ∇) of a Riemannian metric g and a torsion-free affine connection ∇ such that the tensor field ∇g is totally symmetric, i.e.,

$$(\nabla_X g)(Y, Z) = (\nabla_Y g)(X, Z) = (\nabla_Z g)(X, Y)$$

for any vector fields $X, Y, Z \in \mathfrak{X}(M)$ (see [14]).

In the framework of Riemannian geometry, the notion of dual connection was firstly introduced by Amari: Let (M, g) be a Riemannian manifold and denote by ∇^g the Levi–Civita connection of g. Two affine connections ∇ and ∇^* on M are said to be *dual connections* with respect to g [14] if the following equation is satisfied:

$$X(g(Y, Z)) = g(\nabla_X Y, Z) + g(Y, \nabla^*_X Z),$$

for any $X, Y, Z \in \mathfrak{X}(M)$, and we call $(g; \nabla, \nabla^*)$ a *dualistic structure*.

We remark that $(\nabla^*)^* = \nabla$ and $\nabla + \nabla^* = 2\nabla^g$. If (g, ∇) is a statistical structure on M, then (g, ∇^*) is a statistical structure, too [14, 204].

We denote by R and R^* the curvature tensor fields of (g, ∇) and (g, ∇^*), respectively. They are connected by the relation

$$g(R(X, Y)Z, W) = -g(R^*(X, Y)W, Z).$$

A statistical structure (g, ∇) is said to be *of quasi-constant curvature* $c \in \mathbb{R}$ [26] if its curvature tensor is of the form

$$\begin{aligned} R(X, Y)Z &= a\,[g(Y, Z)X - g(X, Z)Y] \\ &+ b\,[\omega(Y)\omega(Z)X - g(X, Z)\omega(Y)\zeta + g(Y, Z)\omega(X)\zeta - \omega(X)\omega(Y)\zeta]\,, \end{aligned}$$

where a and b are smooth functions and ω is the dual 1-form of a unit vector field ζ. In particular, a statistical structure (g, ∇) is said to be *of constant curvature* $c \in \mathbb{R}$ if

$$R(X, Y)Z = c\,[g(Y, Z)X - g(X, Z)Y]\,.$$

Moreover, if $c = 0$, then it is called a *Hessian structure*.

If (g, ∇) is a statistical structure of quasi-constant curvature (in particular, of constant curvature c), then (g, ∇^*) is a statistical structure of quasi-constant curvature (in particular, of constant curvature c), too. Moreover, if (g, ∇) is a Hessian structure, then (g, ∇^*) is a Hessian structure, too [80].

On a Hessian manifold, let $\gamma := \nabla - \nabla^g$ and we define the *Hessian curvature* of (g, ∇) by [172]

$$Q(X, Y) := [\gamma_X, \gamma_Y],$$

for $X, Y \in \mathfrak{X}(M)$. Remark that

$$Q = \frac{1}{2}(R + R^*) - R^g.$$

A corresponding notion of sectional curvature can be defined on Hessian manifolds, too. Let π be a plane field and let $\{X, Y\}$ be an orthonormal basis of π. Define the *Hessian sectional curvature* by [172]

$$K(\pi) := g(Q(X, Y)Y, X).$$

A statistical structure (g, ∇) is said to be *of constant Hessian curvature* $c \in \mathbb{R}$ if its Hessian sectional curvature satisfies

$$Q(X, Y)Z = \frac{c}{2}\,[g(Y, Z)X + g(X, Y)Z]\,.$$

Remark that a manifold of constant Hessian curvature c is a space form of constant sectional curvature $-\frac{c}{4}$.

Let M be an n-dimensional submanifold of an $(n+k)$-manifold Riemannian $(\widetilde{M}, \tilde{g})$. If $(\tilde{g}; \widetilde{\nabla}, \widetilde{\nabla}^*)$ is a dualistic structure on $\widetilde{M}$, then, for any $X, Y \in \mathfrak{X}(M)$, according to [204], the corresponding Gauss formulas are:

$$\widetilde{\nabla}_X Y = \nabla_X Y + h(X, Y), \quad \widetilde{\nabla}^*_X Y = \nabla^*_X Y + h^*(X, Y),$$

where h and h^* are symmetric and bilinear, called the *imbedding curvature tensors* of M in $\widetilde{M}$ for ∇ and ∇^*, respectively, and ∇ and ∇^* are called the *induced connections*. In [204], it is also proved that the induced structure $(g; \nabla, \nabla^*)$ on M is a dualistic structure, too.

We define the *mean curvature vector fields* by

$$H = \text{Trace}(h) = \frac{1}{n}\sum_{i=1}^{n} h(E_i, E_i) = \frac{1}{n}\sum_{\alpha=1}^{k}\sum_{i=1}^{n} h_{ii}^{\alpha} E_{n+\alpha},$$

$$h_{ij}^{\alpha} = \tilde{g}(h(E_i, E_j), E_{n+\alpha})$$

and

$$H^* = \text{Trace}(h^*) = \frac{1}{n}\sum_{i=1}^{n} h^*(E_i, E_i) = \frac{1}{n}\sum_{\alpha=1}^{k}\sum_{i=1}^{n} h_{ii}^{*\alpha} E_{n+\alpha},$$

$$h_{ij}^{*\alpha} = \tilde{g}(h^*(E_i, E_j), E_{n+\alpha}).$$

Let Ric and Ric^g denote the Ricci tensors of the connections ∇ and ∇^g induced by $\widetilde{\nabla}$ and $\widetilde{\nabla}^{\tilde{g}}$, respectively. For $X \in \mathfrak{X}(M)$, let $\widetilde{K}^g(X \wedge \cdot)$ be the sectional curvature function of $\widetilde{M}$ with respect to the Levi–Civita connection $\widetilde{\nabla}^{\tilde{g}}$, restricted to 2-plane sections of TM tangent to X.

19 Chen–Ricci Inequalities for Submanifolds in Statistical Manifolds

19.1 Submanifolds of Statistical Space Forms

Theorem 19.1 ([26]) *Let $(\widetilde{M}, \tilde{g}, \widetilde{\nabla})$ be a statistical manifold of quasi-constant curvature and (M, g, ∇) an n-dimensional statistical submanifold of $\widetilde{M}$. Then, for any $X \in T^1M$, we have:*

$$\begin{aligned}\operatorname{Ric}(X) \geq & 2\operatorname{Ric}^g(X) - \frac{n^2}{8}(\|H\|^2 + \|H^*\|^2) \\ & + (n-1)a + b + (n-2)b\omega(X)^2 - 2\sum_{i=2}^{n} \widetilde{K}^g(X \wedge E_i).\end{aligned}$$

Theorem 19.2 ([21]) *Let $(\widetilde{M}, \tilde{g}, \widetilde{\nabla})$ be a statistical manifold of constant curvature c and (M, g, ∇) an n-dimensional statistical submanifold of $\widetilde{M}$. Then, for any $X \in T^1M$, we have:*

$$\begin{aligned}\operatorname{Ric}(X) \geq & 2\operatorname{Ric}^g(X) - \frac{n^2}{8}(\|H\|^2 + \|H^*\|^2) \\ & + (n-1)c - 2(n-1)\max \widetilde{K}^g(X \wedge E_i).\end{aligned}$$

In particular, for a minimal submanifold, we get the following.

Corollary 19.3 ([21]) *If $(\widetilde{M}, \tilde{g}, \widetilde{\nabla})$ is a statistical manifold of constant curvature c and (M, g, ∇) is a minimal n-dimensional statistical submanifold of $\widetilde{M}$, then, for any $X \in T^1M$, we have:*

$$\operatorname{Ric}(X) \geq 2\operatorname{Ric}^g(X) - \frac{n^2}{4}\tilde{g}(H, H^*) + (n-1)c - 2(n-1)\max \widetilde{K}^g(X \wedge E_i).$$

19.2 Submanifolds of Hessian Space Forms

Theorem 19.4 ([132]) *Let $(\widetilde{M}, \tilde{g}, \widetilde{\nabla})$ be a statistical manifold of constant Hessian curvature c and (M, g, ∇) an n-dimensional statistical submanifold of $\widetilde{M}$. Then, for any $X \in T^1M$, we have:*

$$\operatorname{Ric}(X) \geq \operatorname{Ric}^g(X) - \frac{n^2}{8}(\|H\|^2 + \|H^*\|^2) + \frac{(n-1)c}{2}.$$

20 Chen–Ricci Inequalities for Submanifolds in Almost Complex Statistical Manifolds

20.1 Almost Complex Statistical Manifolds

(M, J, g, ∇) is called a *holomorphic statistical manifold* [81] if it is a Kähler and a statistical manifold and the fundamental 2-form is parallel with respect to the Levi–Civita connection ∇.

A holomorphic statistical manifold is said to be *of constant holomorphic sectional curvature c* if its curvature is of the form [81]

$$R(X,Y)Z = \frac{c}{4}[g(Y,Z)X - g(X,Z)Y + g(JY,Z)JX - g(JX,Z)JY - 2g(JX,Y)JZ].$$

A statistical submanifold M in a holomorphic statistical manifold is said to be a *CR-statistical submanifold* [81] if there exist a holomorphic distribution $\mathcal{D}$ (i.e., $J\mathcal{D} = \mathcal{D}$) and a totally real distribution $\mathcal{D}^\perp$ (i.e., $J\mathcal{D}^\perp \subset T^\perp M$) such that

$$TM = \mathcal{D} \oplus \mathcal{D}^\perp.$$

For any vector X tangent to a submanifold M, we put

$$JX = PX + FX,$$

where PX and FX denote the tangential component and normal component of JX, respectively.

20.2 CR-Statistical Submanifolds of Holomorphic Statistical Manifolds

Theorem 20.1 ([178]) *Let $(\widetilde{M}, J, \tilde{g}, \widetilde{\nabla})$ be a holomorphic statistical manifold of constant holomorphic sectional curvature c and (M, g, ∇) an n-dimensional statistical submanifold of $\widetilde{M}$. Then, for any $X \in T^1M$, we have:*

$$\mathrm{Ric}(X) \geq 2\mathrm{Ric}^g(X) - \frac{n^2}{8}(\|H\|^2 + \|H^*\|^2) - \frac{3c}{4}\|PX\|^2 - \frac{(n-1)c}{4}.$$

In particular, for a CR-statistical submanifold, we get the following.

Corollary 20.2 ([176]) *If $(\widetilde{M}, J, \tilde{g}, \widetilde{\nabla})$ is a holomorphic statistical manifold of constant holomorphic sectional curvature c and (M, g, ∇) is an n-dimensional CR-statistical submanifold of $\widetilde{M}$, then*

(1) *for any unit vector $X \in \mathcal{D}$, we have:*

$$\mathrm{Ric}(X) \geq 2\mathrm{Ric}^g(X) - \frac{n^2}{8}(\|H\|^2 + \|H^*\|^2) - \frac{(n+2)c}{4};$$

(2) *for any unit vector $X \in \mathcal{D}^\perp$, we have:*

$$\mathrm{Ric}(X) \geq 2\mathrm{Ric}^g(X) - \frac{n^2}{8}(\|H\|^2 + \|H^*\|^2) - \frac{(n-1)c}{4}.$$

20.3 Submanifolds of Kähler-Like Statistical Space Forms

Motivated by one of the problems proposed by Vîlcu and Vîlcu [200], Aquib proved the Chen–Ricci inequality in [15] for totally real statistical submanifold in quaternionic Kähler-like statistical space form.

21 Chen–Ricci Inequalities for Submanifolds in Almost Contact Metric Statistical Manifolds

21.1 Almost Contact Metric Statistical Manifolds

In an almost contact metric manifold, Takano [188] introduced an additional (1, 1)-tensor field ϕ^* satisfying

$$g(\phi^* X, Y) = -g(X, \phi Y),$$

for $X, Y \in \mathfrak{X}(M)$, and called (M, ϕ, ξ, η, g) an *almost contact metric manifold of certain kind* [189]. Moreover, in [188], he defined the notion of *Sasaki-like statistical manifold* $(M, \phi, \xi, \eta, g, \nabla)$ if it is an almost contact metric and a statistical manifold and it satisfies $\nabla_X \xi = -\phi X$ and $(\nabla_X \phi)Y = g(X, Y)\xi - \eta(Y)X$, for any $X, Y \in \mathfrak{X}(M)$.

In a Sasaki-like statistical manifold, the curvature of the statistical structure (g, ∇) is given by

$$\begin{aligned} R(X, Y)Z &= \frac{c+3}{4}[g(Y, Z)X - g(X, Z)Y] \\ &+ \frac{c-1}{4}\{\eta(X)\eta(Z)Y - \eta(Y)\eta(Z)X + g(X, Z)\eta(Y)\xi - g(Y, Z)\eta(X)\xi \\ &+ g(X, \phi Z)\phi Y - g(Y, \phi Z)\phi X + \big[g(X, \phi Y) - g(\phi X, Y)\big]\phi Z\}, \end{aligned}$$

where c is a constant.

Furuhata defined a *Kenmotsu statistical manifold* $(M, \phi, \xi, \eta, g, \nabla)$ [82] if it is a Kenmotsu and a statistical manifold and it satisfies $\gamma_X \phi Y = -\phi(\gamma_X Y)$, for any $X, Y \in \mathfrak{X}(M)$, where $\gamma := \nabla - \nabla^g$.

A Kenmotsu statistical manifold is said to be *of constant ϕ-sectional curvature c* if its curvature is of the form

$$\begin{aligned} R(X, Y)Z = &\frac{c-3}{4}[g(Y, Z)X - g(X, Z)Y] \\ &+ \frac{c+1}{4}[g(\phi Y, Z)\phi X - g(\phi X, Z)\phi Y - 2g(\phi X, Y)\phi Z \\ &- \eta(Y)\eta(Z)X + \eta(X)\eta(Z)Y + \eta(Y)g(X, Z)\xi]. \end{aligned}$$

For any vector X tangent to a submanifold M of an almost contact metric statistical manifold, we put

$$\phi X = PX + FX, \quad \phi^* X = P^* X + F^* X,$$

where PX, P^*X are the tangential and FX, F^*X are the normal components of ϕX and $\phi^* X$, respectively.

21.2 Submanifolds of Sasaki-Like Statistical Manifolds

Theorem 21.1 ([25]) *Let $(\widetilde{M}, \tilde{g}, \widetilde{\nabla})$ be a Sasaki-like statistical manifold and (M, g, ∇) an n-dimensional statistical submanifold of $\widetilde{M}$ such that ξ is tangent to M. Then*

(1) *If M is an invariant submanifold, then, for any $X \in T^1M$, we have:*

$$\begin{aligned}\mathrm{Ric}(X) \geq 2\mathrm{Ric}^g(X) - \frac{n^2}{8}(\|H\|^2 + \|H^*\|^2) + \frac{(n-1)(c+2)}{4}\\ - \frac{c-1}{4}\Big[1 + (n-1)\eta(X)^2 + \mathrm{Trace}(P)g(PX, X) - \|X\|^2 - \|PX\|^2 - \|P^*X\|^2\Big]\\ - 2\sum_{i=2}^{n} \widetilde{K}^g(X \wedge E_i).\end{aligned}$$

(2) *If M is an anti-invariant submanifold, then, for any $X \in T^1M$, we have:*

$$\begin{aligned}\mathrm{Ric}(X) \geq 2\mathrm{Ric}^g(X) - \frac{n^2}{8}(\|H\|^2 + \|H^*\|^2)\\ + \frac{(n-1)(c+2)}{4} - \frac{c-1}{4}\Big[1 + (n-2)\eta(X)^2\Big] - 2\sum_{i=2}^{n} \widetilde{K}^g(X \wedge E_i).\end{aligned}$$

21.3 Submanifolds of Kenmotsu Statistical Manifolds

Theorem 21.2 ([181]) *Let $(\widetilde{M}, \tilde{g}, \widetilde{\nabla})$ be a Kenmotsu statistical manifold of constant ϕ-sectional curvature c and (M, g, ∇) an n-dimensional statistical submanifold of $\widetilde{M}$. Then, for any $X \in T^1M$, we have:*

$$\begin{aligned}\mathrm{Ric}(X) \geq 2\mathrm{Ric}^g(X) - \frac{n^2}{8}(\|H\|^2 + \|H^*\|^2)\\ + \frac{3(c+2)}{4}\|PX\|^2 + \frac{n-1}{4}\Big[(c+1)(1 - \eta(X)^2) - 4\Big].\end{aligned}$$

22 Chen–Ricci Inequalities for Riemannian Maps

Riemannian maps between Riemannian manifolds were introduced in 1992 by Fischer [79] as a generalization of isometric immersions and Riemannian submersions as follows: Let $F : (M, g_M) \to (N, g_N)$ be a map between two Riemannian manifolds such that $0 < \text{rank } F < \min\{n, m\}$ with $n = \dim M$ and $m = \dim N$. Let $\ker F_*$ denote the kernel space of F_* and let $\mathcal{H} = (\ker F_*)^\perp$ be the orthogonal complementary space of $\ker F_*$. Then we have the decomposition $TM = \ker F_* \oplus \mathcal{H}$.

Let range F_* be the range of F_* and $(\text{range } F_*)^\perp$ the orthogonal complementary space of range F_* in TN. Since $\text{rank } F < \min\{n, m\}$, we have $(\text{range } F_*)^\perp \neq \{0\}$. Hence, TN has the decomposition: $TN = (\text{range } F_*) \oplus (\text{range } F_*)^\perp$.

Now, a smooth map $F : (M, g_M) \to (N, g_N)$ is called a *Riemannian map* at $p \in M$ if the horizontal restriction $\overset{h}{F}_{*p} : (\ker F_{*p})^\perp \to (\text{range } F_{*p})$ is a linear isometry between the inner product spaces $((\ker F_{*p})^\perp, g_M(p)|_{(\ker F_{*p})^\perp})$ and $(\text{range } F_{*p}, g_N(q)|_{(\text{range } F_{*p})})$ with $q = F(p)$. It is easy to verify that isometric immersions and Riemannian submersions are Riemannian maps with $\ker F_* = \{0\}$ and $(\text{range } F_*)^\perp = \{0\}$, respectively.

For a Riemannian map $F : (M, g_M) \to (N, g_N)$, the second fundamental form of F is defined by

$$(\nabla F_*)(X, Y) = \nabla^F_X F_*(Y) - F_*(\nabla^M_X Y)$$

for $X, Y \in \mathfrak{X}(M)$, where ∇ is the induced connection from the Levi–Civita connection ∇^M on M. Let R^M and R^N denote the curvature tensors for (M, g_M) and (N, g_N), respectively.

Let $\{e_1, \ldots, e_r\}$ and $\{u_{r+1}, \ldots, u_n\}$ be orthonormal bases of $\mathcal{H}_p = \ker F^\perp_{*p}$ and $(\text{Im } F_{*p})^\perp$, $p \in M$, respectively. Then the Ricci curvature $\text{Ric}^{\mathcal{H}}$ on the horizontal space $\mathcal{H}_p$ is defined by

$$\text{Ric}^{\mathcal{H}}(v) = \sum_{j=1}^{r} g_M(R^M(e_j, v)v, e_j)$$

for $0 \neq v \in \mathcal{H}_p$. Consider $\zeta^{\mathcal{H}}$ defined by

$$\zeta^{\mathcal{H}\alpha}_{ij} = g_N((\nabla F_*)(e_i, e_j), u_\alpha), \quad 1 \leq i, j \leq r; \quad r + 1 \leq \alpha \leq n.$$

We put $\text{Trace}\, \zeta^{\mathcal{H}} = \sum_{j=1}^{r} (\nabla F_*)(e_j, e_j)$.

22.1 Riemannian Maps into Real Space Forms

In [113], Lee et al., extended Theorem 3.1 to the following.

Theorem 22.1 *Let $F : (M, g_M) \to (\widetilde{M}^m(c), g_N)$ be a Riemannian map from a Riemannian manifold M into a real space form $\widetilde{M}^m(c)$ with* $\text{rank } F = r < m$. *Then*

(1) *For any unit horizontal vector field v on M, we have:*

$$\mathrm{Ric}^{\mathcal{H}}(v) \leq (r-1)c + \frac{1}{4}\|\mathrm{Trace}\, \zeta^{\mathcal{H}}\|^2. \tag{22.1}$$

(2) *The equality case of* (22.1) *holds for $v \in \mathcal{H}_p$ if and only if*

$$\begin{cases} (\nabla F_*)(u, v) = 0 & \text{for all } u \in \mathcal{H}_p \text{ orthogonal to } v. \\ (\nabla F_*)(v, v) = \frac{1}{2}\mathrm{Trace}\, \zeta^{\mathcal{H}}. \end{cases}$$

(3) *The equality case of* (22.1) *holds identically for all $v \in \mathcal{H}_p$ if and only if either $\zeta^{\mathcal{H}} = 0$, or $r = 2$ and $\zeta^{\mathcal{H}\alpha}_{11} = \zeta^{\mathcal{H}\alpha}_{22}$, $\alpha = r+1, \ldots, m$.*

The following is an immediate consequence of Theorem 22.1.

Corollary 22.2 *Let $F : (M, g_M) \to \mathbb{E}^m$ be a Riemannian map from a Riemannian manifold M into a Euclidean space such that* rank $F = r < m$. *Then we have:*

$$\|\mathrm{Trace}\, \zeta^{\mathcal{H}}\|^2 \geq 4 \max_v \mathrm{Ric}^{\mathcal{H}}(v),$$

where v runs over all unit horizontal vectors at $p \in M$.

22.2 Anti-invariant and Lagrangian–Riemannian Maps

A Riemannian map $F : M \to \widetilde{M}^m$ from a Riemannian manifold M into an almost Hermitian manifold $\widetilde{M}^m$ is called *anti-invariant* at a point $p \in M$ if $J(\mathrm{range}\, F_{*p}) \subset (\mathrm{range}\, F_{*p})^{\perp}$. If F is anti-invariant for every $p \in M$, then F is called an *anti-invariant Riemannian map*. In particular, if $n = m$, then M becomes a Lagrangian submanifold. An anti-invariant Riemannian map $F : M \to \widetilde{M}^m$ is called a *Lagrangian–Riemannian map* if rank $F = \dim\, (\mathrm{range}\, F_*)^{\perp}$. Obviously, every Lagrangian submanifold is a Lagrangian–Riemannian map.

For anti-invariant and Lagrangian–Riemannian maps, Lee et al., proved in [113] the following two results.

Theorem 22.3 *Let $F : (M, g_M) \to (\widetilde{M}^m(c), g_N)$ be an anti-invariant Riemannian map from a Riemannian manifold M into a complex space form $\widetilde{M}^m(c)$ with* rank $F = r$. *Then*

(1) *For any unit horizontal vector field v on M, we have:*

$$\mathrm{Ric}^{\mathcal{H}}(v) \leq \frac{(r-1)c}{4} + \frac{1}{4}\|\mathrm{Trace}\, \zeta^{\mathcal{H}}\|^2. \tag{22.2}$$

(2) *The equality case of* (22.2) *holds for $v \in \mathcal{H}_p$ if and only if*

$$\begin{cases} (\nabla F_*)(u,v) = 0 & \textit{for all } u \in \mathcal{H}_p \textit{ orthogonal to } v. \\ (\nabla F_*)(v,v) = \frac{1}{2}\mathrm{Trace}\,\zeta^{\mathcal{H}}. \end{cases}$$

(3) *The equality case of* (22.2) *holds identically for all* $v \in \mathcal{H}_p$ *if and only if either* $\zeta^{\mathcal{H}} = 0$, *or* $r = 2$ *and* $\zeta^{\mathcal{H}\alpha}_{11} = \zeta^{\mathcal{H}\alpha}_{22}$, $\alpha = r+1, \ldots, m$.

Theorem 22.4 *Let* $F : (M, g_M) \to (R^m(c), g_N)$ *be a Lagrangian–Riemannian map from a Riemannian manifold* M *into a complex space form* $R^m(c)$ *of complex dimension* $m > 1$. *Then, for any unit horizontal vector* v *at* $p \in M$, *we have:*

$$\mathrm{Ric}^{\mathcal{H}}(v) \leq \frac{(m-1)c}{4} + \frac{m-1}{4m}\|\mathrm{Trace}\,\zeta^{\mathcal{H}}\|^2. \tag{22.3}$$

The equality case of (22.3) *holds for any unit horizontal vector at* p *if and only if either* $\zeta^{\mathcal{H}} = 0$, *or* $m = 2$ *and*

$$(\nabla F_*)(e_1, e_1) = 3\mu J e_1, \quad (\nabla F_*)(e_2, e_2) = \mu J e_1, \quad (\nabla F_*)(e_1, e_2) = \mu J e_1, \tag{22.4}$$

for some function μ *with respect to some orthonormal local frame field* $\{e_1, e_2\}$.

Remark 22.5 If M is a Lagrangian submanifold for the Lagrangian–Riemannian map F, then Theorem 22.4 reduces to a result of [68, 153]. Moreover, in this case, condition (22.4) means that M is a Lagrangian H-umbilical submanifolds with ratio 3 in the sense of [39, 40, 61].

23 Chen–Ricci Inequalities for Riemannian Submersions

Let M and B be Riemannian manifolds with $n = \dim M > \dim B = b > 0$. A *Riemannian submersion* $\pi : M \to B$ is a mapping of M onto B satisfying the two axioms: (S1) π *has maximal rank* and (S2) *the differential* π_* *preserves lengths of horizontal vectors*.

For each $p \in B$, $\pi^{-1}(p)$ is an $(n-b)$-dimensional submanifold of M. The submanifolds $\pi^{-1}(p)$, $p \in B$, are called fibers. A vector field on M is called *vertical* if it is always tangent to fibers, and *horizontal* if it is always orthogonal to fibers. We use the corresponding terminology for individual tangent vectors as well. Let $\mathcal{H}$ and $\mathcal{V}$ denote the projections of tangent spaces of M onto the subspaces of horizontal and vertical vectors, respectively.

For a Riemannian submersion $\pi : M \to B$, there exist two natural $(1,2)$-tensors T and A on M defined as follows: For vector fields E, F tangent to M, the tensor T is defined by (see, for instance, [152])

$$T_E F = \mathcal{H}\nabla_{\mathcal{V}E}(\mathcal{V}F) + \mathcal{V}\nabla_{\mathcal{H}E}(\mathcal{H}F).$$

The other tensor A, known as the *integrability tensor*, is defined by

$$A_E F = \mathcal{V}\nabla_{\mathcal{H}E}(\mathcal{H}F) + \mathcal{H}\nabla_{\mathcal{H}E}(\mathcal{V}F).$$

23.1 A Submersion Invariant and Its Application

For a Riemannian submersion $\pi : M \to B$, the first author introduced in [49] a submersion invariant on M, denoted by $\breve{A}_\pi$, defined by

$$\breve{A}_\pi = \sum_{i=1}^{b} \sum_{s=b+1}^{m} \|A_{u_i} v_s\|^2, \quad b = \dim B, \ m = \dim M,$$

where $u_1, \ldots, u_b$ are orthonormal basic horizontal vector fields and $v_{b+1}, \ldots, v_m$ are orthonormal vertical vector fields on M.

The first author proved in [49] the following sharp inequality.

Theorem 23.1 *Let $\pi : M \to B$ be a Riemannian submersion with totally geodesic fibers. Then, for any isometric immersion ϕ of M into a Riemannian manifold $\widetilde{M}$, we have:*

$$\breve{A}_\pi \leq \frac{m^2}{4}\|H\|^2 + b(m-b) \max \widetilde{K},$$

where $b = \dim B$, $m = \dim M$, $\|H\|^2$ is the squared mean curvature of ϕ, and $\max \widetilde{K}$ denotes the maximum value of the sectional curvature function of $\widetilde{M}$.

A Riemannian submersion $\pi : M \to B$ is called *trivial* if both horizontal and vertical distributions of the submersion are totally geodesic distributions (see [49]).

The following non-existence result is an immediate consequence of Theorem 23.1.

Theorem 23.2 *If a Riemannian manifold admits a non-trivial Riemannian submersion with totally geodesic fibers, then it cannot be isometrically immersed in any Riemannian manifold of non-positive sectional curvature as a minimal submanifold.*

23.2 Anti-invariant Riemannian Submersion from Space Forms

The Chen–Ricci inequality (3.1) has been extended by several authors for anti-invariant Riemannian submersion from space forms onto Riemannian manifolds.

For anti-invariant Riemannian submersions from a complex space form to a Riemannian manifold, Gündüzalp and Polat proved the following.

Theorem 23.3 ([91]) *Let* $\pi : (\widetilde{M}(c), g) \to (B, g_B)$ *be an anti-invariant Riemannian submersion from a complex space form of constant holomorphic sectional curvature* c *to a Riemannian manifold. Then the Ricci curvature* $\widehat{\text{Ric}}$ *of the vertical distribution satisfies*

$$\widehat{\text{Ric}}(U) \geq \frac{c}{4}(r-1) - \frac{r^2}{4}\|H\|^2, \tag{23.1}$$

for any unit vertical vector field U, *where* r *is the rank of vertical distribution and* H *is the mean curvature vector of fiber.*

Gündüzalp and Polat also discussed in [91] the necessary and sufficient conditions for the inequality (23.1) to be an equality.

In [27], Aytimur and Özgür derived the Chen–Ricci inequalities on the vertical and horizontal distributions for anti-invariant Riemannian submersions from Sasakian space forms. In particular, they proved the following.

Theorem 23.4 ([27]) *Let* $F : (\widetilde{M}(c), g) \to (B, g_B)$ *be an anti-invariant Riemannian submersion from a Sasakian space form* $\widetilde{M}(c)$ *of constant* ϕ*-sectional curvature* c *to a Riemannian manifold. Then the Ricci curvature* Ric^* *of the horizontal distribution satisfies*

$$\text{Ric}^*(X) \geq \frac{c+3}{2}(n-1) - \frac{3}{4}(c-1)\|CX\|^2,$$

for any unit horizontal vector field X, *where* $n = \dim \mathcal{H}$ *and* CX *is the horizontal component of* ϕX.

Very recently, Aytimur [24] studied anti-invariant Riemannian submersions from cosymplectic space forms to Riemannian and derived the Chen–Ricci inequality for such submersions. She proved the following two results.

Theorem 23.5 ([24]) *Let* $\pi : (\widetilde{M}(c), g) \to (B, g_B)$ *be an anti-invariant Riemannian submersion from a cosymplectic space form* $\widetilde{M}(c)$ *of constant* ϕ*-sectional curvature* c *onto a Riemannian manifold such that the characteristic vector field* ξ *is vertical. Then, for any unit vertical vector field* U, *we have:*

$$\widehat{\text{Ric}}(U) \geq \frac{c}{4}(r-2)\left[1-\eta(U)^2\right] - \frac{r^2}{4}\|H\|^2, \tag{23.2}$$

where r *is the rank of the vertical distribution and* H *is the mean curvature vector of fiber.*

Theorem 23.6 ([24]) *Let* $\pi : (\widetilde{M}(c), g) \to (B, g_B)$ *be an anti-invariant Riemannian submersion from a cosymplectic space form* $\widetilde{M}(c)$ *onto a Riemannian manifold such that* ξ *is vertical. Then, for any unit horizontal vector field* X, *we have:*

$$\mathrm{Ric}^*(X) \leq \frac{c}{4}\Big[(n-1) - 3\|CX\|^2\Big], \tag{23.3}$$

where n is the rank of the horizontal distribution.

In the same article, H. Aytimur also obtained the necessary and sufficient conditions for either inequality (23.2) or (23.3) to be an equality.

For further results in this direction, we refer to [24, 27, 85, 91].

23.3 Hemi-slant Submersion from Complex Space Forms

Let π be a Riemannian submersion from an almost Hermitian manifold (M, J, g_M) onto a Riemannian manifold (B, g_B). Then π is called a *hemi-slant submersion* [191] if there is a distribution $\mathcal{D}^\perp \subset \ker \pi_*$ such that

$$\ker \pi_* = \mathcal{D}^\perp \oplus \mathcal{D}^\theta, \quad J\mathcal{D}^\perp \subseteq (\ker \pi_*)^\perp$$

and the angle $\theta = \theta(X)$ between JX and the space $(\mathcal{D}^\theta)_q$ is constant for non-zero vector $X \in (\mathcal{D}^\theta)_q$ and $q \in M$, where $\mathcal{D}^\theta$ is the orthogonal complement of $\mathcal{D}^\perp$ in $\ker \pi_*$. In this case, θ is called the *hemi-slant angle* of π. The hemi-slant submersion is called *proper* if $\mathcal{D}^\perp \neq \{0\}$ and $\theta \neq 0, \frac{\pi}{2}$.

Recently, Akyol et al. [4] established some Chen–Ricci type inequalities involving the Ricci curvature of the vertical and the horizontal distributions for hemi-slant submersions having the total space a complex space form. In particular, they proved the following two results in [4].

Theorem 23.7 *Let $\pi : (\widetilde{M}(c), g) \to (B, g_B)$ be a proper hemi-slant Riemannian submersion from a complex space form $\widetilde{M}(c)$ of constant holomorphic sectional curvature c onto a Riemannian manifold. Suppose U is a unit vertical vector field. Then*

(1) *If U is in $\mathcal{D}^\perp$, we have:*

$$\widehat{\mathrm{Ric}}(U) \geq \frac{c}{4}(r-1) - \frac{r^2}{4}\|H\|^2,$$

where $\widehat{\mathrm{Ric}}$ is the Ricci curvature of the vertical distribution, H the mean curvature vector of any fiber, and r the rank of the vertical distribution.

(2) *If U is in $\mathcal{D}^\theta$, we have:*

$$\widehat{\mathrm{Ric}}(U) \geq \frac{c}{4}(r - 1 + 3\cos^2\theta) - \frac{r^2}{4}\|H\|^2.$$

Theorem 23.8 *Let $\pi : (\widetilde{M}(c), g) \to (B, g_B)$ be a proper hemi-slant Riemannian submersion from a complex space form $\widetilde{M}(c)$ onto a Riemannian manifold. If X is a unit horizontal vector field, then the Ricci curvature* Ric^* *of the horizontal distribution satisfies*

$$\mathrm{Ric}^*(X) \geq \frac{c}{4}\Big[(n-1) - 3\|CX\|^2\Big],$$

where CX denotes the $(\ker \pi_*)$*-component of JX.*

The equality case holds identically for all unit horizontal vector fields if and only if the horizontal distribution is integrable.

23.4 Statistical Submersions Between Statistical Manifolds

Siddiqui et al., defined and investigated statistical submersions between statistical manifolds in [177]. In the same article, they established the Chen–Ricci inequality for the vertical distributions of statistical submersions.

24 Chen–Ricci Inequalities for Light-Like and Null Hypersurfaces

Let (M, g) be a hypersurface of a pseudo-Riemannian $(m+2)$-manifold $(\widetilde{M}, \tilde{g})$ with index q $(1 \leq q \leq m+1)$. For a point $p \in M$, the tangent space T_pM is a hyperplane of the pseudo-Euclidean space $(T_p\widetilde{M}, \tilde{g}_p)$. Let us consider

$$(T_p\widetilde{M})^{\perp} = \{X_p \in T_p\widetilde{M} : \tilde{g}_p(X_p, Y_p) = 0, \forall Y_p \in T_pM\}$$

and let us put

$$\mathrm{Rad}\,(T_pM) = T_pM \cap (T_p\widetilde{M})^{\perp},$$

which is called *radical distribution*. The complementary non-degenerate vector bundle $S(TM)$ of the radical distribution $\mathrm{Rad}\,(TM)$ in TM is called the *screen bundle* of M. The hypersurface M is called a *light-like hypersurface* of $\widetilde{M}$ if $\mathrm{Rad}\,(T_pM) \neq 0$ for every $p \in M$. The pseudo-Riemannian metric $\tilde{g}$ induces a symmetric tensor field g of type $(0, 2)$ on M. And g has a constant rank m and $\mathrm{Rad}\,(T_pM) = (T_p\widetilde{M})^{\perp}$ for each $p \in M$.

The dimension r of fibers of $\mathrm{Rad}\,(TM)$ is called the degree of nullity of M. For $r = 0$, M is non-degenerate. For $r > 0$, M is called *r-null* if rank $(\mathrm{Rad}\,(TM)) = r$ at every point $p \in M$. A *null hypersurface* of $\widetilde{M}$ is a submanifold of codimension one with $r = 1$ (see [73]). A light-like hypersurface (M, g) of a pseudo-Riemannian manifold $(\widetilde{M}, \tilde{g})$ is called *screen locally conformal* if the shape operator A of M and the shape operator A^* of $S(TM)$ are related by $A = \varphi A^*$, where φ is a non-vanishing function on a neighborhood U on M. In particular, M is called *screen homothetic* if φ is a non-zero constant (see [88]).

Remark 24.1 Gülbahar et al., derived in [88] the Chen–Ricci inequality for screen homothetic light-like hypersurfaces in Lorentzian manifolds. In [165], Poyraz et al., investigated Chen-like inequalities for light-like hypersurface of a Lorentzian product manifold endowed with semi-symmetric non-metric connection. In another article [164], Poyraz and Yaşar established the Chen–Ricci inequality for light-like hypersurface of a Lorentzian product manifold endowed with quarter-symmetric non-metric connection.

Let M be a null hypersurface of a Lorentzian manifold. A *rigging* for M is a vector field L defined on some open set containing M such that $L_p \notin T_pM$ for each $p \in M$ (see [92]). An important property of a rigging is that it allows a definition of geometric objects globally on M.

Remark 24.2 For null hypersurfaces of pseudo-Riemannian manifolds, Ménédore derived in [126] a generalized Chen–Ricci inequality for closed normalized null hypersurfaces with rigged vector field in a pseudo-Riemannian manifold. Also, Pal Singh et al., studied in [161] null hypersurfaces of a Lorentzian manifold with a closed rigging for the hypersurface. In particular, they established a generalized Chen–Ricci inequality for a screen homothetic null hypersurface of a Lorentzian manifold with a closed rigging for the null hypersurface.

25 Chen–Ricci Inequalities for Warped Product Submanifolds

In recent years, there are several articles investigating the Ricci curvature of a warped product submanifold in some model spaces in terms of Ricci curvature, the squared mean curvature, and the warping function of the warped product submanifolds. For results in this respect, we refer to [10, 11, 13, 16, 102, 124].

Also, there are several articles which investigated Chen–Ricci inequalities for submanifolds in product manifolds or in warped product manifolds. For such results in this respect, we refer to [89, 107, 163, 190].

References

1. E. Abedi, R.B. Ziabari, M.M. Tripathi, Ricci and scalar curvatures of submanifolds of a conformal Sasakian space form. Arch. Math. (Brno) **52**(2), 113–130 (2016)
2. A. Ahmad, G. He, W. Tang, P. Zhao, Chen's inequalities for submanifolds in (κ, μ)-contact space form with a semi-symmetric metric connection. Open Math. **16**(1), 380–391 (2018)

3. A. Ahmad, F. Shahzad, J. Li, Chen's inequalities for submanifolds in (κ, μ)-contact space form with a semi-symmetric metric connection. J. Appl. Math. Phys. **6**, 389–404 (2018)
4. M.A. Akyol, R. Demir, N.Ö. Poyraz, G.-E. Vîlcu, Optimal inequalities for hemi-slant Riemannian submersions. Mathematics **2022**(10), 3993, 18 pp (2022)
5. I. Al-Dayel, M.A. Khan, Ricci curvature of contact CR-warped product submanifolds in generalized Sasakian space forms admitting nearly Sasakian structure. AIMS Math. **6**(3), 2132–2151 (2021)
6. Al-Khaldi, A. H., Aquib, M., Aslam, M., Khan, M. A.: Chen-Ricci inequalities with a quarter symmetric connection in generalized space forms. Adv. Math. Phys. **2021** (2021), Art. ID 3221643, 10 pp
7. Al-Solamy, Falleh R.; Chen, B.-Y.; Deshmukh, S.: Two optimal inequalities for anti-holomorphic submanifolds and their applications. Taiwanese J. Math. **18**(1) (2014), 199–217
8. Alegre, P., Barrera, J., Carriazo, A.: A new class of slant submanifolds in generalized Sasakian space forms. Mediterr. J. Math. **17**(3) (2020), 76, 18 pp
9. P. Alegre, D.E. Blair, A. Carriazo, Generalized Sasakian space forms. Israel J. Math. **141**, 157–183 (2004)
10. Ali, A., Pişcoran, L.-I., Alkhaldi, A. H.: Ricci curvature on warped product submanifolds in spheres with geometric applications. J. Geom. Phys. **146** (2019), 103510, 17 pp
11. Ali, A., Pişcoran, L.-I., Alkhaldi, A. H., Alqahtani, L. S.: Ricci curvature on warped product submanifolds of complex space forms and its applications. Int. J. Geom. Methods Mod. Phys. **16**(9) (2019), 1950142, 25 pp
12. Ali, A., Mofarreh, F., Othman, W. A. M., Patra, D. S.: Applications of differential equations to characterize the base of warped product submanifolds of cosymplectic space forms. J. Inequal. Appl. **2020** (2020), 241, 17 pp
13. A.H. Alkhaldi, M.A. Khan, S.K. Hui, P. Mandal, Ricci curvature of semi-slant warped product submanifolds in generalized complex space forms. AIMS Math. **7**(4), 7069–7092 (2022)
14. S. Amari, *Differential-Geometrical Methods in Statistics, New York, NY* (Springer-Verlag, USA, 1985)
15. Aquib, M.: Some inequalities for statistical submanifolds of quaternion Kaehler-like statistical space forms. Int. J. Geom. Methods Mod. Phys. **16**(8) (2019), 1950129, 17 pp
16. M. Aquib, M. Aslam, M.H. Shahid, Bounds on Ricci curvature for doubly warped products pointwise bi-slant submanifolds and applications to Physics. Filomat **37**(2), 505–518 (2023)
17. K. Arslan, R. Ezentas, I. Mihai, C. Murathan, C. Özgür, Certain inequalities for submanifolds in (κ, μ)-contact space forms. Bull. Austral. Math. Soc. **64**(2), 201–212 (2001)
18. Arslan, K., Ezentas, R., Mihai, I., Murathan, C., Özgür, C.: B. Y. Chen inequalities for submanifolds in locally conformal almost cosymplectic manifolds. Bull. Inst. Math. Acad. Sinica **29**(3) (2001), 231–242
19. K. Arslan, R. Ezentas, I. Mihai, C. Murathan, C. Özgür, Ricci curvature of submanifolds in Kenmotsu space forms. Int. J. Math. Math. Sci. **29**, 719–726 (2002)
20. K. Arslan, R. Ezentas, I. Mihai, C. Murathan, C. Özgür, Ricci curvature of submanifolds in locally conformal almost cosymplectic manifolds. Math. J. Toyama Univ. **26**, 13–24 (2003)
21. M.E. Aydin, A. Mihai, I. Mihai, Some inequalities on submanifolds in statistical manifolds of constant curvature. Filomat **29**(3), 465–476 (2015)
22. Aydin, M. E., Mihai, A., Özgür, C.: Relations between extrinsic and intrinsic invariants statistical submanifolds in Sasaki-like statistical manifolds. Mathematics **2021**(9), 1285, 13 pp
23. Akgün, A., Gülbahar, M.: Bi-slant submanifolds of an S-manifold. Arab J. Math. Sci. (2021) (in press). DOI: https://doi.org/10.1108/AJMS-04-2021-0073.
24. Aytimur, A.: Curvature invariants for anti-invariant Riemannian submersions from cosymplectic space forms. Mediterr. J. Math. **20**(1) (2023), 38, 17 pp
25. H. Aytimur, C. Özgür, Inequalities for submanifolds of Sasaki-like statistical manifolds. Turkish J. Math. **42**(6), 3149–3163 (2018)
26. H. Aytimur, C. Özgür, Inequalities for submanifolds in statistical manifolds of quasi-constant curvature. Ann. Polon. Math. **121**(3), 197–215 (2018)

27. Aytimur, H., Özgür, C.: Sharp inequalities for anti-invariant Riemannian submersions from Sasakian space forms. J. Geom. Phys. **166** (2021), 104251, 12 pp
28. P. Bansal, S. Uddin, M.H. Shahid, Optimal inequalities for submanifolds in statistical manifolds of quasi constant curvature. Filomat **35**(10), 3319–3330 (2021)
29. M. Barros, B.-Y. Chen, F. Urbano, Quaternion CR-submanifolds of quaternion manifolds. Kodai Math. J. **4**, 399–417 (1981)
30. A. Bejancu, CR submanifolds of a Kaehler manifold I. Proc. Am. Math. Soc. **69**, 135–142 (1978)
31. Blair, D. E.: Riemannian Geometry of Contact and Symplectic Manifolds. Progress in Mathematics, 203. Boston, MA, USA, Birkhäuser, 2002
32. A. Carriazo, Y.H. Kim, D.W. Yoon, Some inequalities on totally real submanifolds in locally conformal Kaehler space forms. J. Korean Math. Soc. **41**(4), 795–808 (2004)
33. A. Carriazo, V. Martín-Molina, M.M. Tripathi, Generalized (κ, μ)-space forms. Mediterr. J. Math. **10**(1), 475–496 (2013)
34. B.-Y. Chen, Some topological obstructions to Bochner-Kaehler metrics and their applications. J. Differential Geom. **13**(4), 547–558 (1978)
35. B.-Y. Chen, Slant immersions. Bull. Austra. Math. Soc. **41**, 135–147 (1990)
36. B.-Y. Chen, *Geometry of Slant Submanifolds* (Katholieke Universiteit Leuven, Belgium, 1990)
37. B.-Y. Chen, Some pinching and classification theorems for minimal submanifolds. Arch. Math. **60**, 568–578 (1993)
38. B.-Y. Chen, Mean curvature and shape operator of isometric immersions in real-space-forms. Glasgow Math. J. **38**, 87–97 (1996)
39. B.-Y. Chen, Complex extensors and Lagrangian submanifolds in complex Euclidean spaces. Tohoku Math. J. **49**(2), 277–297 (1997)
40. B.-Y. Chen, Interaction of Legendre curves and Lagrangian submanifolds. Israel J. Math. **99**, 69–108 (1997)
41. Chen, B.-Y.: Strings of Riemannian invariants, inequalities, ideal immersions and their applications, The Third Pacific Rim Geometry Conference, pp. 7-60, Int. Press, Cambridge, MA, 1998
42. B.-Y. Chen, Relations between Ricci curvature and shape operator for submanifolds with arbitrary codimension. Glasgow Math. J. **41**, 33–41 (1999)
43. B.-Y. Chen, On Ricci curvature of isotropic and Lagrangian submanifolds in complex space forms. Arch. Math. **74**(2), 154–160 (2000)
44. B.-Y. Chen, Some new obstructions to minimal and Lagrangian isometric immersions. Japanese J. Math. **26**, 105–127 (2000)
45. B.-Y. Chen, Ideal Lagrangian immersions in complex space forms. Math. Proc. Cambridge Philo. Soc. **128**, 511–533 (2000)
46. B.-Y. Chen, Riemannian geometry of Lagrangian submanifolds. Taiwanese J. Math. **5**(4), 681–723 (2001)
47. B.-Y. Chen, Ricci curvature of real hypersurfaces in complex hyperbolic space. Arch. Math. (Brno) **38**(1), 73–80 (2002)
48. Chen, B.-Y.: A general optimal inequality for arbitrary Riemannian submanifolds. J. Inequal. Pure Appl. Math. **6**(3) (2005), 77, 10 pp
49. Chen, B.-Y.: Riemannian submersions, minimal immersions and cohomology class. Proc. Japan Acad. Ser. A Math. Sci. **81**(10) (2005), 162–167
50. B.-Y. Chen, Examples and classification of Riemannian submersions satisfying a basic equality. Bull. Austral. Math. Soc. **72**(3), 391–402 (2005)
51. Chen, B.-Y.: δ-invariants, inequalities of submanifolds and their applications. in: Topics in differential geometry 29–155, Ed. Acad. Române (Bucharest), 2008. DOI: https://doi.org/10.48550/arXiv.1307.1877
52. B.-Y. Chen, *Pseudo-Riemannian Geometry, δ-invariants and Applications* (World Scientific, Hackensack, NJ, USA, 2011)
53. B.-Y. Chen, A tour through δ-invariants: from Nash's embedding theorem to ideal immersions, best ways of living and beyond. Inst. Math. Publ. **94**(108), 67–80 (2013)

54. B.-Y. Chen, *Differential Geometry of Warped Product Manifolds and Submanifolds* (World Scientific, Hackensack, NJ, USA, 2017)
55. B.-Y. Chen, *Geometry of Submanifolds* (Dover Publications, Mineola, New York, 2019)
56. Chen, B.-Y., Blaga, A. M., Vîlcu, G.-E.: Differential geometry of submanifolds in complex space forms involving δ-invariants. Mathematics **2022**(10) (2022), 591, 38 pp
57. B.-Y. Chen, F. Dillen, J. Van der Veken, L. Vrancken, Curvature inequalities for Lagrangian submanifolds: the final solution. Differential Geom. Appl. **31**, 808–819 (2013)
58. B.-Y. Chen, F. Dillen, L. Verstraelen, δ-invariants and their applications to centroaffine geometry. Differential Geom. Appl. **22**(3), 341–354 (2005)
59. Chen, B.-Y., Mihai, A, Mihai, I.: A Chen first inequality for statistical submanifolds in Hessian manifolds of constant Hessian curvature. Results Math. **74**(4) (2019), 165, 11 pp
60. B.-Y. Chen, K. Ogiue, On totally real submanifolds. Trans. Amer. Math. Soc. **193**, 257–266 (1974)
61. B.-Y. Chen, A. Prieto-Martín, X. Wang, Lagrangian submanifolds in complex space forms satisfying an improved equality involving $\delta(2, 2)$. Publ. Math. Debrecen **82**(1), 193–217 (2013)
62. B.-Y. Chen, K. Yano, Hypersurfaces of a conformally flat space. Tensor (New Series) **26**, 318–322 (1972)
63. B.-Y. Chen, K. Yano, Manifolds with vanishing Weyl or Bochner curvature tensor. J. Math. Soc. Japan **27**, 106–112 (1975)
64. Cioroboiu, D.: Some inequalities for Ricci curvature of certain submanifolds in Sasakian space forms. Acta Math. Acad. Paedagog. Nyházi. (N.S.) **19**(2) (2003), 233–243
65. Costache, S.: B.-Y. Chen inequalities for slant submanifolds in Kenmosu space forms. Bull. Transilv. Univ. Brasov Ser. III **1**(50) (2008), 87–92
66. S. Costache, I. Zamfir, An improved Chen-Ricci inequality for special slant submanifolds in Kenmotsu space forms. Ann. Polon. Math. **110**(1), 81–89 (2014)
67. U.C. De, A.K. Gazi, On the existence of nearly quasi-Einstein manifolds. Novi Sad J. Math. **39**, 111–117 (2009)
68. S. Deng, An improved Chen-Ricci inequality. Int. Electron. J. Geom. **2**(2), 39–45 (2009)
69. S. Deng, Improved Chen-Ricci inequality for Lagrangian submanifolds in quaternion space forms. Int. Electron. J. Geom. **5**(1), 163–170 (2012)
70. S. Decu, Optimal inequalities for submanifolds in quaternion-space-forms with semi-symmetric metric connection. Bull. Transilv. Univ. Braşov Ser. III **2**(51), 175–184 (2009)
71. Decu-Marinescu, S.: Chen inequalities for submanifolds in quaternion-space-forms with semi-symmetric non-metric connection. Riemannian geometry and applications-Proceedings RIGA 2011, 115–126, Ed. Univ. Bucuresti, Bucharest, 2011
72. S. Dragomir, L. Ornea, *Locally conformal Kähler geometry, Progress in Mathematicsi 155* (Birkhäuser Boston Inc, Boston, MA, 1998)
73. Duggal, K. L., Bejancu, A.: Lightlike Submanifolds of Semi-Riemannian Manifolds and Applications, Kluwer Academics Publishers, 1996
74. Dwivedi, M. K., Kim, Je.-S.: Chen-Tripathi inequality for warped product submanifolds of S-space forms. An. Stiint. Univ. Al. I. Cuza Iasi. Mat. (N.S.) **58**(1) (2012), 195–208
75. Faghfouri, M., Ghaffarzadeh, N.: Chen's inequality for invariant submanifolds in a generalized (κ, μ)-space form. Global J. Adv. Research Class. Mod. Geom. **4**(2), (2015), 86–101
76. Faghfouri, M., Ghaffarzadeh, N.: Chen's inequality for C-totally real submanifolds in a generalized (κ, μ)-space form. Global J. Adv. Research Class. Mod. Geom. **10**(2), (2021), 137–152
77. L.M. Fernández, A.M. Fuentes, Some relationships between intrinsic and extrinsic invariants of submanifolds in generalized S-space-forms. Hacet. J. Math. Stat. **44**(1), 59–74 (2015)
78. L.M. Fernández, M.B. Hans-Uber, New relationships involving the mean curvature of slant submanifolds in S-space-forms. J. Korean Math. Soc. **44**(3), 647–659 (2007)
79. A.E. Fischer, Riemannian maps between Riemannian manifolds. Contemporary Math. **132**, 331–366 (1992)
80. H. Furuhata, Hypersurfaces in statistical manifolds. Differential Geom. Appl. **27**, 420–429 (2009)

81. Furuhata, H., Hasegawa, I.: Submanifold theory in holomorphic statistical manifolds. In: Dragomir, S., Shahid, M.H., Al-Solamy, F.R. (eds.) Geometry of Cauchy-Riemann Submanifolds, 179–215 (2016), Springer, Singapore
82. H. Furuhata, I. Hasegawa, Y. Okuyama, K. Sato, Kenmotsu statistical manifolds and warped product. J. Geom. **108**, 1175–1191 (2017)
83. V. Ghişoiu, New results for submanifolds of generalized complex space forms with a semi-symmetric metric connection. Riemannian Geometry and Applications - Proceedings RIGA **2014**, 97–116 (2014)
84. Gromov, M.: Isometric immersions of Riemannian manifolds. in: The mathematical heritage of Élie Cartan (Lyon, 1984). Astérisque 1985, Numéro Hors Série, pp. 129–133
85. M. Gülbahar, Ş Eken Meriç, E. Kiliç, Sharp inequalities involving the Ricci curvature for Riemannian submersions. Kragujevac J. Math. **41**(2), 279–293 (2017)
86. Gülbahar, M., Eken Meriç, Ş., Kiliç, E.: Chen invariants for Riemannian submersions and their applications. Commun. Fac. Sci. Univ. Ank. Ser. A1. Math. Stat. **71**(4) (2022), 1007–1022
87. Gülbahar, M., Kiliç, E., Keleş, S.: Chen-like inequalities on lightlike hypersurfaces of a Lorentzian manifold. J. Inequal. Appl. **2013** (2013), 266, 18 pp
88. M. Gülbahar, E. Kiliç, S. Keleş, Some inequalities on screen homothetic lightlike hypersurfaces of a Lorentzian manifold. Taiwanese J. Math. **17**(6), 2083–2100 (2013)
89. Gülbahar, M., Tripathi, M. M., Kiliç, E.: Inequalities involving k-Chen invariants for submanifolds of Riemannian product manifolds. Commun. Fac. Sci. Univ. Ank. Ser. A1. Math. Stat. **68**(1) (2019), 466–483
90. M. Gülbahar, E. Kiliç, S. Keleş, M.M. Tripathi, Some basic inequalities for submanifolds of nearly quasi-constant curvature manifolds. Differ. Geom. Dyn. Syst. **16**, 156–167 (2014)
91. Y. Gündüzalp, M. Polat, Some inequalities of anti-invariant Riemannian submersions in complex space forms. Miskolc Math. Notes **23**(2), 703–714 (2022)
92. M. Gutiérrez, B. Olea, Induced Riemannian structures on null hypersurfaces. Math. Nachr. **289**, 1219–1236 (2015)
93. H.A. Hayden, Subspaces of a space with torsion. Proc. London Math. Soc. **34**, 27–50 (1932)
94. G.Q. He, P. Cheng, Chen inequalities for submanifolds of generalized space forms with semi-symmetric metric connections. J. Anhui Normal U. (Natural Sci.) **37**(5), 418–424 (2014)
95. He, G. Q., Zhang, L., Liu, H. R.: Chen inequalities for submanifolds of generalized Sasakian space forms with a semi-symmetric metric connection. (Chinese) J. Jiin Univ. (Sci. Ed.) **54**(6), (2016), 1248–1254
96. He, G. Q., Zhang, L., Liu, H. R.: Chen-Ricci inequalities for submanifolds of generalized Sasakian space forms with a semi-symmetric metric connection. (Chinese) J. Shandong Univ. Nat. Sci. **52**(10) (2017), 56–63
97. G. He, J. Zhang, P. Zhao, Geometric inequalities for non-integrable distributions in statistical manifolds with constant curvature. Filomat **35**(11), 3585–3596 (2021)
98. S. Hineva, Submanifolds for which a lower bound of the Ricci curvature is achieved. J. Geom. **88**, 53–69 (2008)
99. S. Hong, K. Matsumoto, M.M. Tripathi, Certain basic inequalities for submanifolds of locally conformal Kaehler space forms. SUT J. Math. **41**(1), 75–94 (2005)
100. S. Hong, M.M. Tripathi, On Ricci curvature of submanifolds. Int. J. Pure Appl. Math. Sci. **2**(2), 227–245 (2005)
101. S. Hong, M.M. Tripathi, On Ricci curvature of submanifolds of generalized Sasakian space forms. Int. J. Pure Appl. Math. Sci. **2**(2), 173–201 (2005)
102. A. Ishan, M.A. Khan, Chen-Ricci inequality for biwarped product submanifolds in complex space forms. AIMS Math. **6**(5), 5256–5274 (2021)
103. Jin, D. H., Lee, J. W.: A basic inequality for the Tanaka-Webster connection. J. Appl. Math. **2012**(11) (2012), 231416, 11 pp
104. Khan, M. A., Aldayel I.: Ricci curvature inequalities for skew CR-warped product submanifolds in complex space forms. Mathematics **2020**(8) (2020), 1317, 19 pp
105. M.A. Khan, C. Ozel, Ricci curvature of contact CR-warped product submanifolds in generalized Sasakian space forms admitting a trans-Sasakian structure. Filomat **35**(1), 125–146 (2021)

106. M.A. Khan, C. Ozel, K. Khan, Ricci curvature for biwarped product submanifolds in Kenmotsu space forms. Differ. Geom. Dyn. Syst. **23**, 105–126 (2021)
107. E. Kiliç, M.M. Tripathi, M. Gülbahar, Chen-Ricci inequalities for submanifolds of Riemannian and Kaehlerian product manifolds. Ann. Polon. Math. **116**(1), 37–56 (2016)
108. D.-S. Kim, Y.H. Kim, C.W. Lee, Shape operator A_H for slant submanifolds in generalized complex space forms. Bull. Korean Math. Soc. **42**(1), 189–201 (2005)
109. J.-S. Kim, M.K. Dwivedi, M.M. Tripathi, Ricci curvature of integral submanifolds of an S-space form. Bull. Korean Math. Soc. **44**(3), 395–406 (2007)
110. J.-S. Kim, M.K. Dwivedi, M.M. Tripathi, Ricci curvature of submanifolds of an S-space form. Bull. Korean Math. Soc. **46**(5), 979–998 (2009)
111. J.-S. Kim, M.M. Tripathi, J. Choi, Ricci curvature of submanifolds in locally conformal almost cosymplectic manifolds. Indian J. Pure Appl. Math. **35**(3), 259–271 (2004)
112. Kobayashi, S., Nomizu, K.: Foundations of differential geometry. Vols I & II. Interscience Publishers, New York–London, 1963 & 1969
113. J.W. Lee, C.W. Lee, B. Şahin, G.-E. Vilcu, Chen-Ricci inequalities for Riemannian maps and their applications. Contemp. Math. **777**, 137–152 (2022)
114. C.W. Lee, J.W. Lee, D.W. Yoon, Improved Chen inequality of Sasakian space forms with the Tanaka-Webster connection. Filomat **29**(7), 1525–1533 (2015)
115. Li, Y., Khatri, M., Singh, J. P., Chaubey, S. K.: Improved Chen's inequalities for submanifolds of generalized Sasakian-space-forms. Axioms **2022**(11) (2022), 324, 14 pp
116. X. Liu, On Ricci curvature of C-totally real submanifolds in Sasakian space forms. Proc. Indian Acad. Sci. Math. Sci. **111**(4), 399–405 (2001)
117. X. Liu, On Ricci curvature of totally real submanifolds in a quaternion projective space. Arch. Math. (Brno) **38**(4), 297–305 (2002)
118. X. Liu, W. Su, Shape operator of slant submanifold in cosympectic space forms. Studia Sci. Math. Hungar. **42**(4), 387–400 (2005)
119. X. Liu, J. Zhou, On Ricci curvature of certain submanifolds in a cosymplectic space form. Sarajevo J. Math. **2**(1), 95–106 (2006)
120. Lone, M. A.: Basic inequalities for submanifolds of quaternionic space forms with a quarter-symmetric connection. J. Geom. Phys. **159** (2021), 103927, 15 pp
121. M.A. Lone, Y. Matsuyama, F.R. Al-Solamy, M.H. Shahid, M. Jamali, Upper bounds for Ricci curvatures for submanifolds in Bochner-Kaehler manifolds. Tamkang J. Math. **51**(1), 53–67 (2020)
122. F. Malek, M.B.K. Balgeshir, Slant submanifolds of almost contact metric 3-structure manifolds. Mediterr. J. Math. **10**(2), 1023–1033 (2013)
123. F. Malek, V. Nejadakbary, A lower bound for the Ricci curvature of submanifolds in generalized Sasakian space forms. Adv. Geom. **13**(4), 695–711 (2013)
124. P. Mandal, T. Pal, S. Hui, K,: Ricci curvature on warped product submanifolds of Sasakian-space-forms. Filomat **34**(12), 3917–3930 (2020)
125. K. Matsumoto, I. Mihai, Y. Tazawa, Ricci tensor of slant submanifolds in complex space forms. Kodai Math. J. **26**(1), 85–94 (2003)
126. Ménédore, K.: Chen's Ricci inequalities and topological obstructions on null hypersurfaces of a Lorentzian manifold. J. Inequal. Appl. **2018** (2018), 126, 27 pp
127. Matehkolaee, M. J.: Ricci curvature of integral submanifolds of an f.p.k.-space form. Bull. Belg. Math. Soc. Simon Stevin **21**(3) (2014), 437–453
128. K. Matsumoto, I. Mihai, Ricci tensor of C-totally real submanifolds in Sasakian space forms. Nihonkai Math. J. **13**(2), 191–198 (2002)
129. A. Mihai, Shape operator A_H for slant submanifolds in generalized complex space forms. Turkish J. Math. **27**(4), 509–523 (2003)
130. A. Mihai, *Modern topics in submanifold theory* (Editura Universităţii din Bucureşti, Bucharest, 2006)
131. A. Mihai, Inequalities on the Ricci curvature. J. Math. Inequal. **9**(3), 811–822 (2015)
132. Mihai, A., Mihai, I.: Curvature invariants for statistical submanifolds of Hessian manifolds of constant Hessian curvature. Mathematics **2018**(6) (2018), 44, 8 pp

133. Mihai, A., Mihai, I.: Some basic inequalities on slant submanifolds in space forms. Contact geometry of slant submanifolds, 39–89, Springer, Singapore (2022)
134. A. Mihai, C. Özgür, Chen inequalities for submanifolds of complex space forms and Sasakian space forms endowed with semi-symmetric metric connections. Rocky Mountain J. Math. **41**(5), 1653–1673 (2011)
135. A. Mihai, C. Özgür, Chen inequalities for submanifolds of real space forms with a semi-symmetric metric connection. Taiwanese J. Math. **14**(4), 1465–1477 (2010)
136. A. Mihai, I.N. Radulescu, An improved Chen-Ricci inequality for Kaehlerian slant submanifolds in complex space forms. Taiwanese J. Math. **16**(2), 761–770 (2012)
137. A. Mihai, I.N. Radulescu, Scalar and Ricci curvatures of special contact slant submanifolds in Sasakian space forms. Adv. Geom. **14**(1), 147–159 (2014)
138. I. Mihai, Ricci curvature of submanifolds in Sasakian space forms. J. Aust. Math. Soc. **72**(2), 247–256 (2002)
139. Mihai, I.: Slant submanifolds in complex space forms. Topics in differential geometry, 157–182, Ed. Acad. Romane, Bucharest, 2008
140. I. Mihai, F. Al-Solamy, M.H. Shahid, On Ricci curvature of a quaternion CR-submanifold in a quaternion space form. Rad. Mat. **12**(1), 91–98 (2003)
141. I. Mihai, V. Ghişoiu, Minimality of certain contact slant submanifolds in Sasakian space forms. Int. J. Pure Appl. Math. Sci. **1**, 95–99 (2004)
142. Mihai, I., Mihai, R.-I.: A new algebraic inequality and some applications in submanifold theory. Mathematics **2021**(9) (2021), 1175, 10 pp
143. Mihai, I., Mihai, R.-I.: General Chen inequalities for statistical submanifolds in Hessian manifolds of constant Hessian curvature. Mathematics **2022**(10) (2022), 3061, 9 pp
144. I. Mihai, I.N. Rădulescu, An improved Chen-Ricci inequality for Legendrian submanifolds in Sasakian space forms. J. Adv. Math. Stud. **4**(2), 51–58 (2011)
145. Mihai, I., Siddiqui, A. N., Shahid, M. H.: Geometry of pointwise slant immersions in almost Hermitian manifolds. Complex geometry of slant submanifolds, 281–325, Springer, Singapore (2022)
146. Mustafa, S., Uddin, S.: Chen–Ricci inequality for CR-warped products and related open problems. Mediterr. J. Math. **18** (2021), 67, 24 pp
147. Mustafa, S., Uddin, S., Al-Solamy, F. R.: Chen-Ricci inequality for warped products in Kenmotsu space forms and its applications. Rev. R. Acad. Cienc. Exactas Fis. Nat. Ser. A Mat. **113**(4) (2019), 3585–3602
148. J.F. Nash, The imbedding problem for Riemannian manifolds. Ann. Math. **63**, 20–63 (1956)
149. V. Nejadakbary Mehraban, S. Azami, A. Rezaei, Some results for slant submanifolds in generalized Sasakian space forms. Balkan J. Geom. Appl. **21**(2), 55–72 (2016)
150. A. Oiagă, Ricci curvature of totally real submanifolds in locally conformal Kaehler space forms. An. Univ. Bucureşti Mat. Inform. **49**(2), 69–76 (2000)
151. Z. Olszak, Locally conformal almost cosymplectic manifolds. Colloq. Math. **57**(1), 73–87 (1989)
152. B. O'Neill, The fundamental equations of a submersion. Michigan Math. J. **13**, 459–469 (1966)
153. T. Oprea, Ricci curvature of Lagrangian submanifolds in complex space forms. Math. Inequal. Appl. **13**(4), 851–858 (2010)
154. Y.-L. Ou, B.-Y. Chen, *Biharmonic Submanifolds and Biharmonic Maps in Riemannian Geometry* (Hackensack, World Scientific, NJ, USA, 2020)
155. J.A. Oubina, New classes of almost contact metric structures. Publ. Math. Debrecen **32**, 187–193 (1985)
156. Özgür, C.: B. Y. Chen inequalities for submanifolds of a Riemannian manifold of quasi-constant curvature. Turkish J. Math. **35**(3) (2011), 501–509
157. C. Özgür, A. Mihai, Chen inequalities for submanifolds of real space forms with a semi-symmetric non-metric connection. Canad. Math. Bull. **55**(3), 611–622 (2012)
158. C. Özgür, A. Mihai, Corrigendum to "Chen inequalities for submanifolds of real space forms with a semi-symmetric non-metric connection." Canad. Math. Bull. **58**(1), 158–159 (2015)

159. Özgür, C., Murathan, C.: Chen inequalities for submanifolds of a cosymplectic space form with a semi-symmetric metric connection. An. Stiint. Univ. Al. I. Cuza Iasi. Mat. (N.S.) **58**(2) (2012), 395–408
160. Özgür, C., Murathan, C.: Chen inequalities for submanifolds of a locally conformal almost cosymplectic manifold with a semi-symmetric metric connection. An. Stiint. Univ. "Ovidius" Constanta Ser. Mat. **18**(1) (2010), 239–253
161. Pal Singh, A., Atindogbe, C., Kumar, R., Jain, V.: Chen-like inequalities on null hypersurfaces with closed rigging of a Lorentzian manifold. Int. J. Geom. Methods Mod. Phys. **18**(8) (2021), 2150125, 23 pp
162. G. Pitiş, Integral submanifolds with closed conformal vector field in Sasakian manifolds. New York J. Math. **11**, 157–170 (2005)
163. Poyraz, N.: Chen inequalities on spacelike hypersurfaces of a GRW spacetime. Differential Geom. Appl. **81** (2022), 101863, 11 pp
164. N. Poyraz, E. Yaşar, Chen-like inequalities on lightlike hypersurface of a Lorentzian product manifold with quarter-symmetric nonmetric connection. Kragujevac J. Math. **40**(2), 146–164 (2016)
165. N. Poyraz, B. Doğan, E. Yaşar, Chen inequalities on lightlike hypersurface of a Lorentzian manifold with semi-symmetric metric connection. Int. Electron. J. Geom. **10**(1), 1–14 (2017)
166. N. Poyraz, H.I. Yoldaş, Chen inequalities for submanifolds of real space forms with a Ricci quarter-symmetric metric connection. Int. Electron. J. Geom. **12**(1), 102–110 (2019)
167. B. Şahin, *Riemannian Submersions, Riemannian Maps in Hermitian Geometry, and their Applications* (Elsevier/Academic Press, London, UK, 2017)
168. T. Sasahara, On Ricci curvature of CR-submanifolds with rank one totally real distribution. Nihonkai Math. J. **12**, 47–58 (2001)
169. S. Sasaki, On differentiable manifolds with certain structures which are closely related to almost contact structure I. Tohoku Math. J. **12**(3), 459–476 (1960)
170. N. Satoh, H. Furuhata, I. Hasegawa, T. Nakane, Y. Okuyama, K. Sato, M.H. Shahid, A.N. Siddiqui, Statistical submanifolds from a viewpoint of the Euler inequality. Inform. Geom. **4**, 189–213 (2021)
171. Sayar, C., Akyol, M. A., Prasad, R.: Bi-slant submersions in complex geometry. Int. J. Geom. Methods Mod. Phys. **17**(4) (2020), 2050055, 17 pp
172. H. Shima, *The Geometry of Hessian Structures* (World Scientific Publishing, Singapore, 2007)
173. Shukla, S. S., Rao, P. K.: Ricci curvature of quaternion slant submanifolds in quaternion space forms. Acta Math. Acad. Paedagog. Nyhazi. (N.S.) **28**(1) (2012), 69–81
174. S.S. Shukla, S.K. Tiwari, Ricci curvature of slant submanifolds in generalized Sasakian space forms. Bull. Allahabad Math. Soc. **23**(2), 405–417 (2008)
175. S.S. Shukla, S.K. Tiwari, Ricci curvature of slant submanifolds in (κ, μ)-contact space forms. Kuwait J. Sci. Engrg. **37**(1A), 31–49 (2010)
176. A.N. Siddiqui, F.R. Al-Solamy, M.H. Shahid, I. Mihai, On CR-statistical submanifolds of holomorphic statistical manifolds. Filomat **35**(11), 3571–3584 (2021)
177. Siddiqui, A. N., Chen, B.-Y., Siddiqi, M. D.: Chen inequalities for statistical submersions between statistical manifolds. Int. J. Geom. Methods Mod. Phys. **18**(4) (2021), 2150049, 17 pp
178. A.N. Siddiqui, M.H. Shahid, J.W. Lee, On Ricci curvature of submanifolds in statistical manifolds of constant (quasi-constant) curvature. AIMS Math. **5**(4), 3495–3510 (2020)
179. Siddiqui, A. N., Murathan, C., Siddiqi, M. D.: The Chen's first inequality for submanifolds of statistical warped product manifolds. J. Geom. Phys. **169** (2021), Paper No. 104344, 13 pp
180. A.N. Siddiqui, A.N. Siddiqui, Y.J. Suh, O. Bahadir, Extremities for statistical submanifolds in Kenmotsu statistical manifolds. Filomat **35**(2), 591–603 (2021)
181. A.N. Siddiqui, Y.J. Suh, O. Bahadir, Extremities for statistical submanifolds in Kenmotsu statistical manifolds. Filomat **35**(2), 591–603 (2021)
182. Siddiqui, A. N., Uddin, S., Shahid, M. H.: B.-Y. Chen's inequality for Kähler-like statistical submersions. Int. Electron. J. Geom. **15**(2) (2022), 277–286

183. A. Song, X. Liu, Some inequalities of slant submanifolds in generalized complex space forms. Tamkang J. Math. **36**(3), 223–229 (2005)
184. Su, M., Zhang, L.: Inequalities for doubly warped product submanifold in a Riemannian manifold of nearly quasi-constant curvature. (Chinese) J. Jilin Univ. (Sci. Ed.) **54**(5) (2016), 952–968
185. Su, M., Zhang, L.: Two results on Chen's inequalities for spacelike submanifolds of a pseudo-Riemannian space form. (Chinese) J. Shandong Univ. (Natural Sci.) **51**(10) (2016), 59–64
186. Suceavă, B. D.: Remarks on B. Y. Chen's inequality involving classical invariants. An. Ştiinţ. Univ. Al. I. Cuza Iaşi. Mat. (N.S.) **45**(2) (1999), 405–412
187. Suceavă, B. D., Vajiac, M. B.: Estimates of B. -Y. Chen's $\hat{\delta}$-invariant in terms of Casorati curvature and mean curvature for strictly convex Euclidean hypersurfaces. Int. Electron. J. Geom. **12**(1) (2019), 26–31
188. K. Takano, Statistical manifolds with almost contact structures and its statistical submersions. J. Geom. **85**, 171–187 (2006)
189. Takano, K.: Statistical manifolds with almost complex structures and its statistical submersions. Proceedings of the 45-th Symposium on Finsler Geometry. Tokyo, 2010-09-05/10. Society of Finsler Geometry (2011), 54–57
190. H.M. Taştan, F. Özdemir, The geometry of hemi-slant submanifolds of a locally product Riemannian manifold. Turkish J. Math. **39**(2), 268–284 (2015)
191. H.M. Taştan, B. Şahin, Ş Yanan, Hemi-slant submersions. Mediterr. J. Math. **13**, 2171–2184 (2016)
192. Tripathi, M. M.: Certain basic inequalities for submanifolds. Proc. Tenth Intern. Workshop on Diff. Geom. **10** (2006), 99–145
193. M.M. Tripathi, Certain inequalities for submanifolds in locally conformal almost cosymplectic manifolds. Trends in Mathematics Information Center for Mathematical Sciences **5**(2), 87–96 (2002)
194. M.M. Tripathi, Chen-Ricci inequality for submanifolds of contact metric manifolds. J. Adv. Math. Stud. **1**(1–2), 111–134 (2008)
195. M.M. Tripathi, Improved Chen-Ricci inequality for curvature-like tensors and its applications. Differential Geom. Appl. **29**, 685–698 (2011)
196. Tripathi, M. M., Ziabari, R. B.: Slant submanifolds of conformal Sasakian space forms. Contact geometry of slant submanifolds, 183–198, Springer, Singapore, 2022
197. M.M. Tripathi, J.-S. Kim, C-totally real submanifolds in (κ, μ)-contact space forms. Bull. Austral. Math. Soc. **67**(1), 51–65 (2003)
198. M.M. Tripathi, J.-S. Kim, S.-B. Kim, Mean curvature and shape operator of slant immersions in a Sasakian space form. Balkan J. Geom. Appl. **7**(1), 101–111 (2002)
199. Uddin, S., Chen, B.-Y., Al-Solamy, F. R.: Warped product bi-slant immersions in Kaehler manifolds. Mediterr. J. Math. **14** (2017), Paper No. 95, 11 pp
200. A.D. Vîlcu, G.E. Vîlcu, Statistical manifolds with almost quaternionic structures and quaternionic Kaehler-like statistical submersions. Entropy **17**, 6213–6228 (2015)
201. Vîlcu, G. E.: On Chen invariants and inequalities in quaternionic geometry. J. Inequal. Appl. **2013** (2013), 66, 14 pp
202. G.E. Vîlcu, Slant submanifolds of quarternionic space forms. Publ. Math. (Debrecen) **81**(3–4), 397–413 (2012)
203. Vilcu, G. E.: B.-Y. Chen inequalities for slant submanifolds in quaternionic space forms. Turkish J. Math. **34**(1) (2010), 115–128
204. P.W. Vos, Fundamental equations for statistical submanifolds with applications to the Bartlett correction. Ann. Inst. Stat. Math. **41**, 429–450 (1989)
205. A.Q. Wang, X.M. Liu, Ricci curvature of semi-invariant submanifolds in cosymplectic space forms. J. Math. Res. Exposition **27**(1), 195–200 (2007)
206. Wang, Y.: Affine connections of non-integrable distributions. Int. J. Geom. Methods Mod. Phys. **17**(8) (2020), 2050127, 25 pp
207. Wang, Y.: Chen's inequalities for submanifolds in (κ, μ)-contact space form with generalized semi-symmetric non-metric connections. arXiv:2003.00185 (2020)

208. K. Yano, On a structure defined by a tensor field f of type $(1, 1)$ satisfying $f^3 + f = 0$. Tensor **14**, 99–109 (1963)
209. K. Yano, On semi-symmetric metric connections. Rev. Roumaine Math. Pures Appl. **15**, 1579–1586 (1970)
210. D.W. Yoon, Certain inequalities for submanifolds in locally conformal almost cosymplectic manifolds. Bull. Inst. Math. Acad. Sinica **32**(4), 263–283 (2004)
211. D.W. Yoon, Inequality for Ricci curvature of certain submanifolds in locally conformal almost cosymplectic manifolds. Int. J. Math. Math. Sci. **2005**(10), 1621–1632 (2005)
212. D.W. Yoon, Inequality for Ricci curvature of slant submanifolds in cosymplectic space forms. Turkish J. Math. **30**(1), 43–56 (2006)
213. D.W. Yoon, Ricci curvature of submanifolds in quaternionic space forms. Int. Math. J. **4**(4), 377–384 (2003)
214. Zhang, L., Zhang, P.: Notes on Chen's inequalities for submanifolds of real space forms with a semi-symmetric non-metric connection. J. East China Norm. Univ. Natur. Sci. Ed. **2015**(1) (2015), 6–15
215. P. Zhang, Remarks on Chen's inequalities for submanifolds of a Riemannian manifold of nearly quasi-constant curvature. Vietnam J. Math. **43**(3), 557–569 (2015)
216. P. Zhang, X. Pan, L. Zhang, Inequalities for submanifolds of a Riemannian manifold of nearly quasi-constant curvature with a semi-symmetric non-metric connection. Rev. Un. Mat. Argentina **56**(2), 1–19 (2015)
217. P. Zhang, L. Zhang, W. Song, Chen's inequalities for submanifolds of a Riemannian manifold of quasi-constant curvature with a semi-symmetric metric connection. Taiwanese J. Math. **18**(6), 1841–1862 (2014)
218. P. Zhang, L. Zhang, W. Song, Chen's inequalities for totally real submanifolds in complex space forms with a semi-symmetric metric connection. J. Math. Res. Appl. **34**(5), 587–596 (2014)
219. Zhang, P., Zhang, L., Song, W.-D.: Geometric inequalities for spacelike submanifolds of a semi-Riemannian space form. (Chinese) J. Shandong Univ. Nat. Sci. **49**(6) (2014), 91–94
220. Zhang, P., Zhang, L., Song, W. D.: Inequalities of Chen type for Lagrangian submanifolds of an indefinite complex space form. (Chinese) J. Jilin Univ. Sci. **52**(3) (2014), 439–444
221. P. Zhang, L. Zhang, W.-D. Song, Some remarks on geometric inequalities for submanifolds of a Riemannian manifold of quasi-constant curvature. J. Math (PRC) **36**(2), 445–457 (2016)
222. P. Zhang, L. Zhang, M.M. Tripathi, Geometric inequalities for Einstein totally real submanifolds in a complex space form. Rev. Un. Mat. Argentina **58**(2), 189–198 (2017)
223. R.B. Ziabari, M.M. Tripathi, Ricci and scalar curvatures of submanifolds of a conformal Sasakian space form. Arch. Math. (Brno) **52**(2), 113–130 (2016)

Solitons in $f(\mathcal{R}, T)$-Gravity

Uday Chand De and Krishnendu De

2020 Mathematics Subject Classification 83D05 · 83C05 · 53C50

1 Introduction

Some fundamental concerns about the universe, beginning with the Big Bang and making their ways into the formation of dark matter, elements, and anti-matter, are now entering into the scientific community. But there are now several options for dealing with these problems. At the same time, cosmological research showed that the universe seemed to be expanding more quickly. Two aspects may be linked to rapid expansion. The first technique entails modifying the energy-momentum tensor (briefly, EMT) in Einstein's field equations (briefly, EFEs), including the quintessence phase, phantom era, Chaplygin gas, viscosity, and so on. The second strategy involves changing the spacetime geometry in Einstein's equations. These are frequently referred to as modified theories of gravity.

Perfect fluids (briefly, PFs), which are special solutions of EFEs with disappearing shear stresses, heat conduction, and viscosity suitable with Bianchi identities, play a significant role in Einstein's general relativity (briefly, GR) theory. They depict the Hubble flow's behavior in cosmology, from inflation through the period of dark energy. The matter content of the PF model is found in an isotropic universe or a

U. C. De (✉)
Department of Mathematics, University of Calcutta, West Bengal, India
e-mail: uc_de@yahoo.com

K. De
Department of Mathematics, Kabi Sukanta Mahavidyalaya, The University of Burdwan, P.O.-Angus, Hooghly, Bhadreswar 712221, West Bengal, India
e-mail: krishnendu.de@outlook.in

B.-Y. Chen et al. (eds.), *Geometry of Submanifolds and Applications*, Infosys Science Foundation Series, https://doi.org/10.1007/978-981-99-9750-3_2

star's interior. Due to these factors, it is crucial to study the geometric investigations of PFs on modified gravity theories.

Without supposing the continuation of some unobservable components that must provide the explanation for the dark energy and the dark matter, EFEs are unsuited to describe the late-time inflation of the universe. It is the primary justification behind its extension in order to produce a few higher-order field equations for gravity.

Many modifications of EFEs have been introduced and extensively researched ([13, 36]). The GR can be modified in a variety of ways; Somehow or another, Einstein-Hilbert gravitational action, specifically, $f(\mathcal{R})$-gravity, $\mathcal{R}$ stands for the Ricci scalar ([1, 11, 12], [18–20, 33, 34]) which generalizes Einstein's GR, Gauss-Bonnet gravity [14] is a modification of the Einstein-Hilbert action to include the Gauss-Bonnet invariant G, $f(G)$-gravity [37], $f(Q)$-gravity where the non-metricity Q's arbitrary function f determines the gravitational action [32], $f(T)$ theory, where T is the trace of the EMT [8], $f(\mathcal{R}, T)$-gravity and $f(\mathcal{R}, L_m)$-gravity, L_m indicates Lagrangian density relating to matter, are the extension of $f(\mathcal{R})$-gravity introduced by Harko et al. ([30, 31]).

Recently, spacetimes were extensively studied in modified gravity theories. In ([19, 20, 34]), the authors have examined weakly Ricci-symmetric spacetimes, conformally flat generalized Ricci recurrent spacetimes and projectively flat spacetimes in $f(\mathcal{R})$-gravity. Pseudo-symmetric spacetimes accepting $f(\mathcal{R})$-gravity have been analyzed by De et al. [18]. Güler and De [27] have studied PF spacetimes in $f(\mathcal{R}, T)$-gravity and weakly Ricci-symmetric spacetimes in $f(\mathcal{R}, T)$-gravity have been discussed by Chen et al. [16].

Physical symmetry in GR is specifically related to spacetime geometry. More specifically, the classification of solutions to EFEs is typically made simpler by the metric of symmetry. Soliton, which is related to the geometrical flow of spacetime geometry, is a significant symmetry. In reality, the concepts of Ricci flow and Yamabe flow are applied to interpret kinematics in GR. Because of the preservation of self-similarity via curvatures, Ricci and Yamabe solitons are in focus.

Our aim is to investigate $f(\mathcal{R}, T)$-gravity theory concerning different solitons.

Let $(\mathcal{N}^n, g)$ be a Lorentzian manifold and g is the Lorentzian metric of signature $(\underbrace{+, +, \cdots, +}_{(n-1)\text{times}}, -)$, that is, g is of index 1. In [2], Alias, Romero, and Sanchez proposed the concept of generalized Robertson-Walker (briefly, GRW) spacetimes in 1995. A GRW spacetime is a Lorentzian manifold $\mathcal{N}^n$ $(n \geq 3)$ that may be expressed as $\mathcal{N} = -I \times f^2 N^*$, where I being the open interval of $\mathbb{R}$, $N^{*(n-1)}$ denotes the Riemannian manifold and $f > 0$ is termed as scale factor or warping function. The above-stated spacetime turns into Robertson-Walker (briefly, RW) spacetime when N^* is of dimension three and the sectional curvature is constant. Therefore, GRW spacetime is a spontaneous extension of the RW spacetime, which is used to explain standard cosmology.

If a Lorentzian manifold's non-vanishing Ricci tensor S fulfills

$$S = \beta \mathcal{A} \otimes \mathcal{A} + \alpha\, g, \tag{1}$$

it is referred to as a PF spacetime, in which α, β (not all zero simultaneously) are scalar. Also for all X_1, $g(X_1, \rho) = \mathcal{A}(X_1)$, ρ is a unit timelike vector field which is named as velocity vector. Hence, we have $g(\rho, \rho) = -1$.

Each RW spacetime is a PF spacetime [38], whereas in 4– dimension, the GRW spacetime represents a PF spacetime if the spacetime is RW and vice-versa.

If T indicates the EMT, then EFEs is described by (without cosmological constant)

$$S - \frac{r}{2}g = kT. \tag{2}$$

where $k = \sqrt{8\pi \mathrm{G}}$, G indicates Newton's gravitational constant.

In PF spacetime if σ indicates the energy density and p stands for isotropic pressure, then T is described as

$$T = pg + (p + \sigma)\mathcal{A} \otimes \mathcal{A}. \tag{3}$$

Combining the Eqs. (1), (2) and (3), we infer that

$$\alpha = \frac{k(\sigma - \ p)}{n - 2}, \quad \beta = \kappa \ (\sigma + \ p). \tag{4}$$

Additionally, the equation $\frac{p}{\sigma} = \omega$, in which ω is known as the equation of state (briefly, EOS) parameter, connects σ and p to characterize the specific type of PF spacetimes. The phrase "isentropic" refers to PF spacetime for the EOS $p = p(\sigma)$. Moreover, the PF spacetime depicts stiff matter fluid if $\sigma - p = 0$ (See [15]). Zeldovich [40] introduced the EOS of stiff matter fluid.

If $p = 0$, PF spacetime describes the dust matter fluid; however, if $p - \frac{\sigma}{3} = 0$, PF spacetime describes radiation era, and $p = -\sigma$ represents dark energy, respectively[15].

Besides, the dark energy era known as phantom energy or phantom era is one that satisfies the EOS with $\omega < -1$. The physical effects are examined in ([9, 10]).

Quintessence is a hypothetical type of dark energy, or more specifically, a scalar field, that has been proposed in physics as a possible explanation for the observation of an accelerating rate of expansion of the universe. Also, dark energy represents quintessence phase if $-1 < \omega < 0$.

Hamilton [29] in order to present the notion of Ricci flow, which is motivated by the challenge to find a canonical metric on a smooth manifold. If the evolution equation $\frac{\partial}{\partial t} g_{ij}(t) = -2S_{ij}$ fulfills by the metric of a semi-Riemannian manifold, then it is named Ricci flow [29]. The Ricci solitons (briefly, RSs) are produced by the self-similar Ricci flow solutions. A RS [28] is a metric of $\mathcal{N}^n$ (semi-Riemannian manifold) if it complies with

$$\mathfrak{L}_{W_1} g + 2S = -2\lambda_1 g, \tag{5}$$

in which λ_1 is a real scalar and W_1 indicates a smooth vector field. The Lie derivative operator is indicated by the symbol $\mathfrak{L}_{W_1}$. On $\mathcal{N}^n$, we denote (g, W_1, λ_1) as a RS which is referred to as shrinking, expanding, or steady depending on whether λ_1 is positive, negative, or zero. In particular, RS is trivial, and $\mathcal{N}^n$ is Einstein if W_1 is Killing or identically zero. Additionally, if W_1 is the gradient of some smooth function Ψ, that is, $W_1 = D\Psi$, then (5) takes the form

$$Hess\,\Psi - S = \lambda_1 g, \tag{6}$$

where $Hess$ and D stand for the gradient operator of g and its Hessian, respectively. A gradient Ricci soliton (briefly, GRS) is the metric obeying Eq. (6). It is claimed that the GRS's potential function is the smooth function Ψ.

In [29] on a Riemannian manifold, Hamilton proposed the concept of Yamabe flow, which was inspired by Yamabe's conjecture. A Yamabe flow is a semi-Riemannian manifold $\mathcal{N}^n$ supplied with a semi-Riemannian metric g that complies with the following conditions:

$$\frac{\partial}{\partial t}g(t) + \mathcal{R}g(t) = 0, \quad g_0 = g(t),$$

where the time is t. A Yamabe soliton (briefly, YS) is a semi-Riemannian manifold $\mathcal{N}$ with a semi-Riemannian metric g that obeys

$$\frac{1}{2}\mathfrak{L}_{W_1}g - (\mathcal{R} - \lambda_1)g = 0. \tag{7}$$

YS with $W_1 = D\Psi$ turns into the gradient Yamabe soliton (briefly, GYS).

Thus, Eq. (7) reduces to the form

$$Hess\Psi - (\mathcal{R} - \lambda_1)g = 0. \tag{8}$$

For $\Psi =$ constant (or W_1 is Killing), the GYS (or YS) becomes trivial.

Some interesting results on YSs have been investigated in ([21, 22, 24]) and also by others. Recently, in [23], the authors have studied YSs and GYSs in PF spacetimes.

Cho and Kimura researched real hypersurfaces in a complex space form, expanding the concept of RS to η-RS [17] and have expanded the notion of RS to η-RS. The η-RS on $(\mathcal{N}, g)$ fulfills

$$\pounds_{W_1}g + 2\beta_2 g + 2\beta_1 \eta \otimes \eta + 2S = 0, \tag{9}$$

β_1 is a constant and η is a 1-form. If $\beta_1 = 0$, then it recovers RS. η-RSs are called proper for $\beta_1 \neq 0$.

η-RS reduces to a gradient η-RS if W_1, the potential function is the gradient of Ψ. Hence the foregoing Eq. (9) transforms to

$$Hess\ \Psi + S + \beta_2 g + \beta_1 \eta \otimes \eta = 0. \tag{10}$$

η-RS is referred to as shrinking, expanding, or steady depending on whether β_2 is positive, negative, or zero.

There have been numerous recent publications on η-RS and gradient η-RS in the context of contact and Riemannian manifolds. The investigations of geometric flows are also started when looking at cosmological models such as PF spacetime. Blaga discovered the Poisson equation by investigating the η-RS in PF spacetime ([6, 7]).

The RSs and GRSs have been studied in the Lorentzian setting in ([3–5]), despite the fact that most solitons research has been conducted in the Riemannian category. Few exciting findings on RSs, GRSs, and GYSs in PF spacetime on GR theory have so far been researched in the available literature ([6, 23]). Also, the authors of this article have investigated different solitons in $f(\mathcal{R})$-gravity [25].

Inspired by the aforementioned studies, we planned to investigate different solitons on a PF spacetime solution of $f(\mathcal{R}, T)$-gravity. We lay out our paper as:

In Sect. 2, we provide some idea of PF spacetime solution of $f(\mathcal{R}, T)$-gravity. Sections 3 and 4 concern with RS and GRS and YS and GYS, respectively fulfilling PF spacetime solution of $f(\mathcal{R}, T)$-gravity. We produce few results of η-RS and gradient η-RS in $f(\mathcal{R}, T)$-gravity in Sect. 5.

2 $f(\mathcal{R}, T)$-Gravity

Harko et al. [31] introduced the modified gravity theory. For a number of specific instances of $f(\mathcal{R}, T)$ gravity, the related field equations have been implemented in metric formalism. Also in [31], the authors pursued possibility of rebuilding the FRW cosmologies by the right value of $f(T)$ for the model $f(\mathcal{R}, T) = \mathcal{R} + 2f(T)$. Here we choose [31]

$$f(\mathcal{R}, T) = 2f(T) + \mathcal{R}, \tag{11}$$

in which the component $2f(T)$ in the gravitational action changes the gravitational interaction between matter and curvature and $f(T)$ is an arbitrary function on the trace of T .

For $G = 1$, we presume a modified form of Einstein- Hilbert action term

$$H = \frac{1}{16\pi} \int [\mathcal{L}_m + f(\mathcal{R}, T)] (-g)^{\frac{1}{2}} d^4x, \tag{12}$$

in which $f(\mathcal{R}, T)$ stands for an arbitrary function and $\mathcal{L}_m$ is the scalar field's matter Lagrangian. The matter's stress energy tensor is written by

$$T_{ij} = \frac{-2\delta(-g)^{\frac{1}{2}}\mathcal{L}_m}{(-g)^{\frac{1}{2}}\delta^{ij}}. \tag{13}$$

Let the scalar field's matter Lagrangian be solely dependent on g_{ij}. The field equations of $f(\mathcal{R}, T)$-gravity are produced by the variation of action (12) with regard to the metric tensor g_{ij}

$$f_{\mathcal{R}}(\mathcal{R}, T)S_{ij} + g_{ij}\nabla_c\nabla^c f_{\mathcal{R}}(\mathcal{R}, T) - \nabla_a\nabla_b f_{\mathcal{R}}(\mathcal{R}, T) - \frac{1}{2}f(\mathcal{R}, T)\, g_{ij} \\ = -f_T(\mathcal{R}, T)T_{ij} + 8\pi T_{ij} - f_T(\mathcal{R}, T)\psi_{ij}, \tag{14}$$

in which $f_{\mathcal{R}}$ and f_T represent, respectively, the partial derivatives of $f(\mathcal{R}, T)$ with regard to $\mathcal{R}$ and T. According to standard notation, the d'Alembert operator is represented by $\Box \equiv \nabla_c\nabla^c$, the covariant derivative is represented by ∇_a, and ψ_{ij} is defined by

$$\psi_{ij} = -2T_{ij} + g_{ij}\mathcal{L}_m - 2g^{lk}\frac{\partial^2\mathcal{L}_m}{\partial g^{ij}\partial g^{lk}}. \tag{15}$$

If $f(\mathcal{R}, T) = f(\mathcal{R})$ is taken into account, then (14) yields the field equations for $f(\mathcal{R})$-gravity.

Choose a PF spacetime with σ, p, and the velocity vector ρ, which is provided by the equation $g(X_1, \rho) = \mu(X_1)$ for all X_1. Additionally, we are aware that there is no unique Lagrangian value, thus we make the assumption that $\mathcal{L}_m = -p$ and making use of (3), we acquire

$$T(X_1, Y_1) = (\sigma + p)\eta(X_1)\eta(Y_1) - pg(X_1, Y_1), \tag{16}$$

where $g(\rho, \rho) = -1$.

We get the variation of stress energy by using (16),

$$\psi(X_1, Y_1) = -pg(X_1, Y_1) - 2\,T(X_1, Y_1). \tag{17}$$

After adopting (11) and (14) we get the form

$$S(X_1, Y_1) = \frac{\mathcal{R}}{2}g(X_1, Y_1) - 2f'(T)T(X_1, Y_1) \\ -2f'(T)\psi(X_1, Y_1) + f(T)g(X_1, Y_1) + 8\pi T(X_1, Y_1). \tag{18}$$

In view of (16), (17) and (18), we acquire

$$S(X_1, Y_1) = \left\{\frac{\mathcal{R}}{2} - 8p\pi + f(T)\}g(X_1, Y_1) + \{(p + \sigma)(2f'(T) + 8\pi)\right\}\eta(X_1)\eta(Y_1). \tag{19}$$

Let

$$\left\{\frac{\mathcal{R}}{2} - 8p\pi + f(T)\} = a, \quad \{(p + \sigma)(2f'(T) + 8\pi)\right\} = b. \tag{20}$$

Hence, we have

$$S(X_1, Y_1) = ag(X_1, Y_1) + b\eta(X_1)\eta(Y_1). \tag{21}$$

Throughout this investigation we consider four-dimensional PF spacetime solution of $f(\mathcal{R}, T)$-gravity and contracting Eq. (19), we infer

$$\mathcal{R} = 4\{8p\pi - f(T)\} - \{(p+\sigma)(2f^{'}(T) + 8\pi)\}. \tag{22}$$

For $f(T) = 0$, the $f(\mathcal{R}, T)$-gravity theory reduces to GR-theory. In this case we can derive

$$a = \frac{8\pi(p-\sigma)}{-2}, \ b = 8\pi(p+\sigma),$$

which have been proved by Mantica et al. [35].

3 Ricci and Gradient Ricci Solitons

Suppose that a PF spacetime solution of $f(\mathcal{R}, T)$-gravity permits a RS and also W_1 is identical with ρ.

Hence, using explicit form of the Lie derivative in (5), we acquire

$$S(X_1, Y_1) = -\lambda_1 g(X_1, Y_1) - \frac{1}{2}[g(\nabla_{X_1}\rho, Y_1) + g(X_1, \nabla_{Y_1}\rho)]. \tag{23}$$

By contracting the aforementioned equation, we get

$$\mathcal{R} = -4\lambda_1 - div\rho, \tag{24}$$

in which 'div' stands for the divergence.

Combining the Eqs. (24) and (22), we obtain

$$4\{8p\pi - f(T)\} + \{(p+\sigma)(2f^{'}(T) + 8\pi)\} = -4\lambda_1 - div\rho. \tag{25}$$

Putting $X_1 = Y_1 = \rho$ in (19) and (23) and then comparing the resulting equations, we infer

$$4\left\{\frac{\mathcal{R}}{2} - 8p\pi + f(T)\} + \{(p+\sigma)(2f^{'}(T) + 8\pi)\right\} = \lambda_1. \tag{26}$$

Using (25) in the above equation, we acquire

$$\lambda_1 + \frac{3}{13}div\rho - \frac{8}{13}[f(T) - 8p\pi] = 0. \tag{27}$$

If ρ is Killing, then $div\rho = 0$. Hence from the above equation we get $\lambda_1 = \frac{8}{13}[f(T) - 8p\pi]$. Therefore, we can state:

Theorem 3.1 *If a PF spacetime obeying $f(\mathcal{R}, T)$-gravity endowed with Killing velocity vector field ρ permits a RS, then the RS is expanding if $p > \frac{f(T)}{8\pi}$, steady if $p = \frac{f(T)}{8\pi}$, and shrinking if $p < \frac{f(T)}{8\pi}$.*

Let in a PF spacetime solution of $f(\mathcal{R}, T)$-gravity, the soliton vector field W_1 of the RS (g, W_1, λ_1) be the gradient of Ψ. Then Eq. (6) turns into

$$\nabla_{X_1} D\Psi = QX_1 + \lambda_1 X_1, \tag{28}$$

for all $X_1 \in \mathfrak{X}_1(N)$.

The Eq. (28) along with the subsequent equation

$$R(X_1, Y_1)D\Psi = \nabla_{X_1}\nabla_{Y_1} D\Psi - \nabla_{Y_1}\nabla_{X_1} D\Psi - \nabla_{[X_1,Y_1]} D\Psi \tag{29}$$

give

$$R(X_1, Y_1)D\Psi = (\nabla_{Y_1} Q)(X_1) - (\nabla_{X_1} Q)(Y_1). \tag{30}$$

Putting the value of a and b in Eq. (19) and then differentiating covariantly we acquire

$$(\nabla_{X_1} Q)(Y_1) = X_1(a)Y_1 + X_1(b)\mathcal{A}(Y_1)\rho + b\{(\nabla_{X_1}\mathcal{A})(Y_1)\rho + \mathcal{A}(Y_1)\nabla_{X_1}\rho\}. \tag{31}$$

Using Eq. (31), we infer

$$\begin{aligned} R(X_1, Y_1)D\Psi &= Y_1(a)X_1 - X_1(a)Y_1 + \{Y_1(b)\mathcal{A}(X_1) - X_1(b)\mathcal{A}(Y_1)\}\rho \\ &+b\{(\nabla_{Y_1}\mathcal{A})X_1\rho + \mathcal{A}(X_1)\nabla_{Y_1}\rho\} - b\{(\nabla_{X_1}\mathcal{A})Y_1\rho + \mathcal{A}(Y_1)\nabla_{X_1}\rho\}. \end{aligned} \tag{32}$$

Making contraction of the Eq. (32), we get

$$S(Y_1, D\Psi) = -Y_1(b) + (n-1)Y_1(a) - \rho(b)\mathcal{A}(Y_1) + b\{(\nabla_{Y_1}\mathcal{A})\rho - (\nabla_{\rho}\mathcal{A})Y_1 - \mathcal{A}(Y_1)div\rho\}. \tag{33}$$

Again, from (19) we acquire

$$S(Y_1, D\Psi) = aY_1(\Psi) + b\eta(Y_1)\rho(\Psi). \tag{34}$$

Setting $Y_1 = \rho$ in Eqs. (33) and (34) and after that comparing the values of $S(\rho, D\Psi)$, we infer

$$(n-1)\{b\ \ div\rho + \rho(a)\} - (a-b)\rho(\Psi) = 0. \tag{35}$$

Assume that under the velocity vector field ρ, Ψ and a are invariant, that is, $\rho(\Psi) = 0$ and $\rho(a) = 0$. Then, from Eq. (35) we get

$$b\ \ div\rho = 0. \tag{36}$$

Hence, we conclude that either $b = 0$, or $div\rho = 0$.

Case I. We assume that $b = 0$ and therefore putting the value of b, we acquire

$$\{(\sigma + p)(4\pi + f^{'}(T))\} = 0,$$

that is, $(\sigma + p) = 0$, if $\{4\pi + f^{'}(T)\} \neq 0$.

Case II. We consider that $div\rho = 0$, then ρ is conservative. A conservative vector field is always irrotational, thus we conclude that the PF has zero vorticity.

Thus, we can state our result as:

Theorem 3.2 *If a PF spacetime obeying $f(\mathcal{R}, T)$-gravity with $\rho(\Psi) = 0$ and $\rho(a) = 0$ permits a GRS, then either the PF spacetime represents dark energy era, provided $\{4\pi + f^{'}(T)\} \neq 0$ or the spacetime has zero vorticity.*

4 Yamabe Solitons and Gradient Yamabe Solitons

Let a PF spacetime obeying $f(\mathcal{R}, T)$-gravity admit a YS. Therefore, we get

$$\mathfrak{L}_{W_1} g - 2(\mathcal{R} - \lambda_1)g = 0. \tag{37}$$

It is identical to

$$g(\nabla_{X_1} W_1, Y_1) + g(X_1, \nabla_{Y_1} W_1) = 2(\mathcal{R} - \lambda_1)g(X_1, Y_1). \tag{38}$$

Setting $W_1 = \rho$ in the previous equation, we acquire

$$g(\nabla_{X_1} \rho, Y_1) + g(X_1, \nabla_{Y_1} \rho) = 2(\mathcal{R} - \lambda_1)g(X_1, Y_1). \tag{39}$$

Again, putting $Y_1 = \rho$ in (39), we infer

$$\nabla_\rho \rho = 2(\mathcal{R} - \lambda_1)\rho, \tag{40}$$

since $g(\nabla_{X_1} \rho, \rho) = 0$.

Again, putting $X_1 = Y_1 = \rho$ in the Eq. (38), we obtain

$$\lambda_1 = \mathcal{R}. \tag{41}$$

Using (41) in (40), we get $\nabla_\rho \rho = 0$. Therefore, the integral curves of ρ are geodesic.

For $\lambda_1 = \mathcal{R}$, from (39), we infer $\mathfrak{L}_\rho g = 0$. It is known that in a PF spacetime if ρ is Killing, then $\mathfrak{L}_\rho p = 0$ and $\mathfrak{L}_\rho \sigma = 0$ [26].

Thus we state:

Theorem 4.1 *If a PF spacetime fulfilling $f(\mathcal{R}, T)$-gravity admits a YS, then the integral curves of the velocity vector field ρ are geodesics and energy density σ and isotropic pressure p remain invariant under ρ.*

From Eq. (8), we find

$$\nabla_{Y_1} D\Psi = (\mathcal{R} - \lambda_1) Y_1. \tag{42}$$

Differentiating (42) covariantly along Y_1, we obtain

$$\nabla_{X_1} \nabla_{Y_1} D\Psi = X_1(\mathcal{R}) Y_1 + (\mathcal{R} - \lambda_1) \nabla_{X_1} Y_1. \tag{43}$$

Interchanging X_1 and Y_1 in the foregoing equation and after that utilizing the above equation, (42) and (43) in $R(X_1, Y_1)D\Psi = \nabla_{X_1}\nabla_{Y_1} D\Psi - \nabla_{Y_1}\nabla_{X_1} D\Psi - \nabla_{[X_1,Y_1]} D\Psi$, we lead

$$R(X_1, Y_1)D\Psi = X_1(\mathcal{R})Y_1 - Y_1(\mathcal{R})X_1.$$

By contracting the aforementioned equation over X_1, we get

$$S(Y_1, D\Psi) = (n-1)Y_1(\mathcal{R}).$$

Also Eq. (19) gives

$$S(Y_1, D\Psi) = aY_1(\Psi) + b\rho(\Psi)\mathcal{A}(Y_1).$$

Combining the last two equations, we infer

$$aY_1(\Psi) + b\rho(\Psi)\mathcal{A}(Y_1) = (n-1)Y_1(\mathcal{R}). \tag{44}$$

Putting $Y_1 = \rho$ in the preceding equation, we acquire

$$(n-1)\rho(\mathcal{R}) - (a-b)\rho(\Psi) = 0. \tag{45}$$

Let ρ be Killing, from which we get $\rho(\mathcal{R}) = 0$. Hence the foregoing equation yields either $a \neq b$, or $a = b$.

Case (i): Let $a \neq b$, then we get $\rho(\Psi) = 0$ which entails $g(D\Psi, \rho) = 0$. Taking the covariant derivative of the last equation, we acquire

$$g(\nabla_{X_1}\rho, D\Psi) + (\mathcal{R} - \lambda_1)\eta(X_1) = 0, \tag{46}$$

in which Eq. (42) is applied. Since ρ is Killing, $g(\nabla_{X_1}\rho, Y_1) + g(X_1, \nabla_{Y_1}\rho) = 0$. Replacing X_1 by ρ, we infer $g(\nabla_\rho \rho, Y_1) = 0$, since $g(\rho, \rho) = -1$ implies $g(\rho, \nabla_{Y_1}\rho) = 0$. Hence, we find $\nabla_\rho \rho = 0$. Putting $X_1 = \rho$ in (46) and using the last result, we infer

$$\lambda_1 = \mathcal{R}. \tag{47}$$

Case (ii): If $a = b$, then from the Eq. (20) we acquire

$$\left[\frac{4\pi + f'(T)}{f'(T) - 4\pi}\right]\sigma + p + \frac{f(T)}{f'(T) - 4\pi} = 0, \tag{48}$$

which yields the EOS.

Thus, we write:

Theorem 4.2 *If a PF spacetime obeying $f(\mathcal{R}, T)$-gravity with Killing velocity vector ρ permits a GYS, then either the PF spacetime represents the EOS (48), or the spacetime has constant Ricci scalar.*

$f(\mathcal{R}, T)$-gravity theory reduces to GR-theory for $f(T) = 0$. Hence, from (48) we acquire $p = \sigma$ which implies that the PF is a stiff matter fluid.

Corollary 4.1 *If a PF spacetime equipped with Killing velocity vector ρ permits a GYS, then either the PF is a stiff matter fluid, or the spacetime has constant Ricci scalar.*

5 η-Ricci and gradient η-Ricci Solitons

Suppose a PF spacetime fulfilling $f(\mathcal{R}, T)$-gravity, admits an η-RS defined by (9), where the 1-form η is identical with the 1-form $\mathcal{A}$ of the PF spacetime, that is, $\eta(X_1) = \mathcal{A}(X_1) = g(X_1, \rho)$, where ρ is a unit timelike vector. Then (9) can be rewritten as

$$\pounds_{W_1} g + 2\beta_2 g + 2\beta_1 \mathcal{A} \otimes \mathcal{A} + 2S = 0.$$

Using the Lie derivative's explicit form, we get

$$S(X_1, Y_1) = -\frac{1}{2}[g(\nabla_{X_1}\rho, Y_1) + g(X_1, \nabla_{Y_1}\rho)] - \beta_2 g(X_1, Y_1) - \beta_1 \mathcal{A}(X_1)\mathcal{A}(Y_1). \tag{49}$$

By contracting the preceding equation, we find

$$\mathcal{R} = -div\rho - 4\beta_2 + \beta_1. \tag{50}$$

Again, contracting (21), we arrive at

$$\mathcal{R} = 4a - b. \tag{51}$$

Using the last two equations, we obtain

$$4a - b = -div\rho - 4\beta_2 + \beta_1. \tag{52}$$

Comparing (21) and (49), we acquire

$$ag(X_1, Y_1) + b\mathcal{A}(X_1)\mathcal{A}(Y_1) = -\frac{1}{2}[g(\nabla_{X_1}\rho, Y_1) + g(X_1, \nabla_{Y_1}\rho)] - \beta_2 g(X_1, Y_1) - \beta_1\mathcal{A}(X_1)\mathcal{A}(Y_1).$$

Since $g(\rho, \rho) = -1$, we have $g(\nabla_\rho\rho, \rho) = 0$. Using this result and putting $X_1 = Y_1 = \rho$ in the previous equation, we get

$$b - a = \beta_2 - \beta_1. \tag{53}$$

Using the preceding equation in (52), we infer

$$\beta_2 = -a - \frac{div\rho}{3}. \tag{54}$$

Assume that the divergence of the potential vector field is vanishing. Therefore, (54) recovers that

$$\beta_2 = -a = -\{\frac{R}{2} + f(T) - 8p\pi\}, \tag{55}$$

where we have used Eq. (21).

Therefore, the soliton is expanding for $p < \frac{\mathcal{R}+2f(T)}{16\pi}$, steady if $p = -\frac{\mathcal{R}+2f(T)}{16\pi}$, and shrinking for $p > \frac{\mathcal{R}+2f(T)}{16\pi}$.

Hence we can write

Theorem 5.1 *If $(g, \rho, \beta_2, \beta_1)$ is an η-RS in a PF spacetime fulfilling $f(\mathcal{R}, T)$-gravity and the 1-form η is identical with the 1-form $\mathcal{A}$ of the PF spacetime and the potential vector field ρ has vanishing divergence, then the soliton is expanding for $p < \frac{\mathcal{R}+2f(T)}{16\pi}$, steady if $p = -\frac{\mathcal{R}+2f(T)}{16\pi}$, and shrinking for $p > \frac{\mathcal{R}+2f(T)}{16\pi}$.*

Let in a PF spacetime fulfilling $f(\mathcal{R}, T)$-gravity, the soliton vector field W_1 of the η-RS be a gradient of Ψ, where the 1-form η is identical with the 1-form $\mathcal{A}$ of the PF spacetime. Then Eq. (10) converts to

$$Hess\ \Psi + S + \beta_2 g + \beta_1\mathcal{A}\otimes\mathcal{A} = 0 \tag{56}$$

which implies

$$\nabla_{X_1} D\Psi = -QX_1 - \beta_2 X_1 - \beta_1\mathcal{A}(X_1)\rho. \tag{57}$$

Due to the previous equation and the relation

$$R(X_1, Y_1)D\Psi = \nabla_{X_1}\nabla_{Y_1}D\Psi - \nabla_{Y_1}\nabla_{X_1}D\Psi - \nabla_{[X_1,Y_1]}D\Psi, \tag{58}$$

we acquire

$$\begin{aligned} R(X_1, Y_1)D\Psi = & -[(\nabla_{X_1}Q)(Y_1) - (\nabla_{Y_1}Q)(X_1)] \\ & +\beta_1[(\nabla_{Y_1}\mathcal{A})(X_1)\rho + \mathcal{A}(X_1)\nabla_{Y_1}\rho - (\nabla_{X_1}\mathcal{A})(Y_1)\rho - \mathcal{A}(Y_1)\nabla_{X_1}\rho]. \end{aligned} \tag{59}$$

Using the Eq. (21), we acquire

$$(\nabla_{X_1} Q)(Y_1) = X_1(a)Y_1 \\ +X_1(b)\mathcal{A}(Y_1)\rho + a\{(\nabla_{X_1}\mathcal{A})Y_1\rho + \mathcal{A}(Y_1)\nabla_{X_1}\rho\}. \quad (60)$$

Utilizing (60) in (59), we obtain

$$\begin{aligned} R(X_1, Y_1)D\Psi = & -[X_1(a)Y_1 - Y_1(a)X_1 + X_1(b)\mathcal{A}(Y_1)\rho - Y_1(b)\mathcal{A}(X_1)\rho \\ & +b\{(\nabla_{X_1}\mathcal{A})Y_1\rho + \mathcal{A}(Y_1)\nabla_{X_1}\rho - (\nabla_{Y_1}\mathcal{A})X_1\rho \\ & -\mathcal{A}(X_1)\nabla_{Y_1}\rho\}] + \beta_1[(\nabla_{Y_1}\mathcal{A})(X_1)\rho \\ & +\mathcal{A}(X_1)\nabla_{Y_1}\rho - (\nabla_{X_1}\mathcal{A})(Y_1)\rho - \mathcal{A}(Y_1)\nabla_{X_1}\rho]. \end{aligned} \quad (61)$$

Taking contraction of the previous equation, we lead

$$\begin{aligned} S(Y_1, D\Psi) = & -[Y_1(b) - 3Y_1(a) + \rho(b)\mathcal{A}(Y_1) + b\{(\nabla_\rho\mathcal{A})Y_1 - \mathcal{A}(Y_1)div\rho\}] \\ & +\beta_1[(\nabla_\rho\mathcal{A})(Y_1) + \mathcal{A}(Y_1)div\rho]. \end{aligned} \quad (62)$$

Also, the Eq. (21) yields

$$S(Y_1, D\Psi) = b\mathcal{A}(Y_1)\mathcal{A}(D\Psi) + ag(Y_1, D\Psi). \quad (63)$$

First put $Y_1 = \rho$ in (62) and (63), then comparing the equations provide

$$[3\rho(a) + bdiv\rho] + \beta_1 div\rho - (a - b)\rho(\Psi) = 0. \quad (64)$$

Let the scalar a and the potential function Ψ remain invariant under ρ, that is, $\rho(a) = 0$ and $\rho(\Psi) = 0$. Hence, Eq. (64) reveals that

$$(b + \beta_1)div\rho = 0, \quad (65)$$

which entails that either $b + \beta_1 = 0$ or $div\rho = 0$.

Case I. Assume that $b + \beta_1 = 0$ and $div\rho \neq 0$. Then putting the value of b from Eq. (20), we conclude that

$$-\{(\sigma + p)(8\pi + 2f^{'}(T))\} = \beta_1, \quad (66)$$

which implies that

$$p + \sigma = -\frac{\beta_1}{(8\pi + 2f^{'}(T))}. \quad (67)$$

Case II. We assume that $b + \beta_1 \neq 0$ and $div\rho = 0$.This indicates that ρ is conservative. A conservative vector field is invariably irrotational, thus we conclude that the PF has zero vorticity.

Therefore we state:

Theorem 5.2 *Let the PF spacetime fulfilling $f(\mathcal{R}, T)$-gravity permit a gradient η-RS. If the 1-form η is identical with the 1-form $\mathcal{A}$ of the PF spacetime and the scalar a and Ψ remain invariant under ρ, then either the spacetime represents the EOS (67), or the PF has vanishing vorticity.*

Remark 5.1 If $\beta_1 = 0$ and $(8\pi + 2f^{'}(T)) \neq 0$, then we have $p + \sigma = 0$, that is, the PF spacetime represents the dark matter era .

Remark 5.2 $div\rho = 0$ plays a significant role in Raychaudhuri's equation [39].

Acknowledgements We would like to express our gratitude to the referees for their comprehensive assessment of the manuscript and for their insightful remarks that helped to improve its quality.

References

1. G. Abbas, M.S. Khan, Z. Ahmad and M. Zubair, Higher-dimensional inhomogeneous perfect fluid collapse in $f(R)$ gravity. Eur. Phys. J. C **77**(2017., Article ID 443
2. L. Alias, A. Romero, M. Sánchez, Uniqueness of complete spacelike hypersurfaces of constant mean curvature in generalized Robertson-Walker spacetimes. Gen. Relat. Gravit. **27**, 71 (1995)
3. E. Barbosa, R. Pina, K. Tenenblat, On gradient Ricci solitons conformal to a pseudo-Euclidean space. Israel J. Math. **200**, 213–224 (2014)
4. W. Batat, M. Brozos-Vazquez, E. Garcia-Rio, S. Gavino-Fernandez, Ricci soliton on Lorentzian manifolds with large isometry groups. Bull. Lond. Math. Soc. **43**, 1219–1227 (2011)
5. M. Brozos-Vazqnez, G. Calvaruso, E. Garcia-Rio, S. Gavino-Fernandez, Three-dimensional Lorentzian homogeneous Ricci solitons. Israel J. Math. **188**, 385–403 (2012)
6. A.M. Blaga, Solitons and geometrical structures in a perfect fluid spacetime. Rocky Mountain J. Math. **50**, 41–53 (2020)
7. A.M. Blaga, η-Ricci solitons on Lorentzian para-Sasakian manifolds. Filomat **30**(2), 489–496 (2016)
8. Y.F. Cai, S. Capozziello, M.D. Laurentis, E.N. Saridakis, *$f(T)$ teleparallel gravity and cosmology*. Rept. Prog. Phys. **79**(2016). Article ID 106901
9. R.R. Caldwell, M. Kamionkowski, N.N. Weinberg, Dark Energy with $w < -1$ causes a cosmic doomsday. Phys. Rev. Lett. **91**, 071301 (2003)
10. R.R. Caldwell, A Phantom Menace? Cosmological consequences of a dark energy component with super-negative equation of state. Phys. Lett. B **545**, 23–29 (2002)
11. S. Capozziello, C.A. Mantica, L.G. Molinari, *Cosmological perfect-fluids in $f(R)$ gravity*. Int. J. Geom. Methods Mod. Phys. **16**(2019., Article ID 1950008
12. S. Capozziello, V.F. Cardone, V. Salzano, *Cosmography of f(R) gravity*. Phys. Rev. D **78**, 063504 (2008). https://doi.org/10.1103/PhysRevD.78.063504
13. S.M. Carroll, V. Duvvuri, M. Trodden, M.S. Turner, Is cosmic speed-up due to new gravitational physics? Phys. Rev. D **70**(2004). Article ID 043528
14. S.M. Carroll, A.D. Felice, V. Duvvuri, D.A. Easson, M. Trodden, M.S. Turner, Cosmology of generalized modified gravity models. Phys. Rev. D **71**(2005). Article ID 063513
15. P.H. Chavanis, Cosmology with a stiff matter era. Phys. Rev. D **92**, 103004 (2015)
16. B.-Y. Chen, U.C. De, F. Mofarreh, Study of Weakly Ricci-symmetric spacetimes under Gray's decomposition and $f(R, T)$-Gravity. Filomat **36**, 2391–2401 (2022)

17. J.T. Cho, M. Kimura, Ricci solitons and real hypersurfaces in a complex space form. Tohoku Math. J. **61**, 205–212 (2009)
18. U.C. De, S.A. Demirbağ and F.Ö. Zengin, Pseudo-symmetric spacetimes admitting $F(R)$-*gravity*. Lett. Math. Phys. **112**(2022). Article ID 17
19. A. De, T.-H. Loo, S. Arora and P.K. Sahoo, Energy conditions for a $(WRS)_4$ *spacetime in* $F(R)$*-gravity*. Eur. Phys. J. Plus **136**(2021). Article ID 218
20. A. De, T.-H. Loo, R. Solanki. and P.K. Sahoo, A conformally flat generalized Ricci recurrent spacetime in $F(R)$-gravity. Phys. Scr. **96**(2021). Article ID 085001
21. K. De, U.C. De, A note on gradient Solitons on para-Kenmotsu manifolds. Int. J.Geom. Methods Mod. Phys. **18**(01), 2150007 (2021). (11 pages)
22. K. De, U.C. De, Almost quasi-Yamabe solitons and gradient almost quasi-Yamabe solitons in paracontact geometry, Quaestiones Mathematicae, https://doi.org/10.2989/16073606.2020.1799882.
23. U.C. De, S.K. Chaubey, S. Shenawy, *Perfect fluid spacetimes and Yamabe solitons*. J. Math. Phys. **62**, 032501 (2021). https://doi.org/10.1063/5.0033967
24. K. De, U. C. De, A. A. Syied, N. B. Turki, S. Alsaeed, Perfect fluid spacetimes and gradient solitons. J. Nonlinear Math. Phys. https://doi.org/10.1007/s44198-022-00066-5
25. K. De, U.C. De, Investigations on solitons in $f(\mathcal{R})$-gravity. Eur. Phys. J. Plus **137**, 180 (2022). https://doi.org/10.1140/epjp/s13360-022-02399-y
26. K.L. Duggal, R. Sharma, *Symmetries of spacetimes and Riemannian manifolds, Mathematics and its Applications 487* (Kluwer Academic Press, Boston, London, 1999)
27. S. Güler, U.C. De, Perfect fluid spacetimes, Gray's decomposition and $f(R, T)$-gravity, Hacet. J. Math. Stat. **51**, 101–112 (2022)
28. R.S. Hamilton, Three-manifolds with positive Ricci curvature. J. Differ. Geom. **17**, 255–306 (1982)
29. R.S. Hamilton, The Ricci flow on surfaces. Contemp. Math. **71**, 237–261 (1988)
30. T. Harko, F.S.N. Lobo, $f(R, L_m)$-gravity. Eur. Phys. J. C **70**, 373–379 (2010)
31. T. Harko, T., F.S.N. Lobo, S. Nojiri, S.D. Odintsov, $f(R, T)$-gravity. Phys. Rev. D **84**(2011). Article ID 024020
32. J.B. Jiménez, L. Heisenberg, T. Koivisto, Coincident general relativity. Phys. Rev. D, **98**(2018). Article ID 044048
33. T.-H. Loo, A. De, S. Arora, P.K. Sahoo, Impact of curvature based geometric constraints on $F(R)$ theory. Eur. Phys. J. C **82**(2022). Article ID 705
34. T.-H. Loo, A. De, S. Mandal, P.K. Sahoo, *How a projectively flat geometry regulates* $F(R)$-gravity theory?. Phys. Scr., **96**(2021). Article ID 125034
35. C.A. Mantica, L.G. Molinari, U.C. De, A condition for a perfect fluid spacetime to be a generalized Robertson-Walker spacetime. J. Math. Phys. **57**(2), 022508 (2016). https://doi.org/10.1063/1.4941942
36. S.Nojiri, S.D.Odintsov, Modified gravity with negative and positive powers of curvature: Unification of inflation and cosmic acceleration. Phys. Rev. D **68**(2003). Article ID 123512
37. S. Nojiri, S.D. Odintsov, Modified Gauss-Bonnet theory as gravitational alternative for dark energy. Phys. Lett. B **631**, 1–6 (2005)
38. B. O'Neill, *Semi-Riemannian Geometry with Applications to Relativity* (Academic, New York, 1983)
39. A.K. Raychaudhuri, S. Banerji, A. Banerjee, *General Relativity, Astrophysics, and Cosmology* (Springer, New York, Inc, 1992)
40. Y.B. Zeldovich, The equation of state of ultrahigh density and its relativistic limitations. Soviet Phys. JETP **14**(5), 1143–1147 (1962)

A Survey on Lagrangian Submanifolds of Nearly Kaehler Six-Sphere

Ramesh Sharma

2010 Subject Class 53B25 · 53C25

1 Introduction

The Cayley multiplication can be used to define a vector cross product on the purely imaginary Octonions $\mathbb{R}^7$ by the formula:

$$u \times v = \frac{1}{2}(uv - vu)$$

and the standard inner product on $\mathbb{R}^7$ is given by

$$(u, v) = -\frac{1}{2}(uv + vu).$$

It turns out that

$$u \times (v \times w) + (u \times v) \times w = 2(u, w)v - (u, v)w - (w, v)u.$$

The Cayley multiplication on the Octonions $\mathcal{O}$ is thus given by

$$(r + u)(s + v) = rs - (u, v) + rv + su + (u \times v)$$

R. Sharma (✉)
Department of Mathematics, University of New Haven, West Haven 06516, USA
e-mail: rsharma@newhaven.edu

B.-Y. Chen et al. (eds.), *Geometry of Submanifolds and Applications*, Infosys Science Foundation Series, https://doi.org/10.1007/978-981-99-9750-3_3

for $r, s \in Re(\mathcal{O})$ and $u, v \in Im(\mathcal{O})$. It is also known that the group G_2 of automorphisms of $\mathcal{O}$ is precisely the group of isometries of $\mathbb{R}^7$ preserving the vector cross product.

An almost complex structure J on the unit six-sphere $S^6 = \{x \in \mathbb{R}^7 : ||x|| = 1\}$ is defined by $Ju = x \times u$, where $x \in S^6$ and $u \in T_x S^6$. It turns out that the Riemannian metric g on S^6 induced from $\mathbb{R}^7$ is a Hermitian metric with respect to J, i.e., $g(JX, JY) = g(X, Y)$ for arbitrary $X, Y \in \mathfrak{X}(S^6)$. It is also known that the Riemannian connection $\bar{\nabla}$ of (S^6, g) satisfies the condition: $(\bar{\nabla}_X J)(Y) + (\bar{\nabla}_Y J)(X) = 0$. An almost Hermitian manifolds satisfying this condition is known as a nearly Kähler (or Tachibana) manifold. Thus, S^6 is nearly Kähler, but not Kähler. Due to this special structure, S^6 has attracted the attention of several researchers. For details, we refer to Dillen and Vrancken [8].

A submanifold M of the nearly Kähler S^6 is said to be Lagrangian (also known as totally real) if J maps the tangent bundle of M onto the normal bundle, and the normal bundle onto the tangent bundle. We denote the metrics of S^6 and M by the same letter g. In [10], Ejiri proved the following results.

Theorem 1 ([10]) *A Lagrangian submanifold of the nearly Kähler S^6 is orientable and minimal.*

Its proof used the following lemma: "Let M be a Lagrangian submanifold of S^6 with nearly Kaehler structure (g, J) and let G be a (1, 2) tensor defined by $G(X, Y) = g(\nabla_X J)Y$ for all $X, Y \in \mathfrak{X}(M)$. Then $G(X, Y)$ is normal to M for all X, Y tangent to M." In [10], the following result was also established.

Theorem 2 ([10]) *Let M be a Lagrangian submanifold of constant curvature c, in nearly Kaehler S^6. Then either $c = 1$ (i.e., M is totally geodesic) or $c = 1/16$ (in which case M is locally equivalent to a minimal immersion: $S^6(1/16) \to S^6$ defined by spherical harmonics of degree 6 [9]).*

Lagrangian submanifolds of nearly Kaehler S^6 were also studied by Dillen, Opozda, Verstraelen and Vrancken [5]. Let us recall the following result of Deshmukh and Hadi [6].

Theorem 3 ([6]) *Let M be a compact three-dimensional Lagrangian submanifold of S^6 with nearly Kähler structure. Then there exists a global unit vector field ξ on M, and if $J\xi$ is parallel in the normal bundle, then M is a Sasakian manifold with ξ as the Reeb vector field.*

Intrigued by this result, Vrancken [16] showed that the second fundamental form of a Sasakian Lagrangian submanifold M of the nearly Kähler S^6 annihilates the Reeb vector field, and provided a complete classification of such submanifolds.

2 Almost Contact Metric Structure

Let us briefly review almost contact metric structures. A $(2n+1)$-dimensional smooth manifold M is said to be an almost contact metric manifold if carries a global 1-form η, a vector field ξ, a (1,1)-tensor field φ, and a Riemannian metric g satisfying

$$\varphi^2 = -I + \eta \otimes \xi, \quad \eta(\xi) = 1,$$
$$g(\varphi X, \varphi Y) = g(X, Y) - \eta(X)\eta(Y) \tag{2.1}$$

Obviously, $\varphi\xi = 0$, $\eta \circ \varphi = 0$, $g(X, \xi) = \eta(X)$, and ξ is unit. The almost contact metric structure (η, ξ, g) on M is called a contact metric structure if

$$(d\eta)(X, Y) = g(X, \varphi Y).$$

For a contact metric manifold, following Blair [1] and denoting the Lie derivative operator by L, we define a (1,1) tensor $h = \frac{1}{2}L_{\xi}\varphi$ which is known to be self-adjoint, trace-free, anti-commutes with φ, and annihilates ξ. We have the following formulas for a contact metric manifold:

$$\nabla_X \xi = -\varphi X - \varphi h X \tag{2.2}$$

$$Ric(\xi, \xi) = 2n - |h|^2. \tag{2.3}$$

The special case when $h = 0$ corresponds to K-contact metrics for which ξ is g-Killing. An almost contact metric is called Sasakian if

$$(\nabla_X \varphi)Y = g(X, Y)\xi - \eta(Y)X \tag{2.4}$$

where ∇ is the Riemannian connection of g. A Sasakian manifold can be characterized by a Riemannian manifold with a unit Killing vector field ξ satisfying the condition

$$R(X, Y)\xi = g(\xi, Y)X - g(\xi, X)Y. \tag{2.5}$$

A contact metric is K-contact if and only if

$$Ric(X, \xi) = 2n\eta(X). \tag{2.6}$$

In dimension three, K-contact condition is equivalent to Sasakian condition.

An almost contact metric structure on M is said to be normal if the almost complex structure $\mathcal{J}$ on $M \times \mathcal{R}$ defined by $\mathcal{J}(X, f\frac{d}{dt}) = (\varphi X - f\xi, \eta(X)\frac{d}{dt})$ is integrable. For a three-dimensional almost contact metric manifold, we have the following formula (Olszak [11])

$$(\nabla_X \varphi)Y = g(\varphi \nabla_X \xi, Y)\xi - \eta(Y)\varphi \nabla_X \xi. \tag{2.7}$$

A three-dimensional normal almost contact structure satisfies [11]

$$(\nabla_X \varphi)Y = a(g(X, Y)\xi - \eta(Y)X) + b(g(\varphi X, Y)\xi - \eta(Y)\varphi X) \tag{2.8}$$

$$\nabla_X \xi = -a\varphi X + b(X - \eta(X)\xi) \tag{2.9}$$

where a, b are smooth functions on M. Using Eq. (2.9) and that φ is anti self-adjoint with respect to g, we find

$$(L_\xi g)(X, Y) = 2b(g(X, Y) - \eta(X)\eta(Y)).$$

It can be easily verified that $\xi a = -2ab$. An almost contact metric structure satisfying the condition (2.8) is known as a trans-Sasakian structure. A cosymplectic manifold is a normal almost contact metric manifold such that η and Φ (the 2-form defined by $\Phi(X, Y) = g(X, \varphi Y)$) are both closed. This definition is equivalent to $\nabla\varphi = 0$ on an almost contact metric manifold. Thus a trans-Sasakian structure includes Sasakian and cosymplectic structures as special cases.

Finally, we recall two more classes of almost contact metric structures. An almost contact metric structure is called a nearly Sasakian structure if

$$(\nabla_X \varphi)Y + (\nabla_Y \varphi)X = 2g(X, Y)\xi - \eta(X)Y - \eta(Y)X.$$

An example of this structure is S^5 as a totally umbilical hypersurface of the nearly Kaehler unit sphere S^6 at a latitude of 45^0 and is non-Sasakian. Another almost contact metric structure that is not necessarily normal is the nearly cosymplectic structure defined by the condition

$$(\nabla_X \varphi)Y + (\nabla_Y \varphi)X = 0.$$

An example of this structure occurs on S^5 as a totally geodesic hypersurface of S^6. This structure is not normal and different from the standard contact metric structure on S^5 which is normal. We note that a normal nearly cosymplectic structure is cosymplectic. For details we refer to Blair [1].

3 Basic Equations for Lagrangian Submanifolds of S^6

On a Lagrangian submanifold M of the nearly Kaehler S^6, the tensor field G of type $(1, 2)$ defined on S^6 by $G(X, Y) = (\bar{\nabla}_X J)(Y)$ satisfies the following properties ([7, 10]):

$$G(X,Y) = -G(Y,X) \tag{3.1}$$

$$G(X,JY) = -JG(X,Y) \tag{3.2}$$

$$g(G(X,Y),Z) = -g(G(X,Z),Y) \tag{3.3}$$

$$\begin{aligned} g(G(X,Y),G(Z,W)) &= g(X,Z)g(Y,W) - g(X,W)g(Y,Z) \\ &+ g(JX,Z)g(Y,JW) - g(JX,W)g(Y,JZ) \end{aligned} \tag{3.4}$$

$$(\bar{\nabla}_X G)(Y,Z) = g(Y,JZ)X + g(X,Z)JY - g(X,Y)JZ \tag{3.5}$$

for arbitrary $X, Y, Z \in \mathfrak{X}(S^6)$. If ∇ and $\nabla^{\perp}$ denote the Riemannian connection of a submanifold M of a Riemannian manifold $\bar{M}$, and the connection in the normal bundle, respectively, then we have the Gauss and Weingarten formulas:

$$\bar{\nabla}_X Y = \nabla_X Y + \sigma(X,Y), \quad \bar{\nabla}_X N = -A_N X + \nabla^{\perp}_X N$$

for arbitrary $X, Y \in \mathfrak{X}(M)$, where A_N denotes the shape operator of M with respect to a normal vector field N, defined by $g(A_N X, Y) = g(\sigma(X,Y), N)$, and $\nabla^{\perp}$ denotes the normal connection. Let R and $\bar{R}$ denote the curvature tensors of M and $\bar{M}$ respectively, X, Y, Z, W any vector fields tangent to M and U, V any vector fields normal to M. The curvature tensor $R^{\perp}$ of the normal bundle (normal curvature operator) is defined by

$$R^{\perp}(X,Y)U = \nabla^{\perp}_X \nabla^{\perp}_Y U - \nabla^{\perp}_Y \nabla^{\perp}_X U - \nabla^{\perp}_{[X,Y]} U. \tag{3.6}$$

It may be noted that the normal curvature operator is conformally invariant (Chen [3]), i.e., if ${}^*g = \Omega^2 g$ for a smooth function Ω on M, and ${}^*R^{\perp}$ is the normal curvature operator corresponding to ${}^*g$, then ${}^*R^{\perp}(X,Y) = R^{\perp}(X,Y)$ for any $X, Y \in \mathfrak{X}(M)$. First we recall the following equations of Gauss, Codazzi and Ricci, respectively, for a general submanifold M of a Riemannian manifold $\bar{M}$:

$$\begin{aligned} g(\bar{R}(X,Y)Z,W) &= g(R(X,Y)Z,W) \\ &- g(\sigma(X,W),\sigma(Y,Z)) + g(\sigma(Y,W),\sigma(X,Z)) \end{aligned} \tag{3.7}$$

$$(\bar{R}(X,Y)Z)^{\perp} = (\nabla_X \sigma)(Y,Z) - (\nabla_Y \sigma)(X,Z) \tag{3.8}$$

$$g(\bar{R}(X,Y)U,V) = g(R^{\perp}(X,Y)U,V) + g(A_V A_U X - A_U A_V X, Y). \tag{3.9}$$

For a Lagrangian submanifold of the nearly Kähler S^6 we have the following formulas (Ejiri [10], Schafer and Smoczyk [15]):

$$\nabla^{\perp}_X JY = J\nabla_X Y + G(X, Y) \tag{3.10}$$

$$\sigma(X, Y) = JA_{JY}X \tag{3.11}$$

$$JG(X, JG(Y, Z)) = g(X, Z)Y - g(X, Y)Z \tag{3.12}$$

$$-\sigma(X, JG(Y, Z)) + JG(\sigma(X, Y), Z) + JG(Y, \sigma(Z, X)) = 0. \tag{3.13}$$

Henceforth M will denote a Lagrangian submanifold of the nearly Kaehler S^6. It is noteworthy that the three-form C defined by $C(X, Y, Z) = g(\sigma(X, Y), JZ)$ is always symmetric. Thus, at each point p of M, this cubic form can be represented by a homogeneous polynomial f_p of degree 3 given by

$$f_p(x, y, z) = g(\sigma(xe_1 + ye_2 + ze_3, xe_1 + ye_2 + ze_3), J(xe_1 + ye_2 + ze_3))$$

where (e_1, e_2, e_3) is an orthonormal basis at p. Minimality of M implies that this cubic form is trace free with respect to the metric.

4 Canonical Almost Contact Structure on *M*

In [14], Sharma, Deshmukh and Al-Solamy proved the following result.

Lemma 1 ([14]) *A unit vector field ξ on a Lagrangian submanifold (M, g) of the nearly Kaehler six-sphere, induces a canonical almost contact metric structure (φ, g, ξ) with structure tensor φ defined by $\varphi X = G(X, J\xi)$.*

Proof We begin with the hypothesis that ξ is a global unit vector field on M with respect to the induced metric g on M, and define a 1-form η by $\eta(X) = g(X, \xi)$. We also note from Eq. (3.2) that, for $X \in \mathfrak{X}(M)$, the vector field $G(X, J\xi) = -JG(X, \xi)$ is tangential to M, because we know from lemma 4.1 of [10] that $G(X, Y)$ is normal to M for all vector fields X, Y tangent to M, and hence $JG(X, Y)$ is tangent to M, as M is Lagrangian. Hence we define a (1,1)-tensor φ on M by

$$\varphi X = G(X, J\xi)$$

which shows, in view of properties (3.1) and (3.2), that $\varphi(\xi) = 0$. We also have that

$$\begin{aligned} g(\varphi X, Y) &= g\,(G(X, J\xi), Y) = -g\,(G(J\xi, X), Y) = g\,(X, G(J\xi, Y)) \\ &= -g\,(X, G(Y, J\xi)) = -g(X, \varphi Y). \end{aligned}$$

Further, we have

$$\begin{aligned} \varphi^2 X &= G(G(X, J\xi), J\xi) = -JG(G(X, J\xi), \xi) \\ &= -JG(-JG(X, \xi), \xi) = -JG(\xi, JG(X, \xi)). \end{aligned}$$

Using Eq. (3.12) in the above equation shows that

$$\varphi^2 X = -X + \eta(X)\xi$$

for any $X \in \mathfrak{X}(M)$. Furthermore,

$$\begin{aligned} g(\varphi X, \varphi Y) &= g(G(X, J\xi), G(Y, J\xi)) = g(G(X, \xi), G(Y, \xi)) \\ &= g(X, Y) - \eta(X)\eta(Y). \end{aligned}$$

Thus (φ, ξ, η, g) is an almost contact metric structure on M.

Definition 1 The structure (φ, ξ, η, g) defined by a unit vector field ξ, as defined in the above Lemma, is called a canonical almost contact metric structure on M.

As M is Lagrangian, setting $\nabla^{\perp}_X J\xi = JFX$, where F is a (1,1)-tensor field on M, the following result was proved in [14].

Proposition 1 ([14]) *Let M be a Lagrangian submanifold of S^6 with nearly Kaehler structure (g, J) and ξ be a global unit vector field on M, with the canonically induced almost contact metric structure (φ, g, ξ) on M. Then the structure is Sasakian if and only if $F = 0$, i.e., $J\xi$ is parallel in the normal bundle.*

Proof Using formulas (3.5), (3.11) and (3.13), we can show that

$$\begin{aligned} (\nabla_X \varphi)(Y) &= \nabla_X G(Y, J\xi) - G(\nabla_X Y, J\xi) \\ &= g(X, Y)\xi - \eta(Y)X + G(Y, \nabla^{\perp}_X J\xi). \end{aligned}$$

As per our setting $\nabla^{\perp}_X J\xi = JFX$, the above equation becomes

$$(\nabla_X \varphi)Y = g(X, Y)\xi - \eta(Y)X + G(Y, JFX). \tag{4.1}$$

If $F = 0$, then obviously the structure is Sasakian. Conversely, if the structure is Sasakian, then (4.1) reduces to $G(Y, JFX) = 0$. Substituting ξ for Y, and using (3.1) and (3.2) gives $\varphi FX = 0$. Operating it by φ provides $F = 0$, because $\eta(FX) = g(FX, \xi) = g(JFX, J\xi) = g(\nabla^{\perp}_X J\xi, J\xi) = 0$. This completes the proof.

Thus the question arises "Is it possible to weaken the condition on F for the canonical structure to reduce to Sasakian?" An answer was provided in [14] by assuming the canonical structure to be contact metric as follows.

Theorem 4 ([14]) *Let M be a Lagrangian submanifold of S^6 with nearly Kaehler structure (g, J) and ξ be a global unit vector field such that the canonically induced almost contact metric structure is contact metric structure on M. Then the structure is Sasakian if and only if F is divergence-free.*

At this point, we present the following generalization of a result of Vrancken stated in the beginning of Sect. 1, as obtained in [14].

Theorem 5 ([14]) *Suppose that the Lagrangian submanifold (M, g) of the nearly Kaehler six-sphere S^6 (g, J) is a contact metric manifold. Then (i) M is Sasakian if and only if its second fundamental form annihilates the Reeb vector field ξ, (ii) for Sasakian M, structure tensor φ is given by $\varphi X = G(X, J\xi)$, (iii) if the Sasakian submanifold M is parallel along ξ, i.e., the second fundamental form σ is parallel along ξ, then it is the totally geodesic three-sphere.*

Next, assuming the canonical almost contact metric structure on M to be normal, we have the following classification result [14].

Theorem 6 ([14]) *Let the canonical almost contact metric structure on the Lagrangian submanifold M of the nearly Kaehler six-sphere be normal. If F is divergence-free, then M is either Sasakian or cosymplectic.*

The following result shows how the nearly Sasakian (respectively, nearly cosymplectic) condition on the canonical almost contact structure on the Lagrangian submanifold of S^6 reduce to Sasakian (respectively, cosymplectic) condition. More precisely, we state

Theorem 7 ([13]) *Let M be a Lagrangian submanifold M of the nearly Kaehler S^6 and (φ, ξ, η, g) be the canonical almost contact metric structure induced on M by a global unit vector field tangent to M. Then, if (φ, ξ, η, g) is nearly Sasakian (resp., nearly cosymplectic), then it reduces to Sasakian (resp., cosymplectic).*

In [13], Sharma and Deshmukh proved the following result.

Theorem 8 ([13]) *If the normal connection of a Lagrangian submanifold M of the nearly Kähler S^6 is flat, then M is the totally geodesic S^3.*

Weakening the flatness of the normal connection, the following result was obtained in [12].

Theorem 9 ([12]) *If the normal curvature operator of a Lagrangian submanifold M of the nearly Kähler S^6 annihilates a vector field V normal to M, then the shape operator with respect to V vanishes and M is quasi-Einstein. Further, M is Sasakian if in addition, JV is Killing on M.*

For a minimal submanifold M in a real space of constant curvature c, Chen's inequality [2]

$$\delta_M \leq \frac{1}{2}(n+1)(n-2)$$

holds, where the invariant δ_M is given at a point p by $\tau(p) - inf(K)(p)$ where $inf(K)$ is the function assigning to each point $p \in M$ the infimum of $K(\pi)$, and where π runs over all planes in T_pM. The equality holds at a point p if and only if the index of relative nullity, i.e., the dimension of the subspace of T_pM annihilated by the second fundamental form σ of M is at least $n-2$ at that point. As the Lagrangian submanifold of the nearly Kähler S^6 is three-dimensional and minimal, Chen's equality $\delta_M = 2$ holds at a point if and only if the index of relative nullity is at least 1 at that point. So, if Chen's equality holds at each point of M, then there exists a unit vector field ξ tangent to M such that $\sigma(\xi, X) = 0$ for an arbitrary tangent vector field X on M, and M is quasi-Einstein (i.e., the two eigenvalues of the Ricci tensor with respect to the eigenspace orthogonal to ξ are equal). For details we refer to Deszcz, Dillen, Verstraelen and Vrancken [4]. As pointed out earlier, for a Sasakian Lagrangian submanifold of nearly Kähler S^6, $\sigma(\xi, X) = 0$, and hence Chen's equality is evidently satisfied [8].

Now let us recall the second result of Ejiri [10], as stated earlier, and note that if a manifold has constant curvature, then it is conformally flat, however, the converse need not be true. Prompted by this fact, the following generalization of Ejiri's second result was obtained in [12].

Theorem 10 ([12]) *If a conformally flat Lagrangian submanifold M of the nearly Kähler S^6 satisfies Chen's equality, then it is the totally geodesic S^3.*

Remark 1 This result holds also when "conformally flat" is replaced by either "locally symmetric" (because a three-dimensional locally symmetric manifold is conformally flat), or "parallel second fundamental form" (this implies, by Gauss equation, that M is locally symmetric).

The next result obtained in [12] is

Proposition 2 ([12]) *Let M be a three-dimensional submanifold (not necessarily Lagrangian) of S^6, and ξ be a unit vector field tangent to M belonging to the relative null space everywhere on M. Then M is quasi-Einstein. Further, (i) if ξ is Killing on M then M is Sasakian, (ii) if M is conformally flat then it is the totally geodesic S^3.*

Finally, we note the fact that a Sasakian structure is normal almost contact, but the converse need not be true. The converse problem was considered in [12] and the following result was established.

Theorem 11 ([12]) *If the almost contact metric structure induced by a unit tangent vector field on a Lagrangian submanifold of the nearly Kähler S^6 is normal, then it is Sasakian.*

Remark 2 This result provides an extrinsic characterization of a three-dimensional Sasakian structure.

5 Concluding Remark

Contracting the Gauss equation (3.7) twice and using the minimality of the Lagrangian submanifold M of the nearly Kaehler S^6, gives $r = 6 - |\sigma|^2$ where r and $|\sigma|$ denote the scalar curvature and the length of the second fundamental form of M. Hence $r \leq 6$, with equality holding if and only if M is the totally geodesic unit three-sphere.

References

1. D.E. Blair, *Riemannian Geometry of Contact and Symplectic Manifolds*, Progress in Mathematics, vol. 203. (Birkhauser, Boston, 2010) , Progress in Mathematics, vol. 203. (Birkhauser, Boston, 2010) , Progress in Mathematics, vol. 203. (Birkhauser, Boston, 2010)
2. B.Y. Chen, *Pseudo-Riemannian Geometry, δ -Invariants and Applications* (World Scientific, Singapore, 2011)
3. B.Y. Chen, in *Handbook of Differential Geometry*, eds. F. Dillen and L. verstraelen, Vol. 1. (Elsevier, Amsterdam, 2000)
4. R. Deszcz, F. Dillen, L. Verstraelen, L. Vrancken, Quasi-Einstein totally real submanifolds of the neraly Kehler 6-sphere. Tohoku Math. J. **51**, 461–478 (1999)
5. F. Dillen, B. Opozda, L. Verstraelen, L. Vrancken, On totally real 3-dimensional submanifold of the nearly Kaehler 6-sphere. Proc. Am. Math. Soc. **99**, 741–749 (1987)
6. S. Deshmukh, K.E. Hadi, A Sasakian structure on a 3-dimensional totally real submanifold of the nearly Kähler 6-sphere. Pan Am. Math. J. **2**, 43–52 (1992)
7. F. Dillen, B. Opozda, L. Verstraelen, L. Vrancken, On totally real 3-dimensional submanifold of the nearly Kähler 6-sphere. Proc. Am. Math. Soc. **99**, 741–749 (1987)
8. F. Dillen, L. Vrancken, Totally real submanifolds in $S^6(1)$ satisfying Chen's inequality. Trans. Am. Math. Soc. **348**, 1633–1646 (1996)
9. M. doCarmo, N.R. Wallach, Minimal immersions of spheres into spheres. Ann. Math. (2) **95**, 43–62 (1971)
10. N. Ejiri, Totally real submanifolds in a 6-sphere. Proc. Am. Math. Soc. **83**, 759–763 (1981)
11. Z. Olszak, Normal almost contact metric manifolds of dimension three. Ann. Pol. Math. **47**, 42–50 (1986)
12. R. Sharma, On Lagrangian submanifolds of the nearly Kähler 6-sphere. Contemp. Math. **756**, 219–228 (2020)
13. R. Sharma, S. Deshmukh, On Lagrangian submanifolds of the nearly Kähler 6-sphere. Contemp. Math. **674**, 153–160 (2016)
14. R. Sharma, S. Deshmukh, F. Al-Solamy, Almost contact Lagrangian submanifolds of nearly Kähler 6-sphere. Results. Math. **65**, 143–153 (2014)
15. L. Schafer, K. Smoczyk, Decomposition and minimality of Lagrangian submanifolds in nearly Kähler manifolds. Ann. Glob. Anal. Geom. **37**, 221–240 (2010)
16. L. Vrancken, Special Lagrangian submanifolds of the nearly Kähler 6-sphere. Glasgow Math. J. **45**, 415–426 (2003)

Pythagorean Submanifolds

Muhittin Evren Aydın, Adela Mihai, and Cihan Özgür

2000 MSC 11C20 · 53B25 · 53C42

1 Introduction

Let a, b, c be three positive integers. We then call (a, b, c) a *Pythagorean triple* (PT), if $a^2 + b^2 = c^2$ [18]. More generally, denote by $\mathbf{A}$, $\mathbf{B}$, $\mathbf{C}$ three matrices of type $n \times n$ ($n \in \{1, 2, \ldots\}$) whose entries are integer. Then $(\mathbf{A}, \mathbf{B}, \mathbf{C})$ is called a *matrix Pythagorean triple* (MPT), if

$$\mathbf{A}^2 + \mathbf{B}^2 = \mathbf{C}^2, \tag{1}$$

where $\mathbf{A}^2$ is the second power of $\mathbf{A}$ [3].

Infinitely many examples of the MPTs can be constructed by diagonal matrices as follows. Let (a_i, b_i, c_i) be a PT ($i \in \{1, 2, \ldots, n\}$) and also

$$\mathbf{A} = \pm\mathrm{diag}[a_1, \ldots, a_n], \quad \mathbf{B} = \pm\mathrm{diag}[b_1, \ldots, b_n], \quad \mathbf{C} = \pm\mathrm{diag}[c_1, \ldots, c_n].$$

M. E. Aydın
Department of Mathematics, Fırat University, 23200 Elazig, Türkiye
e-mail: meaydin@firat.edu.tr

A. Mihai (✉)
Department of Mathematics and Computer Science, Technical University of Civil Engineering Bucharest, 020396 Bucharest, Romania

Interdisciplinary Doctoral School, Transilvania University of Brasov, 500036 Brasov, Romania
e-mail: adela.mihai@utcb.ro; adela.mihai@unitbv.ro

C. Özgür
Department of Mathematics, İzmir Democracy University, 35140 Izmir, Türkiye
e-mail: cihan.ozgur@idu.edu.tr

B.-Y. Chen et al. (eds.), *Geometry of Submanifolds and Applications*, Infosys Science Foundation Series, https://doi.org/10.1007/978-981-99-9750-3_4

Then $(\mathbf{A}, \mathbf{B}, \mathbf{C})$ is an example of MPT. A non-trivial example will be given in the case $n = 2$:

$$\begin{bmatrix} 1 & 4 \\ -2 & 7 \end{bmatrix}^2 + \begin{bmatrix} -4 & 16 \\ -8 & 20 \end{bmatrix}^2 = \begin{bmatrix} -3 & 16 \\ -8 & 21 \end{bmatrix}^2.$$

Although the PTs have a geometrical interpretation (for example, see Pythagorean Theorem [18, Theorem 1.1]), the MPTs are established by pure algebraical arguments. With the motivation of geometrically interpreting for the MPTs, the differential-geometrical surfaces were investigated in [5]. The reason is that three fundamental bilinear forms on a surface are defined and the corresponding three square matrices are of type 2×2.

Explicitly, let $\mathbb{M}^3$ be one of the standard complete simply connected models of real space forms, i.e., the hyperbolic space $\mathbb{H}^3$, the Euclidean space $\mathbb{E}^3$ or the 3-sphere $\mathbb{S}^3$. The existence of a surface M^2 isometrically immersed into $\mathbb{M}^3$ was investigated when $\mathbf{I}, \mathbf{II}, \mathbf{III}$ are a MPT, namely,

$$\mathbf{I}^2 + \mathbf{II}^2 = \mathbf{III}^2, \tag{2}$$

with $\mathbf{I}, \mathbf{II}, \mathbf{III}$ the corresponding matrices to the first, second and third fundamental forms of M^2.

Denote by $\varphi := \left(1 + \sqrt{5}\right)/2$ and $\Phi := \left(-1 + \sqrt{5}\right)/2$ the *Golden Ratio* and the *conjugate Golden Ratio*, respectively. It was proven in [5, 6] that M^2 is isometric to $\mathbb{S}^2(\sqrt{\Phi})$ if $c = 0$, $\mathbb{S}^2(\sqrt{\varphi})$ if $c = -1$, $\mathbb{S}^2(\Phi)$ if $c = 1$. Moreover, in such a case, the Gaussian curvature of M^2 is $\varphi + c$. The present authors called this surface M^2 as a *Pythagorean surface*.

The importance of the notion of a Pythagorean surface comes from the fact that Golden Ratio and matrix Pythagorean triples surprisingly meet in a differential-geometrical result (see the words of Johannes Kepler (1571–1630): "*Golden Ratio and Pythagorean Theorem are two great treasures of Geometry*" [12, p. 62]).

There are two items which we need to underline: the first one is that although a Pythagorean surface is based on the Pythagorean Theorem, it is not related, as usual, to the distance among points. The second one is that the entries of $\mathbf{I}, \mathbf{II}, \mathbf{III}$ are allowed to be any real number, contrary to the original definition of the MPT.

Let $\mathbf{A}$ be the shape operator of M^2 and κ_1, κ_2 the eigenvalues of $\mathbf{A}$. Hence, Eq. (2) is equivalent to

$$\mathbf{I}^2 + (\mathbf{I} \cdot \mathbf{A})^2 = (\mathbf{I} \cdot \mathbf{A}^2)^2, \tag{3}$$

which is more useful because of the following reasons:

1. The first is solvability, that is, we may choose an orthonormal frame at a point $p \in M^2$ such that $\mathbf{A}$ is diagonal. Then, Eq. (3) is now the algebraical equation $1 + \kappa_i(p)^2 = \kappa_i(p)^4$ $(i \in \{1, 2\})$, which means that $\kappa_i = \pm\sqrt{\varphi}$ $(i \in \{1, 2\})$ is constant on M^2. Hence, a Pythagorean surface in $\mathbb{M}^3$ is also an isoparametric surface, that is, a surface whose principal curvatures are constants (see [7]).

2. Equation (3) can be directly satisfied for hypersurfaces in higher dimensions. Hence, we easily conclude that a Pythagorean hypersurface M^n in $\mathbb{M}^{n+1}$ is an isoparametric hypersurface with at most two distinct principal curvatures $\kappa_i = \pm\sqrt{\varphi}$ $(i \in \{1, \ldots, n\})$ when it satisfies Eq. (3). Here we recall that an *isoparametric hypersurface* in $\mathbb{M}^{n+1}$ is defined by the property that all principal curvatures are constant. For the Lorentzian setting a similar result can be found in [6].
3. Since we are able to choose an orthonormal frame on M^n, in some sense Eq. (3) is a polynomial equation on the shape operator. Hence, solving Eq. (3) is a remarkable problem (see [9, 10, 13, 14]).

In this paper, we consider a submanifold M^n in $\mathbb{M}^{m+n}$ satisfying an analogous of Eq. (3), introducing the notion of a Pythagorean submanifold (see Definition 1). We give examples observing the existence of totally umbilical and non-totally umbilical submanifolds. We also classify Pythagorean hypersurfaces.

2 Preliminaries

Let $(\tilde{M}, \langle\cdot,\cdot\rangle)$ be a Riemannian manifold and $\tilde{\nabla}$ the Levi-Civita connection on $\tilde{M}$. *The Riemannian curvature tensor* of $\tilde{M}$ is defined by

$$\tilde{R}(X, Y)Z = \tilde{\nabla}_X\tilde{\nabla}_Y Z - \tilde{\nabla}_Y\tilde{\nabla}_X Z - \tilde{\nabla}_{[X,Y]}Z,$$

where X, Y, Z are tangent vector fields to $\tilde{M}$ and $[X, Y] = XY - YX$.

Denote by $T_p\tilde{M}$ the tangent space of $\tilde{M}$ at $p \in \tilde{M}$ and by Γ a plane section of $T_p\tilde{M}$ with an orthonormal basis $\{e_1, e_2\}$. Then the *sectional curvature* of $\tilde{M}$ is $K(e_1, e_2) = \left\langle \tilde{R}(e_1, e_2)e_2, e_1 \right\rangle$. $\tilde{M}$ is called a *space form*, if it is complete simply connected and K is constant for each plane in $T_p\tilde{M}$ and for all $p \in \tilde{M}$.

Let M be a submanifold of $(\tilde{M}, \langle\cdot,\cdot\rangle)$ and denote by $\langle\cdot,\cdot\rangle$ the *induced metric tensor field* (or *first fundamental form*) on M from $\tilde{M}$. Then the *formula of Gauss* is

$$\tilde{\nabla}_X Y = \nabla_X Y + h(X, Y),$$

where ∇ is the induced Levi-Civita connection on M and h is the *second fundamental form* of the submanifold M.

Let ξ be a normal vector field of M. Then the *formula of Weingarten* is

$$\tilde{\nabla}_X \xi = -A_\xi X + D_X \xi,$$

where D is the metric connection on the normal bundle $T^\perp M$ and A_ξ is the *shape operator* in the direction of the normal vector field ξ of M. We have

$$\langle h(X,Y),\xi\rangle = \left\langle A_\xi X, Y\right\rangle.$$

Let $\{\xi_1,\ldots,\xi_m\}$ be an orthonormal basis of $T^\perp M$, where m is the codimension. The *Obata operator* is defined by

$$W = \sum_{i=1}^{m} A_{\xi_i}^2.$$

The *third fundamental form* III of M in $\tilde{M}$ is defined by

$$\mathrm{III}(X,Y) = \langle WX, Y\rangle,$$

(see [2]).

A normal vector field ξ to M is said to be *D-parallel*, if $D_X\xi = 0$, for some tangent vector X to M.

The *mean curvature vector* of M is $\mathbf{H} = \mathrm{trace}(h)/\dim(M)$; $\|\mathbf{H}\|$ is the *mean curvature*, where $\|\cdot\|$ is the induced norm. A *minimal submanifold* is defined by the property that $\|\mathbf{H}\| = 0$.

We call M *totally geodesic*, if h is identically 0. M^n is said to be *umbilical* with respect to ξ, if $A_\xi = \lambda \mathrm{Id}$, for the identity map Id defined on TM and a smooth function $\lambda : M \to \mathbb{R}$. Here the direction ξ is also said to be *umbilical*. M is called a *totally umbilical* submanifold in $\tilde{M}$, if each normal direction is umbilical. Moreover, we call M *pseudo-umbilical*, if $\mathbf{H}$ is umbilical.

Let X, Y, Z, W and respectively ξ, η be arbitrary tangent and respectively normal vector fields to M. Denote by R, $\tilde{R}$ and R^D the Riemannian curvature tensors associated with ∇, $\tilde{\nabla}$ and D, respectively. Then the *equations of Gauss* and *Ricci* are

$$\langle R(X,Y)Z,W\rangle = \left\langle \tilde{R}(X,Y)Z,W\right\rangle + \langle h(X,W),h(Y,Z)\rangle - \langle h(X,Z),h(Y,W)\rangle$$

and

$$\left\langle \tilde{R}(X,Y)\xi,\eta\right\rangle = \left\langle R^D(X,Y)\xi,\eta\right\rangle + \left\langle [A_\xi, A_\eta]X, Y\right\rangle.$$

We call M as *flat*, if R is identically 0. Analogously, M is said to have *flat normal bundle*, if R^D is identically 0. Also, via the equation of Ricci, if $\tilde{M}$ is of constant curvature, then all the shape operators can be diagonalized, simultaneously. In addition, when R^D is identically 0, the existence of m (m is the codimension of M) orthonormal D-parallel vector fields is assured (see [8, Propositions 3.1 and 3.2, pp. 124–127]).

We say that M has *constant principal curvatures*, if A_ξ has constant eigenvalues for any D-parallel vector field ξ. If, in addition, R^D is identically 0, then M is said to be *isoparametric* [7, 19]. A known result (see [11]) states that a Euclidean submanifold M has constant principal curvatures if and only if M is either isoparametric or a focal manifold to an isoparametric submanifold.

3 Examples of Pythagorean Submanifolds

We recall that $(\mathbf{I}, \mathbf{II}, \mathbf{III})$ is an MPT at every point of a Pythagorean surface $M^2 \subset \mathbb{E}^3$. The other cases to consider are that the triples $(\mathbf{I}, \mathbf{III}, \mathbf{II})$ or $(\mathbf{II}, \mathbf{III}, \mathbf{I})$ are a MPT. These three cases are given by a general formula

$$\lambda \mathbf{I}^2 + \mu \mathbf{II}^2 + \mathbf{III}^2 = \mathbf{0},$$

where $\lambda, \mu \in \{-1, 1\}$ are some parameters such that $\lambda + \mu \neq 2$, i.e., $(\lambda, \mu) \neq (1, 1)$. For example, $(\mathbf{I}, \mathbf{II}, \mathbf{III})$ is assumed to be a MPT provided $(\lambda, \mu) = (-1, -1)$.

When we use the shape operator and choose an orthonormal frame, at some point of M^2 we get

$$\lambda \mathrm{Id} + \mu \mathbf{A}^2 + \mathbf{A}^4 = \mathbf{0}, \tag{4}$$

or equivalently $\lambda + \mu \kappa_i^2 + \kappa_i^4 = 0$ $(i \in \{1, 2\})$. The discriminant of this algebraical equation is negative when $(\lambda, \mu) = (1, -1)$, implying the non-existence of a surface such that $(\mathbf{I}, \mathbf{III}, \mathbf{II})$ is a MPT. Then the only cases to study are $(\lambda, \mu) = (-1, -1)$ and $(\lambda, \mu) = (-1, 1)$ and hence Eq. (4) may be given by using a unique parameter (see Eq. (5)).

Let $\mathbb{M}^{m+n}$ denote the hyperbolic space $\mathbb{H}^{m+n}$ or the Euclidean space $\mathbb{E}^{m+n}$ or the sphere $\mathbb{S}^{m+n}$. Now, we generalize the notion of the Pythagorean surface to the case of arbitrary dimension and codimension.

Definition 1 Let $M^n \subset \mathbb{M}^{m+n}$ be a complete submanifold $(n \geq 2)$. Then M^n is called *ξ-Pythagorean*, if $\mathbf{A}_\xi$ satisfies

$$\mathbf{I}^2 + \epsilon (\mathbf{I} \cdot \mathbf{A}_\xi)^2 = (\mathbf{I} \cdot \mathbf{A}_\xi^2)^2, \quad \epsilon \in \{-1, 1\}, \tag{5}$$

where ξ is a D-parallel normal vector field to M^n, with $\|\xi\| = 1$. In addition, if there is an orthonormal parallel frame $\mathcal{B}$ of $T^\perp M^m$ such that M^m is ξ-Pythagorean for every $\xi \in \mathcal{B}$, then M^n is called *Pythagorean* with respect to $\mathcal{B}$.

Here we point out that the existence of a parallel frame $\mathcal{B}$ assures that all the shape operators can be diagonalized, simultaneously; this holds trivially in the particular case $m = 1$. Also, it is obvious that $\mathbf{A}_\xi \neq \pm \mathrm{Id}$ and $\mathbf{A}_\xi \neq \mathbf{0}$ at every point of M^n.

The notion of a Pythagorean submanifold is valid for the case of dimension ≥ 2, but still the one-dimensional case is meaningful (see Fig. 1). Clearly, in the Euclidean setting, we prove the following.

Proposition 3.1 *A Pythagorean curve in $\mathbb{E}^{m+1}$ is a part of the circle lying in a 2-plane of $\mathbb{E}^{m+1}$ with radius $\sqrt{\Phi/m}$ $(\epsilon = 1)$ or $\sqrt{\varphi/m}$ $(\epsilon = -1)$.*

Proof Let $s \in I \subset \mathbb{R}$ and $s \mapsto \mathbf{x}(s) \in \mathbb{E}^{m+1}$ be a parameterized curve by arc length. According to Definition 1, we choose an orthonormal parallel frame $\{\mathbf{t}, \xi_1, \ldots, \xi_m\}$ on $\mathbf{x}$ such that

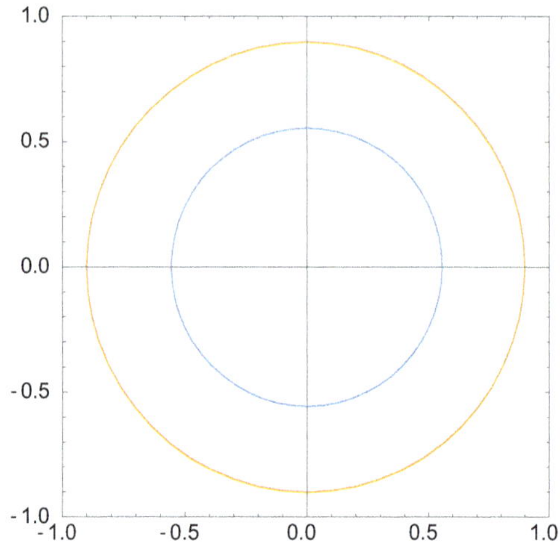

Fig. 1 Pythagorean curves. Blue: a circle of radius $\sqrt{\Phi/2}$. Orange: a circle of radius $\sqrt{\varphi/2}$

$$\mathbf{x}' = \mathbf{t}, \quad \mathbf{t}' = \kappa_1 \xi_1 + \cdots + \kappa_m \xi_m, \quad \xi_i' = -\kappa_i \mathbf{t}, \quad i \in \{1, \ldots, m\}, \tag{6}$$

where κ_i is the i-th curvature with respect to $\{\mathbf{t}, \xi_1, \ldots, \xi_m\}$ and the prime denotes the derivative with respect to the arc length parameter. The role of the shape operator in Definition 1 is played by the i-th curvature κ_i and hence Eq. (5) is now

$$1 + \epsilon \kappa_i^2 = \kappa_i^4, \quad \epsilon \in \{-1, 1\},$$

where $\kappa_i = \pm\sqrt{\varphi}$ $(\epsilon = 1)$ or $\kappa_i = \pm\sqrt{\Phi}$ $(\epsilon = -1)$, for $i \in \{1, \ldots, m\}$. Since κ_i is constant for each $i \in \{1, \ldots, m\}$, we have $\mathbf{x}''' = -(\kappa_1^2 + \cdots + \kappa_m^2)\mathbf{x}'$, implying that, on the interval I, $\mathbf{x}$ lies in a 2-plane of $\mathbb{E}^{m+1}$ (see [17, p. 45]). Denoting by κ the first Frenet curvature of $\mathbf{x}$, then we have from the middle equality in Eq. (6) that $\kappa = \sqrt{m\varphi}$ $(\epsilon = 1)$ or $\kappa = \sqrt{m\Phi}$ $(\epsilon = -1)$. This completes the proof. □

Remark 3.2 In order to be compatible with Definition 1, we use an orthonormal parallel frame on a curve in the proof of Proposition 3.1, which is different from the Frenet frame.

In what follows we will discuss several examples of Pythagorean submanifolds of dimension ≥ 2.

Example 3.3 Let $M^n = \mathbb{S}^n(r)$ be an ordinary hypersphere of radius r (see Fig. 2). Then, for $\epsilon = 1$ in Eq. (5), it is Pythagorean in $\mathbb{E}^{n+1}$ if $r = \sqrt{\Phi}$, in $\mathbb{H}^{n+1}$ if $r = \sqrt{\varphi}$ and in $\mathbb{S}^{n+1}$ if $r = \Phi$. See [6] for the calculating the values of r.

Example 3.4 Let $(x_1, x_2, x_3, x_4, x_5, x_6)$ be the canonical coordinates in $\mathbb{E}^6$ and consider the Clifford torus $M^4 = \mathbb{S}^2(\sqrt{\Phi/2}) \times \mathbb{S}^2(\sqrt{\Phi/2})$ (resp. $\mathbb{S}^2(\sqrt{\varphi/2}) \times \mathbb{S}^2(\sqrt{\varphi/2})$). We may choose a D-parallel orthonormal frame $\{\xi_1, \xi_2\}$ on M^4, where

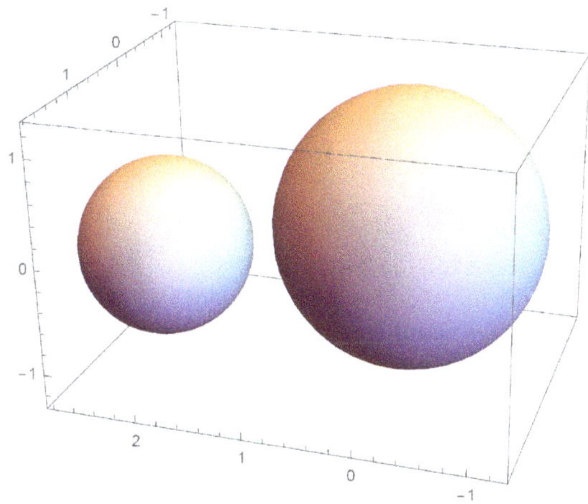

Fig. 2 Pythagorean surfaces. Left: a sphere of radius $\sqrt{\Phi}$ centered at (2, 1, 0). Right: a sphere of radius $\sqrt{\varphi}$ centered at (0, 0, 0)

$$\xi_1 = \sqrt{\varphi}(x_1, x_2, x_3, x_4, x_5, x_6), \quad \xi_2 = \sqrt{\varphi}(-x_1, -x_2, -x_3, x_4, x_5, x_6).$$

A direct calculation follows

$$\mathbf{A}_{\xi_1} = \sqrt{\varphi}\mathrm{Id}, \quad \mathbf{A}_{\xi_2} = \sqrt{\varphi}\mathrm{diag}[1, -1].$$

Then M^4 is a Pythagorean submanifold because $\mathbf{A}_{\xi_1}$ and $\mathbf{A}_{\xi_2}$ satisfy Eq. (5) for $\epsilon = 1$ (resp. $\epsilon = -1$).

Example 3.5 Let $\mathbb{H}_1^{n+1}$ be the anti-de Sitter space and $M^n = \mathbb{S}_1^2(\Phi) \times \mathbb{H}^{n-2}(-\varphi)$. Then, $\mathbf{A} = \sqrt{1+\Phi}\mathrm{Id}_2 \oplus \sqrt{1-\varphi}\mathrm{Id}_{n-2}$, where Id_2 is identity matrix of type 2 and Id_{n-2} is identity matrix of type $n-2$ (see [1]). Then M^n is a Pythagorean hypersurface in $\mathbb{H}_1^{n+1}$ because $\mathbf{A}$ satisfies Eq. (5) with $\epsilon = 1$.

We would like to point out that in Examples 3.4 and 3.5 the Pythagorean submanifolds are not totally umbilical.

4 Pythagorean Euclidean Submanifolds

In this section, we will review the new results obtained by the same authors in [4].

Denote by $M^n \subset \mathbb{E}^{m+n}$ a Pythagorean submanifold throughout the section.

By using Definition 1, it is easy to see that M^n has constant principal curvatures and hence it is isoparametric. An immediate result states that if $m = 1$, then M^n is isometric to $\mathbb{S}^n(\sqrt{\Phi})$ or $\mathbb{S}^n(\sqrt{\varphi})$. In addition, there is a D-parallel frame $\mathcal{B}$ on M^n such that the principal curvatures along every vector field that belongs to $\mathcal{B}$ are $\pm\sqrt{\varphi}$ or $\pm\sqrt{\Phi}$. Hence, since the principal curvatures are non-zero, M^n is full, i.e., it cannot be contained in a hyperplane of $\mathbb{E}^{m+n}$ and, because of [19, Proposition 1.3], it follows $m \leq n$. Furthermore, since the principal curvatures are all non-zero, M^n must be compact, see [19, Theorems 1.18, 1.20].

By the fact that a compact isoparametric manifold of codimension $m = 2$ is also an isoparametric hypersurface in $\mathbb{S}^{n+1}$ [19], we conclude that M^n is isometric to the Clifford torus $\mathbb{S}^k(\rho) \times \mathbb{S}^{n-k}(\rho)$, where ρ is either $\sqrt{\Phi/2}$ or $\sqrt{\varphi/2}$.

On the other hand, if $m \geq 2$, then M^n in $\mathbb{E}^{m+n}$ cannot be totally umbilical. Otherwise, it is contained in $\mathbb{S}^{n+1}$ (see [8]), where the $m-1$ shape operators vanish and in this case Eq. (5) does not satisfy for these shape operators. In addition, denoting by $\mathbf{H}$ the mean curvature vector field of M^n, $\|\mathbf{H}\|$ is constant and hence D-parallel. Since M^n is compact, $\|\mathbf{H}\| \neq 0$.

Assume that $m \geq n \geq 3$ and M^m is of non-negative sectional curvature. Then it is isometric to (see [4])

$$\mathbb{S}^{l_1}(\rho) \times \cdots \times \mathbb{S}^{l_n}(\rho), \quad l_1, \ldots, l_n \geq 1, \quad l_1 + \cdots + l_n = m,$$

where ρ is $\sqrt{\Phi/n}$ ($\epsilon = 1$) or $\sqrt{\varphi/n}$ ($\epsilon = -1$). If in addition $m \geq 2$ (resp. $m = 1$), then M^m is pseudo-umbilical (resp. totally umbilical).

5 Pythagorean Submanifolds of Space Forms

By Definition 1, we have an immediate result.

Proposition 5.1 *Let $M^n \subset \mathbb{M}^{m+n}$ be a Pythagorean submanifold with respect to $\mathcal{B}$. Then, M^n is isoparametric and the principal curvatures along every vector field in $\mathcal{B}$ are $\pm\sqrt{\Phi}$ or $\pm\sqrt{\varphi}$.*

In the Euclidean setting, by means of the important results in the theory of isoparametric submanifolds, we have classified Pythagorean submanifolds in Sect. 4. In a similar way, when the ambient space is hyperbolic, Pythagorean submanifolds can be found by using the results from [20]. In contrast, the problem is complicated, if $\mathbb{M}^{m+n}$ is a sphere.

Now assume that $\mathbb{M}^{n+1}$ is a hyperbolic space or a sphere and $M^n \subset \mathbb{M}^{n+1}$ is a complete hypersurface. If M^n is Pythagorean, then it is an isoparametric hypersurface in $\mathbb{M}^{n+1}$ with at most two distinct principal curvatures $\pm\sqrt{\Phi}$ for the case $\epsilon = 1$ (or $\pm\sqrt{\varphi}$ for the case $\epsilon = -1$).

In what follows, we recall the isoparametric hypersurfaces in $\mathbb{H}^{n+1}$ or $\mathbb{S}^{n+1}$ from [15, 16]. Let $X = (x_1, \ldots, x_{n+2}) \in \mathbb{E}^{n+2}$ and $l \in \{1, \ldots, n+1\}$.

Then, the isoparametric hypersurfaces of $\mathbb{H}^{n+1}$ are

1. $M^n = \{X \in \mathbb{H}^{n+1} : x_l = 0\}$ with $\mathbf{A} = \mathbf{0}$,
2. $M^n = \{X \in \mathbb{H}^{n+1} : x_l = r \in (1, \infty)\}$ with $\mathbf{A} = \sqrt{r^2-1}/r\mathrm{Id}$,
3. $M^n = \{X \in \mathbb{H}^{n+1} : x_{n+2} = x_{n+1} + 1\}$ with $\mathbf{A} = \mathrm{Id}$,
4. $M^n = \mathbb{H}^{n+1} \cap \mathbb{S}^n(r) = \{X \in \mathbb{H}^{n+1} : x_1^2 + \cdots + x_{n+1}^2 = r^2, r > 0\}$ with $\mathbf{A} = \sqrt{r^2+1}/r\mathrm{Id}$,
5. $M^n = \mathbb{H}^{n+1} \cap \mathbb{S}^l(r) = \{X \in \mathbb{H}^{n+1} : x_1^2 + \cdots + x_{l+1}^2 = r^2, r > 0\}$ with $\mathbf{A} = \sqrt{r^2+1}/r\mathrm{Id}_l \oplus r/\sqrt{r^2+1}\mathrm{Id}_{n-l}$.

The isoparametric hypersurfaces of $\mathbb{S}^{n+1}$ having at most two distinct principal curvatures are

1. $M^n = \{X \in \mathbb{S}^{n+1} : x_{n+2} = \sqrt{1-r^2}, r \in (0, 1)\}$ with $\mathbf{A} = \sqrt{1-r^2}/r\mathrm{Id}$,
2. $M^n = \mathbb{S}^l(r_1) \times \mathbb{S}^{n-l}(r_2)$ with $r_1^2 + r_2^2 = 1$, $1 \le l < n$.

Consequently, we obtain the classification of Pythagorean hypersurfaces in $\mathbb{H}^{n+1}$ or $\mathbb{S}^{n+1}$ as follows.

Theorem 5.2 *A Pythagorean hypersurface in $\mathbb{S}^{n+1}$ is isometric to $\mathbb{S}^n(\Phi)$ and in $\mathbb{H}^{n+1}$ is isometric to $\mathbb{S}^n(\sqrt{\varphi})$.*

Proof We refer to [6, Theorems 2 and 3] for details. □

References

1. N. Abe, N. Koike, S. Yamaguchi, Congruence theorems for proper semi-Riemannian hypersurfaces in a real space form. Yokohama Math. J. **35**, 123–136 (1987)
2. J. Álvarez-Vizoso, Curvature of the Gauss map for normally flat submanifolds in space forms. arXiv:2209.06246v2
3. M. Arnold, A. Eydelzon, On matrix Pythagorean triples. Amer. Math. Monthly **126**, 158–160 (2019)
4. M.E. Aydin, A. Mihai, C. Özgür, Pythagorean submanifolds in Euclidean spaces. Results Math. **78**, 211 (2023)
5. M.E. Aydin, A. Mihai, A note on surfaces in space forms with Pythagorean fundamental forms. Mathematics **8**, 444 (2020)
6. M.E. Aydin, A. Mihai, C. Özgür, Pythagorean isoparametric hypersurfaces in Riemannian and Lorentzian space forms. Axioms **11**, 59 (2022)
7. T.E. Cecil, Isoparametric and Dupin hypersurfaces. SIGMA Symmetry Integr. Geom. Methods Appl. **4**, 62 (2008)
8. B.-Y. Chen, *Total Mean Curvature and Submanifolds of Finite Type*, 2nd edn. With a foreword by Leopold Verstraelen. Series in Pure Mathematics, 27 (World Scientific Publishing Co. Pte. Ltd., Hackensack, NJ, 2015)
9. M. Crasmareanu, C.-E. Hretcanu, M.-I. Munteanu, Golden and product-shaped hypersurfaces in real space forms. Int. J. Geom. Methods Mod. Phys. **10**, 1320006 (2013)
10. A.J. Di Scala, On submanifolds whose shape operator is unipotent. J. Geom. **85**, 32–34 (2006)
11. E. Heintze, C. Olmos, G. Thorbergsson, Submanifolds with constant principal curvatures and normal holonomy groups. Internat. J. Math. **2**, 167–175 (1991)

12. M. Livio, *The Golden Ratio: The Story of Phi, the World's Most Astonishing Number* (Broadway Books, New York, USA, 2003)
13. C. Özgür, N.Y. Özgür, Classification of metallic shaped hypersurfaces in real space forms. Turkish J. Math. **39**, 784–794 (2015)
14. C. Özgür, N.Y. Özgür, Metallic shaped hypersurfaces in Lorentzian space forms. Rev. Un. Mat. Argentina **58**, 215–226 (2017)
15. P.J. Ryan, Homogeneity and some curvature conditions for hypersurfaces. Tohoku Math. J. **21**, 363–388 (1969)
16. P.J. Ryan, Hypersurfaces with parallel Ricci tensor. Osaka Math. J. **8**, 251–259 (1971)
17. M. Spivak, *A Comprehensive Introduction to Differential Geometry*, Vol. II, 3rd edn. (Publish or Perish, Inc., Houston, Texas, 1999)
18. R. Takloo-Bighash, *A Pythagorean Introduction to Number Theory: Right Triangles, Sums of Squares, and Arithmetic*. Undergraduate Texts in Mathematics (Springer, Cham, 2018)
19. C.L. Terng, Isoparametric submanifolds and their Coxeter groups. J. Differ. Geom. **21**, 79–107 (1985)
20. B. Wu, Isoparametric submanifolds of hyperbolic spaces. Trans. Amer. Math. Soc. **331**, 609–626 (1992)

On 4-Dimensional Submanifolds of f-Manifolds

Gherici Beldjilali

Mathematics Subject Classification (2010) Primary 53C55 · Secondary 53C25

1 Introduction

The concept of an f-structure on a $(2n + s)$-dimensional manifold which satisfies $f^3 + f = 0$, where f is a (1, 1)-tensor field of constant rank $2n$, was introduced in 1961 by Yano [20]. Soon after Yano's papers, numerous investigations of manifolds endowed with an f-structure, and submanifolds of them, were begun.

An f-structure is a generalization of some structures defined on different types of differentiable manifolds. Almost complex structures ($s = 0$) defined by Weil [22] and almost contact structures ($s = 1$) introduced by Sasaki [18] are well-known examples of f-structures. Later, some authors continued to study these kinds of structures. For example, Goldberg and Yano [13] introduced and studied globally framed metric f-structures on $(2n + s)$-dimensional differentiable manifolds. In [4], Blair defined K-structures (and particular cases of S-structures and C-structures). We note that K-structures are the analogue of Kählerian structures in the almost complex geometry and also S-structures (resp., C-structures) of Sasakian structures (resp. cosymplectic structures) in the almost contact geometry. In [1], Alegre, Fernandez and Prieto-Martin introduced a new class of metric f-manifolds, which are called trans-S-manifolds because they play the same role as trans-Sasakian manifolds in metric f-geometry. The class of trans-S-manifolds includes S-manifolds, C-manifolds and

G. Beldjilali (✉)
Laboratory of Quantum Physics and Mathematical Modeling (LPQ3M), University of Mascara, Mascara, Algeria
e-mail: gherici.beldjilali@univ-mascara.dz

B.-Y. Chen et al. (eds.), *Geometry of Submanifolds and Applications*, Infosys Science Foundation Series, https://doi.org/10.1007/978-981-99-9750-3_5

f-Kenmotsu manifolds [19] as special cases. They prove some properties of trans-S-manifolds and show that several examples constructed by other authors are included in this class.

An f-structure with $s = 2$ has arisen in the study of hypersurfaces in almost contact spaces [6]. The structure has been firstly studied by Goldberg and Yano [12]. Inspired by this structure, Debnath and Konar [9] constructed a new kind of structure on an even-dimensional Riemannian manifold with dimension $2n + 2$ ($s = 2$), named an almost pseudo-contact structure with two associated vector fields. They presented some basic properties of this structure. Recently, Beldjilali and Aydin [2] introduced a new kind of this type which combines the Sasaki and Kenmotsu structures and, at the same time, is different from the trans-S-manifolds.

By motivating these studies, in this paper we study a 4-dimensional almost pseudo contact structure, or, more precisely, we call it an almost bi-contact metric structure to indicate the presence of two global vector fields, as a class of f-structure on a 4-dimensional Riemannian manifold, and we deduce some properties from the complex structures associated with it.

The paper is organized as follows.

In Sect. 2, we review the basic definitions and results that are needed to state and prove our results. In Sect. 3, we derive certain necessary and sufficient conditions for almost bi-contact metric structures to be normal or integrable, and we point out some of their consequences. We give a concrete example. Section 4 is intended to study an interesting class of almost bi-contact metric manifolds such as the "bi-C_{12}-manifolds", and we construct an example to show the existence of such manifolds.

2 Review Definitions and Needed Results

Through this paper, (M, g) always denotes a Riemannian manifold and $\mathfrak{X}(M)$ the Lie algebra of smooth vector fields on M.

2.1 Almost Hermitian Manifolds

An almost complex manifold M is a differentiable manifold equipped with a $(1, 1)$-tensor field J which satisfies $J^2 = -I$, where I is the identity. Such a manifold is even-dimensional. M^{2n} is an almost Hermitian manifold provided it is almost complex and has a Riemannian metric g for which

$$g(JX, JY) = g(X, Y)$$

for all X and Y vector fields on M. To describe the geometry of an almost Hermitian manifold (M^{2n}, J, g), it is useful to consider two special tensors. The first, called the Nijenhuis tensor, is a $(1, 2)$-tensor field N_J defined by

$$N_J(X, Y) = J^2[X, Y] + [JX, JY] - J[X, JY] - J[JX, Y]. \qquad (2.1)$$

An almost complex structure J is integrable if its Nijenhuis tensor N_J vanishes. It is also shown [15] that J is integrable if and only if

$$(\nabla_{JX} J)Y = J(\nabla_X J)Y,$$

where ∇ denote the Levi–Civita connection corresponding to g. If J is integrable, then the almost Hermitian manifold (M^{2n}, J, g) is a Hermitian manifold.

The second tensor is a differential 2-form Ω, called the Kähler form, and it is defined by

$$\Omega(X, Y) = g(X, JY),$$

for all $X, Y \in \mathfrak{X}(M)$. The almost Hermitian manifold M^{2n} is an almost Kähler manifold if Ω is closed, i.e. $d\Omega = 0$. If both $d\Omega = 0$ and $N_J = 0$ are satisfied, then (M^{2n}, J, g) is called a Kähler manifold. Recall that $d\Omega = 0$ and $N_J = 0$ are equivalent to

$$\nabla J = 0.$$

Definition 2.1 ([11, 16]) A Hermitian manifold (M, J, g) is called locally conformal Kähler (shortly, l.c.K) manifold if there exists a closed one-form θ (called the Lee form) such that

$$d\Omega = \theta \wedge \Omega.$$

For more background on almost complex manifolds, we refer to [11, 21].

2.2 Almost Contact Metric Manifolds

An odd-dimensional Riemannian manifold (M, g) is said to be an almost contact metric manifold if there exist on M a $(1, 1)$-tensor field φ, a vector field ξ (called the structure vector field) and a 1-form η such that

$$\begin{cases} \eta(\xi) = 1, \\ \varphi^2(X) = -X + \eta(X)\xi, \\ g(\varphi X, \varphi Y) = g(X, Y) - \eta(X)\eta(Y), \end{cases} \qquad (2.2)$$

for any $X, Y \in \mathfrak{X}(M)$. On an almost contact metric manifold, we immediately have $\varphi\xi = 0$ and $\eta \circ \varphi = 0$.

The fundamental 2-form ϕ is defined by $\phi(X, Y) = g(X, \varphi Y)$. It is known that the almost contact structure (φ, ξ, η) is said to be normal if and only if

$$N^{(1)}(X,Y) = N_\varphi(X,Y) + 2\mathrm{d}\eta(X,Y)\xi = 0, \tag{2.3}$$

for all X and Y vector fields on M, where N_φ denotes the Nijenhuis tensor of φ given by

$$N_\varphi(X,Y) = \varphi^2[X,Y] + [\varphi X, \varphi Y] - \varphi[\varphi X, Y] - \varphi[X, \varphi Y]. \tag{2.4}$$

Some special cases are worthy of attention. A contact metric structure is an almost contact metric structure with $\mathrm{d}\eta = \phi$ and a Sasakian manifold is a normal contact metric manifold or, equivalently

$$(\nabla_X\varphi)Y = g(X,Y)\xi - \eta(Y)X, \tag{2.5}$$

for all $X, Y \in \mathfrak{X}(M)$ where ∇ denotes the Levi–Civita connection of g.

An almost contact metric structure is said to be almost cosymplectic if both η and ϕ are closed. In addition, If the structure (φ, ξ, η) is normal, then it is said to be cosymplectic, equivalently $(\nabla_X\varphi)Y = 0$.

An almost contact metric manifold is an almost Kenmotsu manifold if $\mathrm{d}\eta = 0$ and $\mathrm{d}\phi = 2\eta \wedge \phi$. Moreover, if the structure (φ, ξ, η) is normal, the manifold is said to be Kenmotsu. This manifold can also be characterized by

$$(\nabla_X\varphi)Y = g(\varphi X, Y)\xi - \eta(Y)\varphi X, \tag{2.6}$$

for all $X, Y \in \mathfrak{X}(M)$.

In [17], Oubiña proves that (φ, ξ, η, g) is a trans-Sasakian structure of type (α, β) if and only if it is normal and

$$\mathrm{d}\eta = \alpha\phi, \qquad \mathrm{d}\phi = 2\beta\eta \wedge \phi, \tag{2.7}$$

where $\alpha = \frac{1}{2n}\delta\phi(\xi)$, $\beta = \frac{1}{2n}\mathrm{div}\xi$, δ is the codifferential of g and $\mathrm{div}\xi$ is the divergence of ξ defined by $\mathrm{div}\xi = trace\{X \to \nabla\xi\}$. It is well known that the trans-Sasakian structure may be mentioned as an almost contact metric structure satisfying

$$(\nabla_X\varphi)Y = \alpha\big(g(X,Y)\xi - \eta(Y)X\big) + \beta\big(g(\varphi X, Y)\xi - \eta(Y)\varphi X\big). \tag{2.8}$$

Recently, in [6, 7], the authors showed that (φ, ξ, η, g) is a C_{12}-structure if and only if there exist a 1-form ω such that

$$\mathrm{d}\eta = \omega \wedge \eta, \qquad \mathrm{d}\phi = 0, \qquad N_\varphi = 0, \tag{2.9}$$

and they characterized it by

$$(\nabla_X\varphi)Y = \eta(X)\big(\omega(\varphi Y)\xi + \eta(Y)\varphi\psi\big), \tag{2.10}$$

where $\omega(X) = g(\psi, X)$ for all X vector fields on M.

For more background on almost contact metric manifolds, we refer to [5, 7, 10, 14, 17, 21].

2.3 Almost Bi-contact Metric Manifolds

A $(2n+2)$-dimensional differentiable manifold M is said to have an almost bi-contact metric structure if it admits a field φ of endomorphisms of the tangent spaces, two global vector fields ξ_1 and ξ_2, and two 1-forms η_1 and η_2 satisfying

$$\eta_1(\xi_1) = \eta_2(\xi_2) = 1, \qquad \varphi^2 = -I + \eta_1 \otimes \xi_1 + \eta_2 \otimes \xi_2, \tag{2.11}$$

$$\eta_1(\xi_2) = \eta_2(\xi_1) = \varphi\xi_1 = \varphi\xi_2 = 0 \quad \textit{and} \quad \eta_1 \circ \varphi = \eta_2 \circ \varphi = 0. \tag{2.12}$$

Here, the endomorphism φ has rank $2n$.

If the manifold M with an almost bi-contact metric structure $(\varphi, \xi_1, \xi_2, \eta_1, \eta_2)$ admits a Riemannian metric g satisfying

$$g(\varphi X, \varphi Y) = g(X, Y) - \eta_1(X)\eta_1(Y) - \eta_2(X)\eta_2(Y) \tag{2.13}$$

for all X and Y vector fields on M, then $(\varphi, \xi_1, \xi_2, \eta_1, \eta_2, g)$ is said to be an almost bi-contact metric structure and $(M, \varphi, \xi_1, \xi_2, \eta_1, \eta_2, g)$ is called an almost bi-contact metric manifold. An immediate consequence is

$$\eta_1(X) = g(X, \xi_1) \qquad \textit{and} \qquad \eta_2(X) = g(X, \xi_2).$$

The 2-form ϕ on M^{2n+2} defined by

$$\phi(X, Y) = g(X, \varphi Y), \tag{2.14}$$

for all X and Y vector fields on M, is called the fundamental 2-form of the almost bi-contact metric structure. If ∇ is the Riemannian connection of g, it is easy to prove

$$(\nabla_X \eta_1)Y = g(\nabla_X \xi_1, Y) \qquad \textit{and} \qquad (\nabla_X \eta_2)Y = g(\nabla_X \xi_2, Y), \tag{2.15}$$

and

$$(\nabla_X \phi)(Y, Z) = g\big(Y, (\nabla_X \varphi)Z\big) = -g\big((\nabla_X \varphi)Y, Z\big). \tag{2.16}$$

The exterior derivatives of η_1, η_2 and ϕ are given by

$$2d\eta_1(X, Y) = (\nabla_X \eta_1)Y - (\nabla_Y \eta_1)X, \tag{2.17}$$

$$2\mathrm{d}\eta_2(X, Y) = (\nabla_X \eta_2)Y - (\nabla_Y \eta_2)X, \tag{2.18}$$

$$3\mathrm{d}\phi(X, Y, Z) = \sigma(\nabla_X \phi)(Y, Z), \tag{2.19}$$

where σ denotes the cyclic sum over $X, Y, Z \in \mathfrak{X}(M)$.

The almost bi-contact metric structure is said to be normal if

$$N_\varphi + 2\sum_{i=1}^{2} \mathrm{d}\eta_i \otimes \xi_i = 0. \tag{2.20}$$

If the almost bi-contact metric structure is normal, then [12]

$$[\xi_1, \xi_2] = 0. \tag{2.21}$$

Next, we will establish a formula for the covariant derivative of φ.

Proposition 2.2 *For an almost bi-contact metric structure* $(\varphi, \xi_1, \xi_2, \eta_1, \eta_2, g)$, *the covariant derivative of* φ *is given by*

$$\begin{aligned} 2g\big((\nabla_X\varphi)Y, Z\big) &= 3\mathrm{d}\phi(X, \varphi Y, \varphi Z) - 3\mathrm{d}\phi(X, Y, Z) + g\big(N^{(1)}(Y, Z), \varphi X\big) \\ &+ \sum_{i=1}^{2}\Big(N_i^{(2)}(Y, Z)\eta_i(X) + 2\mathrm{d}\eta_i(\varphi Y, X)\eta_i(Z) - 2\mathrm{d}\eta_i(\varphi Z, X)\eta_i(Y)\Big). \end{aligned} \tag{2.22}$$

Proof Using Koszul's formula

$$\begin{aligned} 2g(\nabla_X Y, Z) &= Xg(Y, Z) + Yg(X, Z) - Zg(X, Y) \\ &+ g([X, Y], Z) + g([Z, X], Y) - g([Y, Z], X) \end{aligned}$$

and the following formulas

$$2(\mathrm{d}\eta_i)(X, Y) = X\big(\eta_i(Y)\big) - Y\big(\eta_i(X)\big) - \eta_i\big([X, Y]\big)$$

$$\begin{aligned} 3(\mathrm{d}\phi)(X, Y, Z) &= X\big(\phi(Y, Z)\big) + Y\big(\phi(Z, X)\big) + Z\big(\phi(X, Y)\big) \\ &- \phi\big([X, Y], Z\big) - \phi\big([Y, Z], X\big) - \phi\big([Z, X], Y\big) \end{aligned}$$

with

$$(\nabla_X\varphi)Y = \nabla_X \varphi Y + \varphi \nabla_X Y,$$

the result is easily obtained. Here, we omit standard calculations. □

3 4-Dimensional Almost Bi-contact Metric Manifolds

Through the rest of this paper M always denotes the 4-dimensional differentiable manifolds, and the indices i, j and k take on values from 1 to 2.

Theorem 3.1 *For an almost bi-contact metric structure* $(\varphi, \xi_1, \xi_2, \eta_1, \eta_2, g)$ *on* M*, we have*

$$(\nabla_X\varphi)Y = \sum_{i=1}^{2}\Big(g(\varphi\nabla_X\xi_i, Y)\xi_i - \eta_i(Y)\varphi\nabla_X\xi_i\Big), \tag{3.1}$$

$$d\phi = 2\phi \wedge \sum_{i=1}^{2}\beta_i\eta_i, \tag{3.2}$$

$$d\eta_i = \omega_i \wedge \eta_i + \alpha_i\phi, \tag{3.3}$$

where

$$\alpha_i = \frac{1}{2}tr(\varphi\nabla\xi_i), \qquad \beta_i = \frac{1}{2}\big(\mathrm{div}\xi_i + \eta_i(\nabla_{\xi_j}\xi_j)\big), \qquad \omega_i = -\nabla_{\xi_i}\eta_i,$$

with $i \neq j$ *and for* $\{e_i\}_{1\leq i\leq 4}$ *is a local orthonormal basis with respect to* g*,* $tr(\varphi\nabla\xi_i) = \sum_{j=1}^{4} g(\varphi\nabla_{e_j}\xi_i, e_j)$.

Proof Since $\eta_1 \wedge \eta_2 \wedge \phi$ is up to a constant factor the volume element on M, it is parallel with respect to ∇, i.e.

$$\nabla_X(\eta_1 \wedge \eta_2 \wedge \phi) = 0.$$

Knowing that

$$\begin{aligned} 4(\eta_1 \wedge \eta_2 \wedge \phi)(X, Y, U, V) = {} & (\eta_1 \wedge \eta_2)(X, Y)\phi(U, V) \\ & - (\eta_1 \wedge \eta_2)(X, U)\phi(Y, V) \\ & + (\eta_1 \wedge \eta_2)(X, V)\phi(Y, U) \\ & + (\eta_1 \wedge \eta_2)(Y, U)\phi(X, V) \\ & - (\eta_1 \wedge \eta_2)(Y, V)\phi(X, U) \\ & + (\eta_1 \wedge \eta_2)(U, V)\phi(X, Y), \end{aligned}$$

and

$$3(d\phi)(X, Y, Z) = (\nabla_X\phi)(Y, Z) + (\nabla_Y\phi)(Z, X) + (\nabla_Z\phi)(X, Y),$$

then, the equation

$$\big(\nabla_X(\eta_1 \wedge \eta_2 \wedge \phi)\big)(X, Y, \xi_1, \xi_2) = 0$$

gives

$$(\nabla_Z\phi)(X, Y) - \eta_1(X)\phi(Y, \nabla_Z\xi_1) + \eta_1(Y)\phi(X, \nabla_Z\xi_1) \\ -\eta_2(X)\phi(Y, \nabla_Z\xi_2) + \eta_2(Y)\phi(X, \nabla_Z\xi_2) = 0,$$

hence

$$(\nabla_Z\phi)(X, Y) = \eta_1(X)\phi(Y, \nabla_Z\xi_1) - \eta_1(Y)\phi(X, \nabla_Z\xi_1) \\ + \eta_2(X)\phi(Y, \nabla_Z\xi_2) - \eta_2(Y)\phi(X, \nabla_Z\xi_2),$$

therefore

$$(\nabla_Z\varphi)Y = g(\varphi\nabla_Z\xi_1, Y)\xi_1 - \eta_1(Y)\varphi\nabla_Z\xi_1 \\ + g(\varphi\nabla_Z\xi_2, Y)\xi_2 - \eta_2(Y)\varphi\nabla_Z\xi_2,$$

which leads to (3.1). To prove (3.2), we note that

$$\mathrm{d}\phi = (\rho_1\eta_1 + \rho_2\eta_2) \wedge \phi$$

for certain functions ρ_1 and ρ_2 on M. Taking a local φ-basis $\{\xi_1, \xi_2, e, \varphi e\}$ we find

$$3\big((\rho_1\eta_1 + \rho_2\eta_2) \wedge \phi\big)(\xi_i, e, \varphi e) = -\rho_i,$$

and

$$\begin{aligned} 3\mathrm{d}\phi(\xi_i, e, \varphi e) &= (\nabla_{\xi_i}\phi)(e, \varphi e) + (\nabla_e\phi)(\varphi e, \xi_i) + (\nabla_{\varphi e}\phi)(\xi_i, e) \\ &= -g(\nabla_e\xi_i, e) - g(\nabla_{\varphi e}\xi_i, \varphi e) \\ &= -\mathrm{div}\xi_i + g(\nabla_{\xi_j}\xi_i, \xi_j) \\ &= -\mathrm{div}\xi_i - \eta_i(\nabla_{\xi_j}\xi_j), \end{aligned}$$

where $i \neq j$. This implies $\rho_i = \mathrm{div}\xi_i + \eta_i(\nabla_{\xi_j}\xi_j)$. Finally, by elementary argumentations we get

$$\mathrm{d}\eta_i = \omega_i \wedge \eta_i + \sigma\phi,$$

where ω_i is a 1-form orthogonal to η_i, i.e. $\omega_i(\xi_i) = 0$, and σ is a scalar function. Consequently,

$$\omega_i(X) = 2(\omega_i \wedge \eta_i + \sigma\phi)(X, \xi_i) = 2\mathrm{d}\eta_i(X, \xi_i) = -(\nabla_{\xi_i}\eta_i)(X),$$

$$\begin{aligned} 2\sigma &= 2(\omega_i \wedge \eta_i + \sigma\phi)(\varphi e, e) = 2\mathrm{d}\eta_i(\varphi e, e) \\ &= g(\nabla_{\varphi e}\xi_i, e) - g(\nabla_e \xi_i, \varphi e) \\ &= g(\varphi\nabla_{\varphi e}\xi_i, \varphi e) + g(\varphi\nabla_e \xi_i, e) \\ &= tr(\varphi\nabla\xi_i), \end{aligned}$$

completing the proof of (3.3). □

Proposition 3.2 *For a normal almost bi-contact metric manifold, the covariant derivative of* φ *is given by*

$$\begin{aligned} (\nabla_X\varphi)Y = \sum_{i=1}^{2}\Big(&\alpha_i\big(g(\varphi X, \varphi Y)\xi_i + \eta_i(Y)\varphi^2 X\big) \\ &+ \beta_i\big(g(\varphi X, Y)\xi_i - \eta_i(Y)\varphi X\big)\Big). \end{aligned} \tag{3.4}$$

Proof If the structure $(\varphi, \xi_1, \xi_2, \eta_1, \eta_2, g)$ is normal then $\nabla_{\xi_i}\xi_i = 0$. Therefore, Eqs. (3.2) and (3.3) become of the form

$$\mathrm{d}\phi = 2\phi \wedge \sum_{i=1}^{2}\beta_i\eta_i, \tag{3.5}$$

$$\mathrm{d}\eta_i = \alpha_i\phi, \tag{3.6}$$

where $2\beta_i = \mathrm{div}\xi_i$ and $2\alpha_i = tr(\varphi\nabla\xi_1)$.

To get our formula, we just substitute (3.5) and (3.6) into (2.22) with $N^{(1)} = 0$ and $N_i^{(2)} = 0$.

Conversely, assume

$$\begin{aligned} (\nabla_X\varphi)Y = \sum_{i=1}^{2}\Big(&\alpha_i\big(g(\varphi X, \varphi Y)\xi_i + \eta_i(Y)\varphi^2 X\big) \\ &+ \beta_i\big(g(\varphi X, Y)\xi_i - \eta_i(Y)\varphi X\big)\Big), \end{aligned}$$

knowing that

$$N^{(1)}(X, Y) = N_\varphi(X, Y) + 2\mathrm{d}\eta_i(X, Y)\xi_i,$$

where N_φ denotes the Nijenhuis torsion of φ given by

$$\begin{aligned} N_\varphi(X, Y) = \varphi^2[X, Y] &+ [\varphi X, \varphi Y] - \varphi[\varphi X, Y] - \varphi[X, \varphi Y] \\ &(\varphi\nabla_Y\varphi - \nabla_{\varphi Y}\varphi)X - (\varphi\nabla_X\varphi - \nabla_{\varphi X}\varphi)Y. \end{aligned}$$

By a direct computation using the hypothesis, one can get $N^{(1)} = 0$. This completes the proof. □

This case was treated in [1] with dimension $2n + s$ and it was called “trans-S-manifold”. Here, we give an interesting result for the 4-dimensional case.

Theorem 3.3 *An almost bi-contact metric structure* $(\varphi, \xi_1, \xi_2, \eta_1, \eta_2, g)$ *on* M *is normal (i.e. trans-S-manifold of dimension* 4*) if and only if*

$$\nabla_X \xi_i = -\alpha_i \varphi X - \beta_i \varphi^2 X. \tag{3.7}$$

Proof Suppose that $\nabla_X \xi_i = -\alpha_i \varphi X - \beta_i \varphi^2 X$, for all X vector fields on M. From (3.1) we get (3.4) then the structure is normal.

Conversely, assuming that $(\varphi, \xi, \psi, \eta, \omega, g)$ is normal, this is equivalent to

$$\begin{aligned}(\nabla_X \varphi)Y = \sum_{i=1}^{2} \Big(&\alpha_i \big(g(\varphi X, \varphi Y)\xi_i - \eta_i(Y)\varphi^2 X\big) \\ &+\beta_i \big(g(\varphi X, Y)\xi_i - \eta_i(Y)\varphi X\big)\Big).\end{aligned}$$

Setting $Y = \xi_i$ gives

$$-\varphi \nabla_X \xi_i = -\alpha_i \varphi^2 X - \beta_i \varphi X,$$

and hence

$$\begin{aligned}\nabla_X \xi_i &= \alpha_i \varphi X - \beta_i \varphi^2 X \\ &= -\alpha_i \varphi X - \beta_i \varphi^2 X.\end{aligned}$$

Notice that $\eta_i(\nabla_{\xi_i} \xi_j) = 0$ because $[\xi_1, \xi_2] = 0$. □

Since in the previous paragraph the normality of an almost bi-contact metric structure was the main condition in some remarkable results, let us now treat in detail with the structures which can be integrable but never normal (i.e. $N_\varphi = 0$ and $N^{(1)} \neq 0$). Let’s start with this important proposition.

Proposition 3.4 *For an integrable non-normal almost bi-contact metric structure* $(\varphi, \xi_1, \xi_2, \eta_1, \eta_2, g)$ *on* M*, the covariant derivative of* φ *is given by*

$$\begin{aligned}(\nabla_X \varphi)Y = \sum_{i=1}^{2} \Big(&\alpha_i \big(g(\varphi X, \varphi Y)\xi_i - \eta_i(Y)\varphi^2 X\big) \\ &+ \beta_i \big(g(\varphi X, Y)\xi_i - \eta_i(Y)\varphi X\big) \\ &+ \eta_i(X)\big(\omega_i(\varphi Y)\xi_i + \eta_i(Y)\varphi \psi_i\big)\Big),\end{aligned} \tag{3.8}$$

where ω_i *is the* g*-dual of* ψ_i*, i.e.* $\omega_i(X) = g(X, \psi_i)$ *and* $i \in \{1, 2\}$.

Proof Assume that $(\varphi, \xi_1, \xi_2, \eta_1, \eta_2, g)$ is integrable structure. So, using formula (2.22), with

$$g\big(N^{(1)}(Y, Z), \varphi X\big) = 2g\big(\mathrm{d}\eta_i(Y, Z)\xi_i, \varphi X\big) = 0,$$

$$\begin{aligned} N_i^{(2)}(Y, Z) &= \big(\mathcal{L}_{\varphi Y}\eta_i\big)(Z) - \big(\mathcal{L}_{\varphi Z}\eta_i\big)(Y) \\ &= \varphi Y\big(\eta_i(Z)\big) - \eta_i\big([\varphi Y, Z]\big) - \varphi Z\big(\eta_i(Y)\big) + \eta_i\big([\varphi Z, Y]\big) \\ &= 2\mathrm{d}\eta_i(\varphi Y, Z) - 2\mathrm{d}\eta_i(\varphi Z, Y), \end{aligned}$$

and

$$\mathrm{d}\phi = 2\sum_{i=1}^{2} \beta_i \eta_i \wedge \phi, \tag{3.9}$$

$$\mathrm{d}\eta_i = \omega_i \wedge \eta_i + \alpha_i \phi, \tag{3.10}$$

where $2\beta_i = \mathrm{div}\xi_i + \eta_i(\nabla_{\xi_j}\xi_j)$, $2\alpha = tr(\varphi\nabla\xi_1)$ and $\omega_i = \nabla_{\xi_i}\eta_i \perp \eta_i$, we obtain

$$\begin{aligned} (\nabla_X\varphi)Y = \sum_{i=1}^{2} \Big(&\alpha_i\big(g(\varphi X, \varphi Y)\xi_i - \eta_i(Y)\varphi^2 X\big) \\ &+ \beta_i\big(g(\varphi X, Y)\xi_i - \eta_i(Y)\varphi X\big) \\ &+ \eta_i(X)\big(\omega_i(\varphi Y)\xi_i + \eta_i(Y)\varphi\psi_i\big)\Big). \end{aligned}$$

□

Example 3.5 Let (x, y, z, t) denote the Cartesian coordinates in $M = \mathbb{R}^4$. Let (θ^i) be the dual frame of differential 1-forms on $\mathbb{R}^4$ given by

$$\theta^1 = \mathrm{e}^{\rho} dx, \quad \theta^2 = \mathrm{e}^{\rho} dy, \quad \theta^3 = \mathrm{e}^{2\rho+\tau}(-2x dy + dz), \quad \theta^4 = \mathrm{e}^{\rho-\tau} dt,$$

and let (e_i) be the frame of vector fields on M defined by

$$e_1 = \mathrm{e}^{-\rho}\frac{\partial}{\partial x}, \quad e_2 = \mathrm{e}^{-\rho}\left(\frac{\partial}{\partial y} + 2x\frac{\partial}{\partial z}\right), \quad e_3 = \mathrm{e}^{-2\rho-\tau}\frac{\partial}{\partial z}, \quad e_4 = \mathrm{e}^{\tau-\rho}\frac{\partial}{\partial t}.$$

Define an almost bi-contact metric structure $(\varphi, \xi_1, \xi_2, \eta_1, \eta_2, g)$ on M by

$$\varphi\frac{\partial}{\partial x} = \frac{\partial}{\partial y} + 2x\frac{\partial}{\partial z}, \quad \varphi\frac{\partial}{\partial y} = -\frac{\partial}{\partial x}, \qquad \varphi\frac{\partial}{\partial z} = \varphi\frac{\partial}{\partial t} = 0,$$

$$\xi_1 = e_3, \qquad \xi_2 = e_4, \qquad \eta_1 = \theta^3, \qquad \eta_2 = \theta^4,$$

and

$$g = \sum \theta^i \otimes \theta^i,$$

where ρ and τ are two functions on M.

For the fundamental form ϕ, $\phi(X, Y) = g(X, \varphi Y)$, we have

$$\phi = -2\mathrm{e}^{2\rho} dx \wedge dy.$$

Then, we have

$$\begin{cases} \eta_1 = \mathrm{e}^{2\rho+\tau}(-2x dy + dz) \\ \eta_2 = \mathrm{e}^{\rho-\tau} dt \\ \phi = -2\mathrm{e}^{2\rho} dx \wedge dy \end{cases} \Rightarrow \begin{cases} \mathrm{d}\eta_1 = \mathrm{d}(2\rho + \tau) \wedge \eta_1 + \mathrm{e}^{\tau}\phi \\ \mathrm{d}\eta_2 = \mathrm{d}(\rho - \tau) \wedge \eta_2 \\ \mathrm{d}\phi = 2(\rho_3 dz + \rho_4 dt) \wedge \phi \end{cases}$$

$$\Rightarrow \begin{cases} \mathrm{d}\eta_1 = \mathrm{d}(2\rho + \tau) \wedge \eta_1 + \mathrm{e}^{\tau}\phi \\ \mathrm{d}\eta_2 = \mathrm{d}(\rho - \tau) \wedge \eta_2 \\ \mathrm{d}\phi = 2\mathrm{e}^{\tau-\rho}\eta_2 \wedge \phi, \end{cases}$$

where $\beta_1 = \rho_3 = 0$ and $\rho_3 = \frac{\partial \rho}{\partial z}$.

By a direct computation, one can get $N^{(1)} = 0$ if and only if ρ and τ are constant in which case, we have

$$\alpha_1 = \mathrm{e}^{\tau}, \qquad \alpha_2 = \beta_1 = \beta_2 = \omega_1 = \omega_2 = 0.$$

But, for all $\rho, \tau \in \mathcal{C}^{\infty}(M)$ with $\rho_3 = 0$ we have $N_{\varphi} = 0$ and in this case

$$\alpha_1 = \mathrm{e}^{\tau}, \quad \alpha_2 = \beta_1 = 0, \quad \beta_2 = \mathrm{e}^{\tau-\rho}, \quad \omega_1 = \mathrm{d}(2\rho + \tau), \quad \omega_2 = \mathrm{d}(\rho - \tau).$$

To verify result in formulas (3.4) and (3.8), the components of the Levi–Civita connection corresponding to g are given by

$$\nabla_{e_1} e_1 = -\mathrm{e}^{-\rho}(\rho_2 + 2x\rho_3)e_2, \quad \nabla_{e_1} e_2 = \mathrm{e}^{-\rho}(\rho_2 + 2x\rho_3)e_1 + \mathrm{e}^{\tau} e_3,$$

$$\nabla_{e_1} e_3 = \rho_3 \mathrm{e}^{-(2\rho+\tau)} e_1 - \mathrm{e}^{\tau} e_2, \qquad \nabla_{e_1} e_4 = \rho_4 \mathrm{e}^{(\tau-\rho)} e_1,$$

$$\nabla_{e_2} e_1 = \rho_1 \mathrm{e}^{-\rho} e_2 + \mathrm{e}^{\tau} e_3, \qquad \nabla_{e_2} e_2 = -\mathrm{e}^{-\rho}\big((\rho_1 e_1 + \rho_3 \mathrm{e}^{-(\rho+\tau)} e_3 + \rho_4 \mathrm{e}^{\tau} e_4\big),$$

$$\nabla_{e_2} e_3 = \mathrm{e}^{-\tau}\big(e_1 + \rho_3 \mathrm{e}^{-2(\rho+\tau)} e_2\big), \qquad \nabla_{e_2} e_4 = \rho_4 \mathrm{e}^{\tau-\rho} e_2$$

$$\nabla_{e_3} e_1 = \mathrm{e}^{-\tau} e_2 + \mathrm{e}^{-\rho}(2\rho_1 + \tau_1)e_3,$$

$$\nabla_{e_3} e_2 = \mathrm{e}^{\tau} e_1 + \mathrm{e}^{-\rho}\big(2x(2\rho_3 + \tau_3) + 2\rho_2 + \tau_2\big)e_3,$$

$$\nabla_{e_3} e_3 = -\mathrm{e}^{-\rho}(2\rho_2 + \tau_2)e_1 - \mathrm{e}^{-\rho}\big(2x(2\rho_3 + \tau_3) + 2\rho_2 + \tau_2)e_2\big) - e^{\tau-\rho}(2\rho_4 + \tau_4)e_4,$$

$$\nabla_{e_3} e_4 = e^{\tau-\rho}(2\rho_2 + \tau_4)e_3, \qquad \nabla_{e_4} e_1 = e^{-\rho}(\rho_1 - \tau_4)e_4,$$

$$\nabla_{e_4} e_2 = e^{-\rho}\big(\rho_2 - \tau_2 + 2x(\rho_3 - \tau_3)\big)e_4, \qquad \nabla_{e_4} e_3 = e^{-(2\rho+\tau)}\big(\rho_3 - \tau_3)\big)e_4,$$

$$\nabla_{e_4} e_4 = -e^{-\rho}\big((\rho_1 - \tau_1)e_1 + (\rho_2 - \tau_2 + 2x(\rho_3 - \tau_3))e_2 + e^{-(\tau+\rho)}(\rho_3 - +\tau_3)e_3\big).$$

Let $\mathcal{D}_1 := \{X \in \mathfrak{X}(M) \;/\; \eta_1(X) = 0\}$ and $\mathcal{D}_2 := \{X \in \mathfrak{X}(M) \;/\; \eta_2(X) = 0\}$ be two distributions of rank 3 transversal to the characteristic vector fields ξ_1 and ξ_2, respectively. So, we notice that we have two 3-dimensional almost contact metric structures, $(\varphi, \xi_1, \eta_1, g)$ on $\mathcal{D}_2$ and $(\varphi, \xi_2, \eta_2, g)$ on $\mathcal{D}_1$. Depending on the two Theorems 3.7 and 4.2, one can define several classes of almost bi-contact metric structures of dimension 4, by discussing the nature of both factors $(\varphi, \xi_1, \eta_1, g)$ and $(\varphi, \xi_2, \eta_2, g)$, more precisely, the parameters α_i, β_i and ω_i. As examples,

(1) For $\alpha_i = \beta_i = \omega_i = 0$ we get a C-manifold [4].
(2) For $\alpha_i \neq 0$, $\beta_i = \omega_i = 0$ we get an S-manifold [4].
(3) For $\alpha_i = \omega_i = 0$, $\beta_i \neq 0$ we obtain an f-Kenmotsu [19].
(4) For $\alpha_i \neq 0$, $\beta_i \neq 0$, $\omega_i = 0$ we obtain an trans-S-manifold [1].
(5) For $\alpha_1 \neq 0$, $\beta_2 \neq 0$, $\alpha_2 = \beta_1 = \omega_i = 0$ we get a nice case [2], where Sasakian structure and Kenmotsu structure are combined together.

We draw attention to some cases that have not yet been studied; we cite, for example,

(1) For $\alpha_i \neq 0$, $\beta_i = 0$ and $\omega_i \neq 0$.
(2) For $\alpha_i = 0$, $\beta_i \neq 0$ and $\omega_i \neq 0$.
(3) For $\alpha_i \neq 0$, $\beta_i \neq 0$ and $\omega_i \neq 0$.

While this is an area of possible future research, we focus in the next section on the case where $\alpha_i = \beta_i = 0$ and $\omega_i \neq 0$. Before that, we give an important result about the Hermitian structures extracted from the normal almost bi-contact metric structures.

The almost bi-contact 4-dimensional metric manifold carries an almost complex natural structure J as follows.

Theorem 3.6 *For any almost bi-contact metric structure $(\varphi, \xi_1, \xi_2, \eta_1, \eta_2)$ on a Riemannian manifold (M^{2n+2}, g), there exists an almost Hermitian structure J on (M^{2n+2}, g) given by*

$$J = \varphi - \xi_1 \wedge \xi_2, \tag{3.11}$$

where the operator $\wedge$ is defined by

$$(X \wedge Y)Z = g(Y, Z)X - g(X, Z)Y, \tag{3.12}$$

for every tangent vector field X, Y and Z on M^{2n+2}.

Proof Let $(M^{2n+2}, \varphi, \xi_1, \xi_2, \eta_1, \eta_2, g)$ be an almost bi-contact metric manifold. For all $X \in \mathfrak{X}(M)$, we have

$$JX = \varphi X + \eta_1(X)\xi_2 - \eta_2(X)\xi_1.$$

It follows that $J\xi_1 = \xi_2$ and $J\xi_2 = -\xi_1$. So, using formulas (2.11) and (2.12) we get

$$\begin{aligned} J^2X &= J\big(\varphi X + \eta_1(X)\xi_2 - \eta_2(X)\xi_1\big) \\ &= \varphi^2 X + \eta_1(X)\xi_1 + \eta_2(X)\xi_2 \\ &= -X. \end{aligned}$$

Moreover, we have

$$\begin{aligned} g(JX, JY) &= g\big(\varphi X + \eta_1(X)\xi_2 - \eta_2(X)\xi_1, \varphi Y + \eta_1(Y)\xi_2 - \eta_2(Y)\xi_1\big) \\ &= g(X, Y) \end{aligned}$$

for every tangent vector field X and Y on M^{2n+2} which completes the proof. □

This allows us to understand the almost Hermitian structures built from an almost bi-contact metric ones. Among the important information, the fundamental 2-form Ω of (J, g) is given by

$$\begin{aligned} \Omega(X, Y) &= g(X, JY) \\ &= g\big(X, \varphi Y - (\xi_1 \wedge \xi_2)Y\big) \\ &= \phi(X, Y) - 2(\eta_1 \wedge \eta_2)(X, Y), \end{aligned}$$

hence

$$\Omega = \phi - 2\eta_1 \wedge \eta_2, \tag{3.13}$$

and we have immediately that

$$\mathrm{d}\Omega = \mathrm{d}\phi - 2\mathrm{d}\eta_1 \wedge \eta_2 + 2\eta_1 \wedge \mathrm{d}\eta_2,$$

with the help of (3.2) and (3.3), and we obtain

$$\mathrm{d}\Omega = 2\Big(\sum_{i=1}^{2} \beta_i \eta_i - \alpha_1 \eta_2 + \alpha_2 \eta_1\Big) \wedge \Omega - 2(\omega_1 + \omega_2) \wedge \eta_1 \wedge \eta_2. \tag{3.14}$$

On the other hand, we know that when the structure $(\varphi, \xi_1, \xi_2, \eta_1, \eta_2)$ is normal the structure (J, g) is said to be integrable in which case the formula (3.14) becomes

$$\mathrm{d}\Omega = 2\Big(\sum_{i=1}^{2} \beta_i \eta_i - \alpha_1 \eta_2 + \alpha_2 \eta_1\Big) \wedge \Omega. \tag{3.15}$$

So, for the normal cases we have the following:

$$\begin{cases} (1): C - manifolds \ : \ d\Omega = 0, \\ (2): S - manifolds \ : \ d\Omega = 2(\alpha_2\eta_1 - \alpha_1\eta_2) \wedge \Omega, \\ (3): f - Kenmotsu \ : \ d\Omega = 2\sum_{i=1}^{2} \beta_i\eta_i \wedge \Omega. \end{cases} \tag{3.16}$$

Therefore, summing up the arguments above, we have the following main theorem:

Theorem 3.7 *1. The almost bi-contact metric structure on M is a C-structure if and only if the almost Hermitian structure (J, g) is Kählerian.*

2. The almost bi-contact metric structure on M is an S-structure if and only if the almost Hermitian structure (J, g) is a locally conformal Kählerian with α_i is constant.

3. The almost bi-contact metric structure on M is an f-structure if and only if the almost Hermitian structure (J, g) is a locally conformal Kählerian.

4 New Class of 4-Dimensional Almost Bi-contact Metric Manifolds

In this section, we establish an interesting class of almost bi-contact metric manifolds that fulfill the conditions

$$\alpha_i = \beta_i = 0 \qquad and \qquad \omega_i \neq 0.$$

To begin with, we point out that in the case of almost contact metric manifolds, this class corresponds to C_{12}-manifolds in the classification of Chinea–Gonzalez (see [8]). Several recent works has been recently published about this subject (see [3, 7, 10]). Therefore, we suggest the name "S-C_{12}-manifolds" for this new class.

Definition 4.1 An almost bi-contact metric structure $(\varphi, \xi_i, \eta_i, g)$ on M^{2n+s} where $i \in \{1, ..., s\}$ is said to be an almost S-C_{12}-structure if there exists a closed 1-form ω_i such that

$$d\eta_i = \omega_i \wedge \eta_i \qquad \text{and} \qquad d\phi = 0,$$

where $i \in \{1, ..., s\}$. In addition if $N_\varphi = 0$, we say that M is an S-C_{12}-structure.

Note that the definition is correct for dimension $2n + s$. But we will focus on the fourth dimension, i.e. in the case $s = 2$; briefly, we denote such a case by "bi-C_{12}-manifold".

Directly, from formula (3.8), one can get the following.

Corollary 4.2 *For an integrable non-normal almost bi-C_{12}-manifold $(\varphi, \xi_1, \xi_2, \eta_1, \eta_2, g)$ on M, the covariant derivative of φ is given by*

$$(\nabla_X\varphi)Y = \sum_{i=1}^{2} \eta_i(X)\big(\omega_i(\varphi Y)\xi_i + \eta_i(Y)\varphi\psi_i\big). \tag{4.1}$$

Theorem 4.3 *A 4-dimensional almost bi-contact metric manifold $(M, \varphi, \xi_1, \xi_2, \eta_1, \eta_2, g)$ is said to be a bi-C_{12}-manifold if and only if*

$$\nabla_X\xi_k = -\eta_k(X)\psi_k + \sum_{i=1}^{2} \big(\eta_i(X)\mathrm{div}\xi_k + \eta_k(X)\mathrm{div}\xi_i\big)\xi_i, \tag{4.2}$$

where $\mathrm{div}\xi_k = -\eta_k(\nabla_{\xi_i}\xi_i)$ *with* $k \neq i$.

Proof We assume that condition (4.2) is satisfied for all X vector field on M. Substituting into the formula (3.1), we obtain (4.1).

Conversely, assuming that $(M, \varphi, \xi_1, \xi_2, \eta_1, \eta_2, \omega, g)$ is a bi-C_{12}-manifold, this is equivalent to

$$(\nabla_X\varphi)Y = \sum_{i=1}^{2} \eta_i(X)\big(\omega_i(\varphi Y)\xi_i + \eta_i(Y)\varphi\psi_i\big).$$

Setting $Y = \xi_k$ gives

$$\begin{aligned} -\varphi\nabla_X\xi_k &= \sum_{i=1}^{2} \eta_i(X)\eta_i(\xi_k)\varphi\psi_i \\ &= \eta_k(X)\varphi\psi_k. \end{aligned}$$

Applying φ, we get

$$\begin{aligned} \nabla_X\xi_k &= -\eta_k(X)\psi_k + \sum_{i=1}^{2} \big(\eta_i(\nabla_X\xi_k) + \eta_k(X)\eta_i(\psi_k)\big)\xi_i \\ &= -\eta_k(X)\psi_k + \sum_{i=1}^{2} \big(\eta_i(\nabla_X\xi_k) + \eta_k(X)\mathrm{div}\xi_i\big)\xi_i. \end{aligned} \tag{4.3}$$

Again, applying η, we obtain

$$\eta_i(\nabla_X\xi_k) = \mathrm{div}\xi_k\eta_i(X),$$

therefore, Eq. (4.3) becomes

$$\nabla_X \xi_k = -\eta_k(X)\psi_k + \sum_{i=1}^{2} \big(\eta_i(X)\mathrm{div}\xi_k + \eta_k(X)\mathrm{div}\xi_i\big)\xi_i.$$

This ends the proof. □

Example 4.4 Let (x, y, z, t) denote the Cartesian coordinates in $M = \mathbb{R}^4$ and (θ^i) be the dual frame of differential 1-forms on $\mathbb{R}^4$ given by

$$\theta^1 = \mathrm{e}^z dx, \quad \theta^2 = \mathrm{e}^t dy, \quad \theta^3 = \mathrm{e}^{-z} dz, \quad \theta^4 = \mathrm{e}^{-t} dt,$$

and let (e_i) be the frame of vector fields on M defined by

$$e_1 = \mathrm{e}^{-z}\frac{\partial}{\partial x}, \quad e_2 = \mathrm{e}^{-t}\frac{\partial}{\partial y}, \quad e_3 = \mathrm{e}^{z}\frac{\partial}{\partial z}, \quad e_4 = \mathrm{e}^{t}\frac{\partial}{\partial t}.$$

Define an almost bi-contact metric structure $(\varphi, \xi_1, \xi_2, \eta_1, \eta_2, g)$ on M by

$$\varphi e_1 = \varphi e_2 = 0, \qquad \varphi e_3 = e_4 \;\; \textit{and} \;\; \varphi e_4 = -e_3,$$

$$\xi_1 = e_1, \qquad \xi_2 = e_2, \qquad \eta_1 = \theta^1, \qquad \eta_2 = \theta^2,$$

and

$$g = \sum \theta^i \otimes \theta^i.$$

For the fundamental form ϕ, $\phi(X, Y) = g(X, \varphi Y)$, we have

$$\phi = -2\mathrm{e}^{-(z+t)} dz \wedge dt.$$

Then, we have

$$\begin{cases} \eta_1 = \mathrm{e}^z dx \\ \eta_2 = \mathrm{e}^t dy \\ \phi = -2\mathrm{e}^{-(z+t)} dz \wedge dt \end{cases} \Rightarrow \begin{cases} \mathrm{d}\eta_1 = dz \wedge \eta_1 \\ \mathrm{d}\eta_2 = dt \wedge \eta_2 \\ \mathrm{d}\phi = 0. \end{cases}$$

By a direct computation, the non trivial components of $N^{(1)}(e_i, e_j)$ are given by

$$N^{(1)}(e_1, e_3) = \mathrm{e}^z e_1 \;\; \textit{and} \;\; N^{(1)}(e_2, e_4) = \mathrm{e}^t e_2.$$

But, $\forall i, j \in \{1, 2, 3, 4\}$

$$N_\varphi(e_i, e_j) = 0,$$

implying that $(\varphi, \xi_1, \eta_1, \xi_2, \eta_2)$ becomes integrable non-normal.

We have $\omega_1 = dz$ and $\omega_2 = dt$ which implies $\psi_1 = e^z e_3$ and $\psi_2 = e^t e_4$ with $d\omega_1 = d\omega_2 = 0$. Thus, $(M, \varphi, \xi_1, \xi_2, \eta_1, \eta_2, g)$ is a bi-C_{12}-manifold.

Here the formula (4.2) becomes of the form

$$\nabla_{e_j}\xi_k = -\delta_{kj}\psi_k,$$

because $\mathrm{div}\xi_1 = \mathrm{div}\xi_2 = 0$. To check it, the non-zero components of the Levi–Civita connection corresponding to g are given by

$$\nabla_{e_1}e_1 = -e^z e_3, \quad \nabla_{e_1}e_3 = e^z e_1, \quad \nabla_{e_2}e_2 = -e^t e_4, \quad \nabla_{e_2}e_4 = e^t e_2.$$

References

1. P. Alegre, L.M. Fernandez, A. Prieto-Martin, A new class of metric f-manifolds. Carpathian J. Math. **34**(2), 123–134 (2018)
2. G. Beldjilali, A. Gezer, Sasaki-Kenmotsu manifolds. Acta et comm. Univ. Tartuensis de Math. **26**(1), 77–88 (2022)
3. G. Beldjilali, 3-dimensional C_{12}-manifolds. Rev. Uni. Mat. Argentina (Article in press). https://doi.org/10.33044/revuma.3088
4. D.E. Blair, Geometry of manifolds with structural group $U(n) \times O(s)$. J. Diff. Geom. **4**, 155–167 (1970)
5. D.E. Blair, *Riemannian Geometry of Contact and Symplectic Manifolds*. Progress in Mathematics, vol. 203. Birhauser, Boston (2002)
6. D.E. Blair, G.D. Ludden, Hypersurfaces in almost contact manifolds. Tôhoku Math. J. **21**, 354–362 (1969)
7. H. Bouzir, G. Beldjilali, B. Bayour, On three dimensional C12-manifolds. Mediterr. J. Math. **18**, 239 (2021). https://doi.org/10.1007/s00009-021-01921-3
8. D. Chinea, C. Gonzalez, A classification of almost contact metric manifolds. Ann. Mat. Pura Appl. **156**(4), 15–36 (1990)
9. P. Debnath, A. Konar, Almost pseudo contact structure. Commun. Korean Math. Soc. **26**(1), 125–133 (2011)
10. S. de Candia, M. Falcitelli, Curvature of $C_5 \oplus C_{12}$-Manifolds. Mediterr. J. Math. (2019) https://doi.org/10.1007/s00009-019-1382-2
11. S. Dragomir, L. Ornea, *Locally Conformally Kähler Manifolds*. Progress in Mathematics, vol. 55. Birkhäuser (1998)
12. S.I. Goldberg, K. Yano, On normal globally framed f-manifolds. Tôhoku Math. J. **2**(2), 362–370 (1970)
13. S.I. Goldberg, K. Yano, Globally framed f -manifolds. Illinois J. Math. **15**(3), 456–474 (1971)
14. K. Kenmotsu, A class of almost contact Riemannian manifolds. Tôhoku Math. J. **24**, 93–103 (1972)
15. M. Nagao, S. Kotô, Curvature in almost Kähler spaces. J. Math. Soc. Japon **12**, 422–433 (1960)
16. L. Ornea, M. Verbitsky, Locally conformally Kaehler manifolds with potential. Mathematische Annalen **148**, 25–33 (2010)
17. J.A. Oubiña, New classes of almost contact metric structures. Publ. Math. Debrecen **32**, 187–193 (1985)
18. S. Sasaki, On differentiable manifolds with certain structures which are closely related to almost contact structure I. Tohoku Math. J. **12**(3), 459–476 (1960)
19. A. Turgut Vanli, R. Sari, Generalized Kenmotsu manifolds. Com. Math. Appl. **7**(4), 311–328 (2016)

20. K. Yano, On a structure defined by a tensor field f of type (1,1) satisfying $f^3 + f = 0$. Tensor **14**, 99–109 (1963)
21. K. Yano, M. Kon, *Structures on Manifolds*. Series in Pure Mathematics, vol. 3. World Scientific (1984)
22. A. Weil, Sur la théorie des formes différentielles attachées à une variété analytique complexe. Commentarii Mathematici Helvetici **20**(1), 110–116 (1947)

Almost Yamabe Solitons on a Total Space of Almost Hermitian Submersions

Tanveer Fatima, Mehmet Akif Akyol, and Rakesh Kumar

Mathematics Subject Classification (2010) 53C15 · 53C40 · 53C50

1 Introduction

Hamilton [17] defined the notion of Yamabe flow in order to solve the Yamabe problem. The importance of the notion is that Yamabe solitons seem as singularity models which are self-similar solutions for Yamabe flows. Many authors have studied this notion intensively (see [3, 4, 6, 9, 20, 22]). Almost Yamabe solitons, which are the natural generalization of Yamabe solitons, were given as follows [3]:

A Riemannian manifold M is said to be an almost Yamabe soliton (M, g, η, μ) if there exists a vector field η on M which satisfies

$$\frac{1}{2}\mathcal{L}_{\eta}g = (\tau - \mu)g, \tag{1.1}$$

T. Fatima
Department of Mathematics and Statistics, College of Sciences, Taibah University, Yanbu, Kingdom of Saudi Arabia
e-mail: tansari@taibahu.edu.sa

M. A. Akyol (✉)
Department of Mathematics, Bingöl University, 12000 Bingol, Turkey
e-mail: mehmetakifakyol@bingol.edu.tr

R. Kumar
Department of Basic and Applied Sciences, Punjabi University Patiala, Patiala 147 002, Punjabi, India
e-mail: rakesh_bas@pbi.ac.in

B.-Y. Chen et al. (eds.), *Geometry of Submanifolds and Applications*, Infosys Science Foundation Series, https://doi.org/10.1007/978-981-99-9750-3_6

where τ is the scalar curvature of M, μ, a smooth function, η, a soliton field for (M, g) and $\mathcal{L}$ is the Lie-derivative. Moreover, we say that *an almost Yamabe soliton is steady, expanding or shrinking if $\mu = 0$, $\mu < 0$ or $\mu > 0$, respectively*. An almost Yamabe soliton is said to be the Yamabe soliton if μ is a constant in the definition (1.1). It is obvious that Einstein manifolds are almost Yamabe solitons. To have the comprehensive understanding on solitions, one can go through the scientific literature [1, 2, 10, 14, 15, 19, 23, 26] etc.

On the other hand, the theory of Riemannian submersions between Riemannian manifolds was introduced by O'Neill [21] and Gray [13], respectively. It has been extended intensively in the last three decades. (For the interested readers, we refer to [12, 13, 21, 24].) Later, Riemannian submersions were considered between almost complex manifolds by Watson [25] under the name of almost Hermitian submersions. He showed that under almost Hermitian submersions, the base manifold takes the form as total space in most cases. Since then, Riemannian submersions have been used as an effective tool to describe the structure of a Riemannian manifold equipped with a differentiable structure.

Considering the above studies of almost Yamabe solitons and Riemannian submersions, the objective of the present paper is to study the almost Hermitian submersions whose total space is an almost Yamabe soliton which is also verified through an example. Basic tools of the mentioned theory are given in Sect. 2. Section 3 essentially concerns with Kaehler submersions admitting almost Yamabe solitons. Some observations are also mentioned while the potential vector field of almost Yamabe solitons is a torqued, recurrent or concurrent vector field.

2 Preliminaries

This section often involves well-known definitions and concepts of almost Hermitian submersions which are present in subsequent sections and can be found through the index.

Let (M_1, g_1) and (M_2, g_2) be Riemannian manifolds, where $\dim(M_1)$ is greater than $\dim(M_2)$. A surjective mapping $\pi : (M_1, g_1) \to (M_2, g_2)$ is called a *Riemannian submersion* [21] if

(S1) π has maximal rank, and
(S2) π_*, restricted to $\ker \pi_*^{\perp}$, is a linear isometry.

In this case, for each $q \in M_2$, $\pi^{-1}(q)$ is a k-dimensional submanifold of M_1 and called a *fibre*, where $k = \dim(M_1) - \dim(M_2)$. A vector field on M_1 is called *vertical* (resp. *horizontal*) if it is always tangent (resp. orthogonal) to fibres. A vector field X on M_1 is called *basic* if X is horizontal and π-related to a vector field X_* on M_2, i.e., $\pi_* X_p = X_{*\pi(p)}$ for all $p \in M_1$. We will denote by $\mathcal{V}$ and $\mathcal{H}$ the projections on the vertical distribution $\ker \pi_*$, and the horizontal distribution $\ker \pi_*^{\perp}$, respectively.

As usual, the manifold (M_1, g_1) is called *total manifold* and the manifold (M_2, g_2) is called *base manifold* of the submersion $\pi : (M_1, g_1) \to (M_2, g_2)$. The geometry of Riemannian submersions is characterized by O'Neill's tensors $\mathcal{T}$ and $\mathcal{A}$, defined as follows:

$$\mathcal{T}_U V = \mathcal{V}\nabla_{\mathcal{V}U}\mathcal{H}V + \mathcal{H}\nabla_{\mathcal{V}U}\mathcal{V}V, \tag{2.1}$$

$$\mathcal{A}_U V = \mathcal{V}\nabla_{\mathcal{H}U}\mathcal{H}V + \mathcal{H}\nabla_{\mathcal{H}U}\mathcal{V}V \tag{2.2}$$

for any vector fields U and V on M_1, where ∇ is the Levi-Civita connection of g_1. It is easy to see that $\mathcal{T}_U$ and $\mathcal{A}_U$ are skew-symmetric operators on the tangent bundle of M_1 reversing the vertical and the horizontal distributions. We now summarize the properties of the tensor fields $\mathcal{T}$ and $\mathcal{A}$. Let V, W be vertical and X, Y be horizontal vector fields on M_1, then we have

$$\mathcal{T}_V W = \mathcal{T}_W V, \tag{2.3}$$

$$\mathcal{A}_X Y = -\mathcal{A}_Y X = \frac{1}{2}\mathcal{V}[X, Y]. \tag{2.4}$$

On the other hand, from (2.3) and (2.4), we obtain

$$\nabla_V W = \mathcal{T}_V W + \hat{\nabla}_V W, \tag{2.5}$$

$$\nabla_V X = \mathcal{T}_V X + \mathcal{H}\nabla_V X, \tag{2.6}$$

$$\nabla_X V = \mathcal{A}_X V + \mathcal{V}\nabla_X V, \tag{2.7}$$

$$\nabla_X Y = \mathcal{H}\nabla_X Y + \mathcal{A}_X Y, \tag{2.8}$$

where $\hat{\nabla}_V W = \mathcal{V}\nabla_V W$. If X is basic

$$\mathcal{H}\nabla_V X = \mathcal{A}_X V.$$

Remark 1 Throughout the article, we have assumed all horizontal vector fields as basic vector fields.

It is not difficult to observe that $\mathcal{T}$ acts on the fibres as the second fundamental form while $\mathcal{A}$ acts on the horizontal distribution and measures of the obstruction to the integrability of this distribution. For details on Riemannian submersions, we refer to O'Neill's paper [21] and the book [12]. Furthermore, the O'Neill tensors $\mathcal{T}$ and $\mathcal{A}$ satisfy

$$\sum_{j=1}^{r} g_1\left(\mathcal{T}_U U_j, \mathcal{T}_V U_j\right) = \sum_{i=1}^{n} g_1\left(\mathcal{T}_U X_i, \mathcal{T}_V X_i\right)$$

$$\sum_{j=1}^{r} g_1\left(\mathcal{A}_X U_j, \mathcal{A}_Y U_j\right) = \sum_{i=1}^{n} g_1\left(\mathcal{A}_X X_i, \mathcal{A}_Y X_i\right)$$

$$\sum_{j=1}^{r} g_1\left(\mathcal{A}_X U_j, \mathcal{T}_U U_j\right) = \sum_{i=1}^{n} g_1\left(\mathcal{A}_X X_i, \mathcal{T}_U X_i\right),$$

where $\{U_j\}_{1\leq j\leq r}$ and $\{X_i\}_{1\leq i\leq n}$ are orthonormal frames of vertical distribution $\mathcal{V}$ and horizontal distribution $\mathcal{H}$, respectively, for any $X, Y \in \mathcal{H}$ and $U, V \in \mathcal{V}$.

Denote the Riemannian curvature tensors of M_1, M_2 and any fibre by R, R' and $\hat{R}$, respectively. Then, we get

$$R(U, V, F, W) = \hat{R}(U, V, F, W) + g_1\left(\mathcal{T}_V F, \mathcal{T}_U W\right) - g_1\left(\mathcal{T}_U F, \mathcal{T}_V W\right), \quad (2.9)$$

$$\begin{aligned} R(X, Y, G, Z) = {} & R'\left(X', Y', G', Z'\right) \circ \pi - g_1\left(\mathcal{A}_Y G, \mathcal{A}_X Z\right) \\ & + 2g_1\left(\mathcal{A}_X Y, \mathcal{A}_G Z\right) + g_1\left(\mathcal{A}_X G, \mathcal{A}_Y Z\right), \end{aligned} \quad (2.10)$$

for any horizontal vectors X, Y, G, Z and vertical vectors U, V, F, W. Moreover, denote the mean curvature vector of any fibre by H which is given as $N = rH$, such that

$$N = \sum_{j=1}^{r} \mathcal{T}_{V_j} V_j, \quad (2.11)$$

where $\{V_1, V_2, \ldots, V_r\}$ is an orthonormal frame of $\mathcal{V}$. Also, remark that

$$\mathcal{T}_U V = g(U, V)H$$

is satisfied for U, V tangent to $\mathcal{V}$ if and only if the fibre is a totally umbilical submanifold. Indeed, the vector field N is zero on M_1 if and only if any fibre is minimal. Using (2.11), one has

$$g_1\left(\nabla_E N, Z\right) = \sum_{j=1}^{r} g_1\left(\left(\nabla_E \mathcal{T}\right)\left(U_j, U_j\right), Z\right),$$

where Z is any horizontal vector field and E, an arbitrary vector field M_1.

A $(1, 1)$-tensor field J on a $2n$-dimensional smooth manifold M is said to be an almost complex structure if $J^2 = -I$. An almost Hermitian manifold (M, g, J) is a smooth manifold endowed with an almost complex structure J and a Riemannian metric g compatible in the sense that

$$g(JX_1, X_2) + g(X_1, JX_2) = 0, \quad X_1, X_2 \in \chi(M).$$

The Kaehler form of the almost Hermitian manifold is defined by $\Phi(X_1, X_2) = g(X_1, JX_2)$. An almost Hermitian manifold is called Kaehler manifold if $\nabla J = 0$, where ∇ is a Levi–Civita connection with respect to g.

Now, we recall some basic notions about almost Hermitian submersions from [12, 18, 25].

Let (M_1, g_1, J_1) and (M_2, g_2, J_2) be almost Hermitian manifolds. Riemannian submersion $\pi : M_1 \to M_2$ is said to be an almost Hermitian submersion if π is an almost complex map, i.e., if $\pi_* \circ J_1 = J_2 \circ \pi_*$ holds. Moreover, an almost Hermitian submersion is called a Kaehler if the total space is a Kaehler manifold. It is well known that the fibres and the base manifold of an almost Hermitian submersion belong to the same class as the total space.

Finally, for a given almost Hermitian submersion the scalar curvatures of total space, the base space and the fibres are related by

$$\tau = \left(\tau' \circ \pi\right) + \hat{\tau} - \|\mathcal{T}\|^2 \tag{2.12}$$

such that $\|\mathcal{T}\|^2 = \sum_{i,j} g_1\left(\mathcal{T}_{U_j} X_i, \mathcal{T}_{U_j} X_i\right)$, where $\hat{\tau}$ and τ' are the scalar curvatures of any fibre of π and the base manifold, respectively [12].

3 Almost Hermitian Submersions from an Almost Yamabe Soliton

This section includes our investigation on the almost Hermitian submersion, in particular, the Kaehler submersion $\pi : (M_1, g_1, J_1) \to (M_2, g_2, J_2)$ whose total space admits an almost Yamabe soliton.

We start with an example to provide a formal assurance of the existence of such solitons on the total manifold of almost Hermitian submersions.

Example 1 Let $\left(\mathbb{R}^4, J_1, g_1\right)$ be an almost Hermitian manifold endowed with an almost complex structure (J_1, g_1) which is given by

$$g_1 = e^{-2x_3} dx_1^2 + e^{-2x_4} dx_2^2 + dx_3^2 + dx_4^2$$
$$J_1(x_1, x_2, x_3, x_3) = (-x_2, x_1, -x_4, x_3).$$

Let $\left(\mathbb{R}^2, J_2, g_2\right)$ be an almost Hermitian manifold endowed with an almost complex structure (J_2, g_2), given by

$$g_2 = e^{2y_2} dy_1^2 + dy_2^2$$
$$J_2(y_1, y_2) = (-y_2, y_1).$$

Let $\pi : \left(\mathbb{R}^4, J_1, g_1\right) \longrightarrow \left(\mathbb{R}^2, J_2, g_2\right)$ be defined by

$$\pi\left(x_1, x_2, x_3, x_4\right) = \left(\frac{x_1 - x_3}{\sqrt{2}}, \frac{x_2 - x_4}{\sqrt{2}}\right).$$

Then

$$\operatorname{Ker} \pi_* = \operatorname{span}\left\{U_1 = e_1 + e_3, \quad U_2 = e_2 + e_4\right\}$$

and

$$(\operatorname{Ker} \pi_*)^{\perp} = \operatorname{span}\left\{X_1 = e_1 - e_3, \quad X_2 = e_2 - e_4\right\}$$

where $\{e_1 = e^{x_3}\partial x_1, e_2 = e^{x_4}\partial x_2, \quad e_3 = \partial x_3, e_4 = dx_4\}$ is the basis of $T\mathbb{R}^4$. Then

$$\pi_* X_1 = \sqrt{2}e_1^*, \quad \pi_* X_2 = \sqrt{2}e_2^*$$

where $\left\{e_1^* = e^{-y_2}\partial y_1, e_2^* = \partial y_2\right\}$ is the basis of $T\mathbb{R}^2$. Then, clearly

$$g_1\left(X_1, X_1\right) = 2 = g_2\left(\pi_* X_1, \pi_* X_1\right)$$
$$g_1\left(X_2, X_2\right) = 2 = g_2\left(\pi_* X_2, \pi_* X_2\right).$$

Therefore, π is a Riemannian submersion from $\mathbb{R}^4$ to $\mathbb{R}^2$. Moreover

$$J_1 X_1 = -e_3 + e_4 \quad , \quad J_1 X_2 = e_1 - e_3 \quad \text{then}$$
$$\pi_* J_1 X_1 = -\sqrt{2}e_2^*, \qquad \pi_* J_1 X_2 = \sqrt{2}e_1^*.$$

Also,

$$J_2 \pi_* X_1 = -\sqrt{2}e_2^* \quad , \quad J_2 \pi_* X_2 = \sqrt{2}e_1^*.$$

Hence

$$\pi_* J_1 X_1 = J_2 \pi_* X_1 \quad \text{and} \quad \pi_* J_1 X_2 = J_2 \pi_* X_2.$$

Thus, π is an almost Hermitian submersion from $\left(\mathbb{R}^4, J_1, g_1\right)$ to $\left(\mathbb{R}^2, J_2, g_2\right)$. Further

$$g_{1ij} = \begin{bmatrix} e^{-2x_3} & 0 & 0 & 0 \\ 0 & e^{-2x^4} & 0 & 0 \\ 0 & 0 & 1 & 0 \\ 0 & 0 & 0 & 1 \end{bmatrix} \text{then } g_1^{ij} = \begin{bmatrix} e^{2x_3} & 0 & 0 & 0 \\ 0 & e^{2x_4} & 0 & 0 \\ 0 & 0 & 1 & 0 \\ 0 & 0 & 0 & 1 \end{bmatrix}$$

Then

$$
\begin{aligned}
&\Gamma^1_{11} = \Gamma^2_{11} = \Gamma^4_{11} = 0 \quad , \quad \Gamma^3_{11} = e^{-2x_3} \\
&\Gamma^1_{12} = 0 = \Gamma^1_{21}, \quad \Gamma^2_{12} = 0 = \Gamma^2_{21}, \Gamma^3_{12} = 0\Gamma^3_{21}, \quad \Gamma^4_{12} = 0 = \Gamma^4_{21} \\
&\Gamma^1_{13} = -1 = \Gamma^1_{31}, \quad \Gamma^2_{13} = 0 = \Gamma^2_{31}, \quad \Gamma^3_{13} = 0 = \Gamma^3_{31}, \Gamma^4_{13} = 0 = \Gamma^4_{31} \\
&\Gamma^1_{14} = 0 = \Gamma_{41}, \quad \Gamma^2_{14} = 0 = \Gamma^2_{41}, \Gamma^3_{14} = 0 = \Gamma^3_{41}, \Gamma^4_{14} = 0 = \Gamma^4_{41} \\
&\Gamma^1_{22} = \Gamma^2_{22} = \Gamma^3_{22} = 0, \Gamma^4_{22} = e^{-2x_4} \\
&\Gamma^1_{23} = 0 = \Gamma_{32}, \quad {}^2_{23} = 0 = \Gamma^2_{32}, \Gamma^3_{23} = 0 = \Gamma^3_{32}, \Gamma^4_{23} = 0 = \Gamma^4_{23} \\
&\Gamma^1_{24} = -1 = \Gamma^1_{42}, \quad \Gamma^2_{24} = 0 = \Gamma^2_{42}, \quad \Gamma^3_{24} = 0 = \Gamma^3_{42}, \quad \Gamma^4_{24} = 0 = \Gamma^4_{42} \\
&\Gamma^1_{33} = \Gamma^2_{33} = \Gamma^3_{33} = \Gamma^4_{33} = 0 \\
&\Gamma^1_{44} = \Gamma^2_{44} = \Gamma^3_{44} = \Gamma^4_{44} = 0.
\end{aligned}
$$

Using these Christoffel symbols, we obtain

$$\text{Ric}\,(e_1, e_1) = 1, \quad \text{Ric}\,(e_2, e_2) = 1, \quad \text{Ric}\,(e_3, e_3) = 1 \text{ and } \text{Ric}\,(e_4, e_4) = 1.$$

Hence the scalar curvature τ is

$$\tau = \sum_i \text{Ric}\,(e_i, e_i) = 4.$$

We also have

$$
\begin{array}{llll}
\nabla_{U_1} U_1 = -X_1, & \nabla_{U_1} U_2 = 0, & \nabla_{U_1} X_1 = U_1, & \nabla_{U_1} X_2 = 0 \\
\nabla_{U_2} U_1 = 0, & \nabla_{U_2} U_2 = -X_2, & \nabla_{U_2} X_1 = 0, & \nabla_{U_2} X_2 = U_2 \\
\nabla_{X_1} U_1 = -X_1, & \nabla_{X_1} U_2 = 0, & \nabla_{X_1} X_1 = U_1, & \nabla_{X_1} X_2 = 0 \\
\nabla_{X_2} U_1 = 0, & \nabla_{X_2} U_2 = -X_2, & \nabla_{X_2} X_1 = 0, & \nabla_{X_2} X_2 = U_2.
\end{array}
$$

Since $T\mathbb{R}^4 = (\text{Ker}\,\pi_*) \oplus (\text{Ker}\,\pi_*)^{\perp}$, therefore any vectors $Z_1, Z_2, Z_3 \in T\mathbb{R}^4$ can be written as

$$
\begin{aligned}
Z_1 &= a_1 U_1 + b_1 U_2 + c_1 X_1 + d_1 X_2 \\
Z_2 &= a_2 U_1 + b_2 U_2 + c_2 X_1 + d_2 X_2 \\
Z_3 &= a_3 U_1 + b_3 U_2 + c_3 X1 + d_3 X_2
\end{aligned}
$$

where $a_i, b_i, c_i, d_i \in \mathbb{R}$, then

$$g_1\,(Z_2, Z_3) = 2\,(a_2 a_3 + b_2 b_3 + c_2 c_3 + d_2 d_3)\,. \tag{3.1}$$

We know that

$$\frac{1}{2}\left(\mathcal{L}_{Z_1}g_1\right)(Z_2, Z_3) = \frac{1}{2}\left[g_1\left(\nabla_{Z_2}Z_1, Z_3\right) + g_1\left(\nabla_{Z_3}Z_1, Z_2\right)\right]$$
$$= c_1a_2(2a_3 - c_3) + a_1b_2(2b_3 + d_3) - a_1c_2(2c_3 + a_3)$$
$$- b_1d_2(2d_3 + b_3) - a_1a_2c_3 - b_1b_2d_3 + c_1c_2a_3 + d_1d_2b_3. \quad (3.2)$$

Using (3.1) and (3.2), the following expression:

$$\frac{1}{2}\left(\mathcal{L}_{Z_1}g_1\right)(Z_2, Z_3) = (\tau - \lambda)g_1(Z_2, Z_3)$$

becomes

$$c_1a_2(2a_3 - c_3) + d_1b_2(2b_3 + d_3) - a_1c_2(2c_3 + a_3) - b_1d_2(2d_3 + b_3)$$
$$- a_1a_2c_3 - b_1b_2d_3 + c_1c_2a_3 + d_1d_2b_3 = (4 - \lambda)2(a_2a_3 + b_2b_3 + c_2c_3 + d_2d_3).$$

Since $a_i, b_i, c_i, d_i \in \mathbb{R}$, hence λ is constant and $\left(\mathbb{R}^4, J_1, g_1\right)$ becomes Yamabe soliton. Moreover, for some suitable choices of a_i, b_i, c_i and d_i, the Yamabe soliton $\left(\mathbb{R}^4, J_1, g_1\right)$ will be shrinking, steady or expanding according to $\lambda < 0, \lambda = 0$ or $\lambda > 0$, respectively.

Now we mention some theorems and corollaries for any fibre of the submersion and the basc space to be an almost Yamabe soliton and a Yamabe soliton.

First we recall the following.

Lemma 1 ([11]) *Following are the statements which are equivalent to each other for a Riemannian submersion π from a Riemannian manifold (M_1, g_1) onto a Riemannian manifold (M_2, g_2)*

(i) The horizontal distribution $\mathcal{H}$ is parallel with respect to ∇ on M.
(ii) The vertical distribution $\mathcal{V}$ is parallel with respect to ∇ on M.
(iii) The fundamental tensors $\mathcal{T}$ and $\mathcal{A}$ vanish identically.

Theorem 1 *Let (M_1, g_1, J_1) be an almost Yamabe soliton with the potential vector field η and $\pi : (M_1, g_1, J_1) \to (M_2, g_2, J_2)$ be a Kaehler submersion. Then the totally geodesic fibres of the submersion are almost Yamabe solitons if the potential vector field η is vertical.*

Proof Since (M_1, g_1, J_1) is an almost Yamabe soliton, then from (1.1), we have

$$\frac{1}{2}(\mathcal{L}_{\eta}g_1)(U, V) = (\tau - \mu)g_1(U, V)$$

$$\frac{1}{2}\{g_1(\nabla^1_U\eta, V) + g_1(\nabla^1_V\eta, U)\} = (\tau - \mu)g_1(U, V)$$

for any vertical vector fields U, V on M_1. Since the potential vector field η is vertical, by using (2.5) we get

$$\frac{1}{2}\{g_1(\hat{\nabla}_U\eta, V) + g_1(\hat{\nabla}_V\eta, U)\} = (\tau - \mu)g_1(U, V).$$

By using (2.12) and the assumption that the fibres of the submersion are totally geodesic, i.e., the second fundamental form of fibres $\mathcal{T} = 0$

$$\frac{1}{2}\{g_1(\hat{\nabla}_U\eta, V) + g_1(\hat{\nabla}_V\eta, U)\} = (\hat{\tau} - \mu)g_1(U, V).$$

In view of the relation (1.1), the above equation yields that

$$\frac{1}{2}(\mathcal{L}_\eta g_1)(U, V) = (\hat{\tau} - \mu)g_1(U, V),$$

which shows that the fibres of the submersion π are almost Yamabe solitons with the potential vector field η.

We remark that an almost Yamabe soliton is said to be a Yamabe soliton if μ is constant.

As a consequence of the above theorem, we have the following.

Corollary 1 *Let π be a Kaehler submersion from a Kaehler manifold (M_1, g_1, J_1) admitting an almost Yamabe soliton to a Kaehler manifold (M_2, g_2, J_2). Then the totally geodesic fibres of the submersion are Yamabe solitons if the potential vector field is vertical.*

Theorem 2 *Let $\pi : (M_1, g_1, J_1) \to (M_2, g_2, J_2)$ be a Kaehler submersion admitting an almost Yamabe soliton with the potential vector field η and a smooth function μ such that the potential vector field is horizontal. If the second fundamental form of fibres $\mathcal{T}$ vanishes, then the base manifold (M_2, g_2, J_2) is also an almost Yamabe soliton.*

Proof Since (M_1, g_1, J_1) is an almost Yamabe soliton, then from (1.1), we get

$$\frac{1}{2}\{g_1(\nabla_X\eta, Y) + g_1(\nabla_Y\eta, X)\} = (\tau - \mu)g_1(X, Y)$$

for any horizontal vector fields X and Y on M_1. By simply using (2.8) and (2.12), the above equation reaches $\frac{1}{2}\{g_1(\mathcal{H}(\nabla_X\eta), Y) + g(\mathcal{H}(\nabla_Y\eta), X)\} = \{(\tau' \circ \pi) + \|\mathcal{T}\|^2 - \hat{\tau} - \mu\}g(X, Y)$.

Since $\mathcal{T} = 0$ and also we note that $\mathcal{H}(\nabla_X\eta)$ and $\mathcal{H}(\nabla_Y\eta)$ are π -related to $\nabla^2_{X_*}\eta_*$ and $\nabla^2_{Y_*}\eta_*$, respectively, it follows that

$$\frac{1}{2}\left\{g_2(\nabla^2_{X_*}\eta_*, Y_*) + g_2(\nabla^2_{Y_*}\eta_*, X_*)\right\} = \{(\tau' \circ \pi_*) - \mu\}g_2(X_*, Y_*) \circ \pi_*, \tag{3.3}$$

where the vector fields X_*, Y_* are tangent to (M_2, g_2, J_2) and η_* is a potential vector field on M_2, which is π-related to η on M_1. By using (1.1) in (3.3), it shows that the base manifold (M_2, g_2, J_2) is an almost Yamabe soliton with the potential vector field η_* and a function μ.

As a conclusion of the above theorem, we state the following corollary.

Corollary 2 *Let $\pi : (M_1, g_1, J_1) \to (M_2, g_2, J_2)$ be a Kaehler submersion admitting Yamabe soliton with the potential vector field η and a smooth function μ such that the potential vector field is horizontal. If the second fundamental form of fibres $\mathcal{T}$ vanishes, then the base manifold (M_2, g_2, J_2) is also a Yamabe soliton.*

4 Characterizations on Total Space of Almost Hermitian Submersions

This section gives the characterization on almost Yamabe solitons by investigating the relation between extrinsic horizontal curvature and extrinsic vertical curvature with function μ. We also obtain some results on torqued, recurrent and concurrent vector fields on Kaehler submersion.

Theorem 3 *Let π be a Kaehler submersion admitting an almost Yamabe soliton (M_1, g_1, η, μ) such that the potential vector field η is vertical and the horizontal distribution is parallel. Then, the extrinsic horizontal scalar curvature $\tau|_{\mathcal{H}}$ satisfies*

$$\tau|_{\mathcal{H}} - \mu = 0.$$

Proof Since the total space is an almost Yamabe soliton, then from (1.1), we get

$$\frac{1}{2}\{g_1(\nabla_X \eta, Y) + g_1(\nabla_Y \eta, X)\} = (\tau - \mu) g_1(X, Y)$$

for any horizontal vectors X, Y on M_1. Considering (2.7) in the above equation, we have

$$\frac{1}{2}\{g_1(\mathcal{A}_X \eta, Y) + g_1(\mathcal{A}_Y \eta, X)\} = (\tau|_{\mathcal{H}} - \mu) g_1(X, Y).$$

Since the horizontal distribution is parallel, therefore in view of the properties of the tensor field $\mathcal{A}$ the above equality in turn yields

$$(\tau|_{\mathcal{H}} - \mu)\, g_1(X, Y) = 0,$$

for any horizontal vector fields X, Y on M_1, which proves our claim.

As an immediate consequence of Theorem 3, we conclude the followingbreak observations.

Corollary 3 *Let $\pi : (M_1, g_1, J_1) \rightarrow (M_2, g_2, J_2)$ be a Kaehler submersion admitting an almost Yamabe soliton (M_1, g, η, μ) such that η is vertical. If the horizontal distribution is parallel, then the following hold:*

(i) (M_1, g_1, η, μ) is shrinking if and only if the extrinsic horizontal scalar curvature is positive.
(ii) (M_1, g_1, η, μ) is expanding if and only if the extrinsic horizontal scalar curvature is negative.
(iii) (M_1, g_1, η, μ) is steady if and only if the extrinsic horizontal scalar curvature is zero.

Theorem 4 *Let π be a Kaehler submersion admitting an almost Yamabe soliton (M_1, g_1, η, μ) such that the potential vector field η is horizontal and the horizontal distribution is parallel. Then, the extrinsic vertical scalar curvature $\hat{\tau}$ satisfies*

$$\hat{\tau} - \mu = 0.$$

Proof Since (M_1, g_1, J_1) is an almost Yamabe soliton and the potential vector field η is a horizontal vector field then by using (2.6), we arrive at

$$\frac{1}{2}\{g_1(\mathcal{T}_U\eta, V) + g_1(\mathcal{T}_V\eta, U)\} = (\hat{\tau} - \mu)g(U, V)$$

for any vertical vector field U, V on M. Since the horizontal distribution is parallel, by using Lemma 1, the extrinsic vertical scalar curvature satisfies $(\hat{\tau} - \mu) = 0$.

In view of the relation of the extrinsic vertical scalar curvature $\hat{\tau}$ with μ, we mention the following conclusion.

Corollary 4 *Let π be a Kaehler submersion admitting an almost Yamabe soliton (M_1, g_1, η, μ) such that the potential vector field η is horizontal and the horizontal distribution is parallel. Then,*

(i) (M_1, g_1, η, μ) is shrinking if and only if the fibre has positive scalar curvature.
(ii) (M_1, g_1, η, μ) is expanding if and only if the fibre has negative scalar curvature.
(iii) (M_1, g_1, η, μ) is steady if and only if the fibre has zero scalar curvature.

Definition 1 A nowhere vanishing vector field η on a Riemannian manifold M is called torse-forming if

$$\nabla_W\eta = \varphi W + \alpha(W)\eta, \tag{4.1}$$

where φ is a function, and α is a 1-form. The vector field η is called concircular, if the 1-form α vanishes identically. The vector field η is called concurrent if the 1-form α vanishes identically and the function $\varphi = 1$. The vector field η is called recurrent if the function $\varphi = 0$ [5, 7].

Finally if $\varphi = \alpha = 0$, then the vector field η is called a parallel vector field. The nowhere zero vector field η is called a torqued vector field if it satisfies

$$\nabla_W \eta = \varphi W + \alpha(W)\eta, \quad \alpha(\eta) = 0, \tag{4.2}$$

where the function φ is called the torqued function and 1-form α is called the torqued form of η; see [8].

Theorem 5 *Let $\pi : (M_1, g_1, J_1) \to (M_2, g_2, J_2)$ be a Kaehler submersion. Then,*

(i) For any torqued vector field η on $\Gamma\,(\ker \pi_*)^{\perp}$, $J_1\eta$ is not a torqued vector field on $\Gamma(ker\pi_*)^{\perp}$.
(ii) For torqued vector field $\eta \in \Gamma\,(ker\pi_*)^{\perp}$, $[\eta, J\eta]$ does not vanish.

Proof Since M_1 is a Kaehler manifold, then we have

$$\nabla^1_X J_1\eta = J_1\nabla^1_X\eta, \text{ for any } \quad X \in \Gamma\,(ker\pi_*)^{\perp},$$

where η is a torqued vector field which belongs to $\Gamma\,(\ker \pi_*)^{\perp}$ and ∇^1 is a Levi–Civita connection on M_1. Using (2.8), we have

$$\nabla^1_X J_1\eta = J_1\left(\mathcal{H}\nabla^1_X\eta + \mathcal{A}_X\eta\right).$$

From (4.2) and $\mathcal{A} = 0$, we arrive at

$$\nabla^1_X J_1\eta = J_1\left(\mathcal{H}\nabla^1_X\eta\right) = \varphi J_1 X + \alpha(X)J_1\eta \tag{4.3}$$

which implies that $J_1\eta$ is never a torqued vector field. We also have

$$\mathcal{H}\nabla^1_{J_1\eta}\eta = \varphi J_1\eta + \alpha(J_1\eta)\eta. \tag{4.4}$$

Now, by taking η instead of X in (4.3) and taking the difference with (4.4), we obtain (ii).

In particular, if we consider η as a recurrent or concurrent vector field then we have the following corollaries.

Corollary 5 *Let $\pi : (M_1, g_1, J_1) \to (M_2, g_2, J_2)$ be a Kaehler submersion. Then,*

(i) For a recurrent vector field η on $\Gamma\,(\ker \pi_*)^{\perp}$, $J_1\eta$ is also a recurrent vector field on $(\ker \pi_*)^{\perp}$.
(ii) For recurrent vector field $\eta \in \Gamma\,(ker\pi_*)^{\perp}$, $[\eta, J\eta]$ vanishes.

For the concurrent vector field, we have the following result from [16].

Corollary 6 *Let $\pi : (M_1, g_1, J_1) \rightarrow (M_2, g_2, J_2)$ be a Kaehler submersion. Then,*

(i) For a concurrent vector field η on Γ $(\ker \pi_*)^\perp$, $J_1\eta$ is not a concurrent vector field on $(\ker \pi_*)^\perp$.
(ii) For concurrent vector field $\eta \in \Gamma$ $(ker\pi_*)^\perp$, $[\eta, J\eta]$ vanishes.

Theorem 6 *Let $\pi : (M_1, g_1, J_1) \rightarrow (M_2, g_2, J_2)$ be a Kaehler submersion and η be the concurrent potential vector field and μ, a function on M. If the second fundamental form of the fibres are totally geodesic then the extrinsic vertical scalar curvature $\hat{\tau}$ is equal to $\mu + 1$ for $\eta \in \Gamma(ker\pi_*)$.*

Proof As (M_1, g_1, J_1) is an almost Yamabe soliton with potential vector field η, then for any $U, V \in \Gamma(ker\pi_*)$ we have

$$\frac{1}{2}\{g_1(\nabla^1_U\eta, V) + g_1(\nabla^1_V\eta, U)\} = (\tau - \mu)g_1(U, V)$$

.Since $\mathcal{T} = 0$, using (2.12) we get

$$\frac{1}{2}\{\hat{g}(\hat{\nabla}_U\eta, V) + \hat{g}(\hat{\nabla}_V\eta, U)\} = (\hat{\tau} - \mu)\hat{g}(U, V).$$

Since the potential vector field η is a concurrent vector field on M_1, with the help of (4.1) we conclude the theorem.

In view of the relation between the extrinsic vertical scalar curvature $\hat{\tau}$ and μ, we give the following result.

Corollary 7 *Let $\pi : (M_1, g_1, J_1) \rightarrow (M_2, g_2, J_2)$ be a Kaehler submersion and (M_1, g_1) be an almost Yamabe soliton with the concurrent potential vector field η. If $\eta \in \Gamma(ker\pi_*)$, then we have*

(i) (M_1, g_1, η, μ) is shrinking if and only if the extrinsic vertical scalar curvature $\hat{\tau} > 1$.
(ii) (M_1, g_1, η, μ) is expanding if and only if the extrinsic vertical scalar curvature $\hat{\tau} < 1$.
(iii) (M_1, g_1, η, μ) is steady if and only if the extrinsic vertical scalar curvature $\hat{\tau} = 1$.

Data Availability Statement

Data sharing is not applicable to this article as no datasets were generated or analysed during the current study.

References

1. A.M. Blaga, η-Ricci solitons on para-Kenmotsu manifolds. Balkan J. Geom. Appl. **20**(1), 1–13 (2015)
2. A.M. Blaga, S.Y. Perktaş, Remarks on almost -Ricci solitons in ϵ-para Sasakian manifolds. Commun. Fac. Sci. Univ. Ankara Ser. A1 Math. Stat. **68**(2), 1621–1628 (2019)
3. E. Barbosa, E. Ribeiro, On conformal solutions of the Yamabe flow. Archiv der Mathematik **101**, 79–89 (2013)
4. H.D. Cao, X. Sun, Y. Zhang, On the structure of gradient Yamabe solitons. Math. Res. Lett. (19), 767–774 (2012)
5. B.-Y. Chen, Classification of torqued vector fields and its applications to Ricci solitons. Kragujevac J. Math. **41**, 239–250 (2017). https://doi.org/10.5937/KGJMATH1702239C
6. B.-Y. Chen, S. Deshmukh, Yamabe and quasi-Yamabe solitons on Euclidean submanifolds. Mediterrenean J. Math. (15), 194 (2018)
7. B.-Y. Chen, S. Deshmukh, Ricci solitons and concurrent vector fields. Balkan J. Geom. Appl. **20**(1), 14–25 (2015)
8. S. Deshmukh, B.-Y. Chen, *A* Note on Yamabe solitons. Balkan J. Geom. Appl. **23**(1), 37–43 (2018)
9. C. Dey, U.C. De, A note on quasi-Yamabe solitons on contact metric manifolds. J. Geom. **111**, 11 (2020)
10. Ş Eken Meriç, E. Kiliç, Riemannian submersions whose total manifolds admitting a Ricci soliton. Int. J. Geom. Methods Modern Phys. **16**(12), 1950196 (12 pages) (2019)
11. Ş Eken Meriç, Some remarks on Riemannian submersions admitting an almost Yamabe soliton. Adıyaman Univ. J. Sci. **10**(1), 295–306 (2020)
12. M. Falcitelli, S. Ianus, A.M. Pastore, *Riemannian Submersions and Related Topics* (World Scientific Publishing Co. Pte. Ltd, 2004)
13. A. Gray, Pseudo-Riemannian almost product manifolds and submersion. J. Math. Mech. **16**, 715–737 (1967)
14. S. Güler, On a class of gradient almost Ricci solitons. Bull. Malays. Math. Sci. Soc. **43**, 3635–3650 (2020)
15. S. Güler, M. Crasmareanu, Ricci-Yamabe maps for Riemannian flow and their volume variation and volume entropy. Turkish J. Math. **43**, 2631–2641 (2019)
16. Y. Gündüzalp, Almost Hermitian submersions whose total manifolds admit a Ricci soliton. Honam Math. J. **42**(4), 733–745 (2020)
17. R. Hamilton, The Ricci flow on surfaces. Math. Gen. Relat. Contemp. Math. **71**, 237–361 (1988)
18. D.L. Johnson, Kaehler Submersions and holomorphic connections. J. Diffe. Geom. **15**, 71–79 (1980)
19. L. Ma, V. Miquel, Remarks on scalar curvature of Yamabe solitons. Ann. Global Anal. Geom. **42**, 195–205 (2012)
20. B.L. Neto, A note on (anti-)self dual quasi Yamabe gradient solitons. Results Math. **71**, 527–533 (2017)
21. B. O'Neill, The fundamental equations of a Riemannian submersions. Michigan Math. J. **13**, 459–469 (1966)
22. T. Seko, S. Matea, Classification of almost Yamabe solitons in Euclidean spaces. J. Geom. Phys. **136**, 97–103 (2019)
23. M.D. Siddiqui, F. Mofarrah, M. A. Akyol, Ali H. Hakami, *η-Ricci-Yamabe Solitons along Riemannian Submersions*. Axioms. **12**(8), 796 (2023)
24. B. Sahin, Riemannian Submersions, Riemannian Maps in Hermitian Geometry, and their Applications. Elsevier, Academic Press, London (2017)
25. B. Watson, Almost Hermitian submersions. J. Diff. Geom. **11**(1), 147–165 (1976)
26. K. Yano, M. Kon, Structures on Manifolds. World Scientific, Singapore (1984)

Generalized Wintgen Inequalities for (ϵ)-Para Sasakian Manifold

Majid Ali Choudhary, Lovejoy S. Das, Mohd. Danish Siddiqi, and Oğuzhan Bahadır

2010 Mathematics Subject Classification: 53B05 · 53C15 · 53C17 · 53C40 · 53D10

1 Introduction

The Wintgen inequality is a sharp geometric inequality involving intrinsic and extrinsic invariants for the surface M^2 in Euclidean space E^4. The discovery is attributed to Wintgen [30], who used the equation below

$$||\mathcal{H}||^2 \geq \mathcal{K} + |\mathcal{K}^{\perp}|.$$

$\mathcal{K}$ and $\mathcal{K}^{\perp}$ are employed in this equation for the Gauss curvature and normal curvature of M^2, respectively. The mean curvature is expressed as $\mathcal{H}$. The equality in the

M. A. Choudhary (✉)
Department of Mathematics, School of Sciences, Maulana Azad National Urdu University, Hyderabad, India
e-mail: majid_alichoudhary@yahoo.co.in

L. S. Das
Department of Mathematics, Kent State University, New Philadelphia, OH 44663, USA
e-mail: ldas@kent.edu

Mohd. D. Siddiqi
Department of Mathematics, Faculty of Sciences, Jazan University, Jazan, Saudi Arabia
e-mail: msiddiqi@jazanu.edu.sa

O. Bahadır
Department of Mathematics, Kahramanmaras Sutçu Imam University, 46050 Kahramanmaras, Turkey

B.-Y. Chen et al. (eds.), *Geometry of Submanifolds and Applications*, Infosys Science Foundation Series, https://doi.org/10.1007/978-981-99-9750-3_7

preceding connection also holds when surface's ellipse of curvature appears to be a circle in Euclidean space.

The aforementioned inequality was later investigated and expanded independently in [17, 26] for surfaces of arbitrary codimension n in real space form $\widetilde{M}^{n+2}(c)$ as

$$||\mathcal{H}||^2 + c \geq \mathcal{K} + |\mathcal{K}^\perp|.$$

The generalized Wintgen inequality was presented in [27] as a natural extension of the foregoing inequality. They also established that the following inequality is maintained at every point of M^n immersed in m-dimensional real space form $\widetilde{M}^{n+m}(c)$

$$\rho \leq ||\mathcal{H}||^2 - \rho^\perp + c,$$

in which ρ stands for the normalized scalar curvature and $\rho^\perp$ for the normalized normal scalar curvature of M, respectively. Other names for this idea include the DDVV conjecture and the normal scalar curvature conjecture. Keep in mind that the submanifolds for which the equality in the previous relation holds are the Wintgen ideal submanifolds [5]. While evidence for the general case of the DDVV conjecture were provided in [16, 21]. The generalized Wintgen inequality has been widely researched and developed for numerous submanifolds in several ambient spaces [8, 9, 22–24] etc.

Comparatively, Sato [28] began researching virtually para-contact structure on a differentiable manifold in 1976. With $\epsilon = 1$ (ξ will be spacelike) or $\epsilon = -1$ (ξ be timelike), Tripathi et al. [29] created what is called as ϵ-almost para-contact structure. They went on to explain the (ϵ)-Sasakian manifold and (ϵ)-almost para-contact manifold. Dirik et al. [15] recently examined the (ϵ)-para Sasakian manifold (shortly (ϵ)-PSM) and found some nice investigations on the contact pseudo-slant (shortly CPS) submanifolds of the (ϵ)-para Sasakian space forms (ϵ-PSSF).

Inspired by aforementioned developments in the field, the CPS submanifolds of (ϵ)-PSSF are the subject of this investigation. The generalized Wintgen inequalities have been constructed for such submanifolds and equality scenarios are discussed.

Next, recall that Ricci solitons are the self-identical solutions of the Ricci flow introduced in 1982 due to [18]. A Ricci soliton can be viewed as a Riemannian manifold (M, g) with a smooth vector field V on M according to

$$S + \frac{1}{2}\mathfrak{L}_V g + \Lambda g = 0, \tag{1.1}$$

where S is the Ricci tensor of (M, g), Λ is constant, and $\mathfrak{L}_V g$ is the Lie-derivative of the metric tensor g with respect to V. As a result, the Ricci soliton is represented by (M, g, V, Λ). A Ricci soliton (M, g, V, Λ) is considered to be decreasing, stationary, or growing according to $\Lambda < 0$, $\Lambda = 0$, or $\Lambda > 0$. Besides this, if $V = \nabla\psi$ in (1.1), then (M, g) is referred to as a gradient Ricci soliton with a smooth function ψ on M and the Ricci tensor of g having the following [19]

$$S + Hess(\psi) = \Lambda g. \tag{1.2}$$

Many researchers have recently worked on the notion of Ricci soliton for many types of manifolds [12, 13]. Bejan and Crasmareanu [2] examined and presented various characterizations in terms of Ricci solitons for the Eisenhart problem of seeing parallel tensors for the symmetric scenario, which had previously been addressed in the context of quasi-constant curvature manifolds. Also, they examined potential Lorentz–Ricci solitons and the (0, 2)-type family of parallel symmetric tensor fields [6].

A necessary and sufficient criterion for a submanifold to be a Ricci soliton in a Riemannian manifold with a concurrent vector field was also developed by Chen and Deshmukh in [3] along with a classification of Ricci solitons with concurrent potential fields. In-depth classifications were also made for the Ricci solitons on Euclidean hypersurfaces that result from the position vector field of hypersurfaces [4].

In light of this, we compute the scalar curvature of a particular kind of submanifold M of a Ricci soliton (M, g, V, Λ) in the second section of this study. Additionally, we create an inequality for Ricci solitons to discover correlations between intrinsic and extrinsic invariants like scalar curvature, sectional curvature, or mean curvature.

2 Preliminaries

Consider a differentiable manifold $\widetilde{M}^n$ and let (1, 1)-tensor field ϕ, 1-form η, vector field ξ defined on $\widetilde{M}^n$ satisfy [28]

$$\phi^2 = \ell - \eta(\ell)\xi, \tag{2.1}$$

$$\eta(\xi) = 1, \quad \phi\xi = 0, \eta(\ell) = g(\ell, \xi) \tag{2.2}$$

$$\eta(\phi) = 0, \tag{2.3}$$

then $\widetilde{M}^n$ is said to have a para-contact structure (ϕ, ξ, η).

The semi-Riemannian metric on $\widetilde{M}$ is defined in terms of a non-degenerate symmetric tensor field g of type (0, 2). Note that the semi-Riemannian metric for index 1 is a Lorentzian metric [1].

For a semi-Riemannian metric g, let $\widetilde{M}^n$ be an almost para-contact manifold. If $\widetilde{M}$ satisfies the conditions listed below [29]

$$g(\phi\ell_1, \phi\ell_2) = g(\ell_1, \ell_2) - \epsilon\eta(\ell_1)\eta(\ell_2), \qquad \forall \ell_1, \ell_2 \in \Gamma(T\widetilde{M}), \tag{2.4}$$

it is said to be a (ϵ)-almost para-contact metric manifold endowed with an (ϵ)-almost para-contact metric structure $(\phi, \xi, \eta, g, \epsilon)$. In the Eq. (2.4), $\epsilon = +1$, or $\epsilon = -1$.

When index of g is 1, an (ϵ)-almost para-contact metric manifold is known as Lorentzian almost para-contact metric manifold. Moreover, an almost para-contact metric manifold with positive definite metric g is a (ϵ)-almost para-contact metric manifold.

An (ϵ)-almost para-contact metric manifold is referred as a Lorentzian almost para-contact metric manifold if index of g is 1. Also, an (ϵ)-almost para-contact metric manifold is the usual almost para-contact metric manifold with positive definite metric g.

With the help of (2.1), (2.2) and (2.4), we write

$$g(\phi\ell_1, \ell_2) = g(\ell_1, \phi\ell_2), \tag{2.5}$$

$$g(\ell_1, \xi) = \epsilon\eta(\ell_1). \tag{2.6}$$

Again, (2.2) and (2.6) produce

$$g(\xi, \xi) = \epsilon.$$

Assume that $\tilde{\nabla}$ is the Levi–Civita connection on an almost para-contact metric manifold $\widetilde{M}(\phi, \xi, \eta, g)$. Therefore $\widetilde{M}$ is said to be (ϵ)-PSM if and only if

$$(\tilde{\nabla}_{\ell_1}\phi)\ell_2 = -g(\phi\ell_1, \phi\ell_2)\xi - \epsilon\eta(\ell_2)\phi^2\ell_1. \tag{2.7}$$

Replacement of ℓ_2 with ξ in (2.7) results

$$\nabla_{\ell_1}\xi = \epsilon\phi\ell_1. \tag{2.8}$$

For any given smooth functions a and b on an (ϵ)-PSM $\widetilde{M}$, when Ricci tensor of $\widetilde{M}$ can be expressed as

$$S(\ell_1, \ell_2) = ag(\ell_1, \ell_2) + b\eta(\ell_1)\eta(\ell_2),$$

$\widetilde{M}$ is known as η-Einstein and with $b = 0$, it becomes Einstein manifold.

Let $\widetilde{M}(k)$ denotes an (ϵ)-PSSF with constant ϕ-para holomorphic sectional curvature k. One can write

$$\begin{aligned} g(\tilde{R}(\ell_1, \ell_2)\ell_3, \ell_4) = {} & \left(\frac{k-3}{4}\right)\{g(\ell_2, \ell_3)g(\ell_1, \ell_4) - g(\ell_1, \ell_3)g(\ell_2, \ell_4)\} \\ & + \left(\frac{k+1}{4}\right)\{\eta(\ell_1)\eta(\ell_2)g(\ell_2, \ell_4) - \eta(\ell_2)\eta(\ell_3)g(\ell_1, \ell_4) \\ & + g(\ell_1, \ell_3)\eta(\ell_2)\eta(\ell_4) - g(\ell_2, \ell_3)\eta(\ell_1)\eta(\ell_4) \\ & + g(\ell_2, \phi\ell_3)g(\phi\ell_1, \ell_4) - g(\ell_1, \phi\ell_3)g(\phi\ell_2, \ell_4) \\ & + 2g(\phi\ell_1, \ell_2)g(\phi\ell_1, \ell_4)\}. \end{aligned} \tag{2.9}$$

Suppose that M be a submanifold of an (ϵ)-PSM $\widetilde{M}$ and ∇ (resp. $\nabla^{\perp}$) be the induced connections on TM (resp. $T^{\perp}M$) of M. Then, for $\forall \ell_1, \ell_2 \in \Gamma(TM)$ and $V \in \Gamma(T^{\perp}M)$, one can recall

$$\tilde{\nabla}_{\ell_1}\ell_2 = \nabla_{\ell_1}\ell_2 + \sigma(\ell_1, \ell_2), \tag{2.10}$$

$$\tilde{\nabla}_{\ell_1}V = -A_V\ell_1 + \nabla^{\perp}_{\ell_1}V, \tag{2.11}$$

σ represents the second fundamental form and A_V stands for the shape operator (corresponding to the normal vector field V), respectively. The fine relation between σ and A_V is given by

$$g(A_V\ell_1, \ell_2) = g(\sigma(\ell_1, \ell_2), V)). \tag{2.12}$$

A submanifold M of a (ϵ)-PSM $\widetilde{M}$ is said to be totally umbilical if $\sigma(\ell_1, \ell_2) = g(\ell_1, \ell_2)H$, totally geodesic if $\sigma = 0$, and minimum if $H = 0$, H here denotes the mean curvature of M.

Let $\tilde{R}$ and R represent the Riemannian curvature tensors of $\widetilde{M}$ and M, respectively. The equation of Gauss is given by

$$\begin{aligned}\tilde{R}(\ell_1, \ell_2)\ell_3 = {} & R(\ell_1, \ell_2)\ell_3 + A_{\sigma(\ell_1,\ell_3)}\ell_2 - A_{\sigma(\ell_2,\ell_3)}\ell_1 + (\tilde{\nabla}_{\ell_1}\sigma)(\ell_2, \ell_3) \\ & -(\tilde{\nabla}_{\ell_2}\sigma)(\ell_1, \ell_3), \qquad \forall \ell_1, \ell_2, \ell_3 \in \Gamma(TM).\end{aligned} \tag{2.13}$$

Let's identify a local orthonormal tangent frame of TM of M and a local orthonormal normal frame of $T^{\perp}M$ of M in $\widetilde{M}$ using the symbols $\{e_1, \ldots, e_n\}$ and $\{e_{n+1}, \ldots, e_m\}$, respectively. Next, calculate the scalar curvature using

$$\tau = \sum_{1 \le i < j \le n} R(e_i, e_j, e_j, e_i).$$

This equation yields the normalized scalar curvature

$$\rho = \frac{2\tau}{n(n-1)} = \frac{2}{n(n-1)} \sum_{1 \le i < j \le n} \mathcal{K}(e_i \wedge e_j), \tag{2.14}$$

where $\mathcal{K}$ is the sectional curvature function on M. The scalar normal curvature $\mathcal{K}_N$ is described as [25]

$$\mathcal{K}_N = \sum_{1 \le r < s \le m-n} \sum_{1 \le i < j \le n} \left[\sum_{k=1}^{n} (h^r_{jk}h^s_{ik} - h^r_{ik}h^s_{jk}) \right]^2. \tag{2.15}$$

The normalized scalar normal curvature can be defined by [25]

$$\rho_N = \frac{2}{n(n-1)}\sqrt{\mathcal{K}_N}. \tag{2.16}$$

Setting $\sigma_{ij}^r = g(\sigma(e_i, e_j), e_r)$, where $i, j = \{1, \ldots, n\}$ and $r = \{n+1, \ldots, m\}$, we have

$$||H||^2 = \frac{1}{n^2} \sum_{r=n+1}^{m} \left(\sum_{i=1}^{n} \sigma_{ii}^r \right)^2, \tag{2.17}$$

$$||\sigma||^2 = \sum_{r=n+1}^{m} \sum_{i,j=1}^{n} \left(\sigma_{ij}^r\right)^2. \tag{2.18}$$

Making use of (2.9) and (2.13), the Riemannian curvature tensor R of M immersed in an (ϵ)-PSSF $\widetilde{M}(k)$ is given by

$$\begin{aligned} R(\ell_1, \ell_2)\ell_3 = &\left(\frac{k-3}{4}\right)\{g(\ell_2, \ell_3)\ell_1 - g(\ell_1, \ell_3)\ell_2\} + \left(\frac{k+1}{4}\right)\{\eta(\ell_1)\eta(\ell_3)\ell_2 \\ &- \eta(\ell_2)\eta(\ell_3)\ell_1 + \eta(\ell_2)g(\ell_1, \ell_3)\xi - \eta(\ell_1)g(\ell_2, \ell_3)\xi \\ &- g(\ell_1, \phi\ell_3)\phi\ell_2 + g(\ell_2, \phi\ell_3)\phi\ell_1 + 2g(\phi\ell_1, \ell_2)\phi\ell_3\} \\ &+ A_{\sigma(\ell_2,\ell_3)}\ell_1 - A_{\sigma(\ell_1,\ell_3)}\ell_2 + (\tilde{\nabla}_{\ell_2}\sigma)(\ell_1, \ell_3) - (\tilde{\nabla}_{\ell_1}\sigma)(\ell_2, \ell_3). \end{aligned} \tag{2.19}$$

Definition 2.1 If two orthogonal distributions $D^\perp$ and D_θ exist on a submanifold M of an (ϵ)-PSM $\widetilde{M}$ such that [15]:
(i) D_θ is slant,
(ii) $\phi(D^\perp) \subset T^\perp M$,
(iii) Reference [20] TM has thc orthogonal direct decomposition $TM = D^\perp \oplus D_\theta$, $\xi \in \Gamma(D_\theta)$, then M is said to be CPS-submanifold.

One can distinguish the different cases of CPS-submanifold as per [15].
In case of the CPS-submanifold, $T^\perp M$ can be decomposed as follows:

$$T^\perp M = \mu \oplus F(D^\perp) \oplus F(D_\theta).$$

if μ is an invariant subspace of $T^\perp M$.
We have considered the bases of the CPS submanifolds as per [14] and let $dim(M) = 2p + q + 1 = n$.
Recall the following results [15]:

Theorem 2.1 *For any CPS-submanifold M of an (ϵ)-PSSF $\widetilde{M}(k)$, we have*

$$\begin{aligned} \tau = &\{\left(\frac{k-3}{4}\right)(2p+q) + \left(\frac{k+1}{4}\right)(-2\epsilon - \epsilon^2 cos^2\theta)\}(2p+q+1) \\ &+ \left(\frac{k+1}{4}\right)(4 - 2p - q + 2\epsilon^2 - \epsilon - \epsilon cos^2\theta) \\ &+ (2p+q+1)^2\|H\|^2 - \|\sigma\|^2. \end{aligned} \tag{2.20}$$

3 Main Results

Theorem 3.1 *For any CPS-submanifold M of an (ϵ)-PSSF $\widetilde{M}(k)$, we have*

$$\rho_N \leq ||\mathcal{H}||^2 - 2\rho + (k-3) + \frac{k+1}{2p+q}(-2\epsilon - \epsilon^2 cos^2\theta) + \left(\frac{k+1}{(2p+q)(2p+q+1)}\right)(4 - 2p - q + 2\epsilon^2 - \epsilon - \epsilon cos^2\theta). \quad (3.1)$$

Additionally, with some orthonormal frame $\{e_1, \ldots, e_n, e_{n+1}, \ldots, e_m\}$, equality holds in (3.1) iff the shape operators A take the following shape:

$$A_{n+1} = \begin{pmatrix} \mathcal{F}_1 & g & 0 & \ldots & 0 & 0 \\ g & \mathcal{F}_1 & 0 & \ldots & 0 & 0 \\ 0 & 0 & \mathcal{F}_1 & \ldots & 0 & 0 \\ \vdots & \vdots & \vdots & \ddots & \vdots & \vdots \\ 0 & 0 & 0 & \ldots & \mathcal{F}_1 & 0 \\ 0 & 0 & 0 & \ldots & 0 & \mathcal{F}_1 \end{pmatrix}, \quad (3.2)$$

$$A_{n+2} = \begin{pmatrix} \mathcal{F}_2 + \mathcal{G} & 0 & 0 & \ldots & 0 & 0 \\ 0 & \mathcal{F}_2 - \mathcal{G} & 0 & \ldots & 0 & 0 \\ 0 & 0 & \mathcal{F}_2 & \ldots & 0 & 0 \\ \vdots & \vdots & \vdots & \ddots & \vdots & \vdots \\ 0 & 0 & 0 & \ldots & \mathcal{F}_2 & 0 \\ 0 & 0 & 0 & \ldots & 0 & \mathcal{F}_2 \end{pmatrix}, \quad (3.3)$$

$$A_{n+3} = \begin{pmatrix} \mathcal{F}_3 & 0 & 0 & \ldots & 0 & 0 \\ 0 & \mathcal{F}_3 & 0 & \ldots & 0 & 0 \\ 0 & 0 & \mathcal{F}_3 & \ldots & 0 & 0 \\ \vdots & \vdots & \vdots & \ddots & \vdots & \vdots \\ 0 & 0 & 0 & \ldots & \mathcal{F}_3 & 0 \\ 0 & 0 & 0 & \ldots & 0 & \mathcal{F}_3 \end{pmatrix}, \qquad A_{n+4} = \cdots = A_m = 0, \quad (3.4)$$

where $\mathcal{F}_1, \mathcal{F}_2, \mathcal{F}_3$ and $\mathcal{G}$ are real functions on M.

Proof 3.1 Taking view of (2.20), one obtains

$$\sum_{1\le i<j\le n} R(e_i, e_j, e_j, e_i) = \left(\frac{k+1}{4}\right)(4 - 2p - q + 2\epsilon^2 - \epsilon - \epsilon cos^2\theta) + \left(\frac{k-3}{4}\right)(2p+q)(2p+q+1) + \left(\frac{k+1}{4}\right)(-2\epsilon - \epsilon^2 cos^2\theta)(2p+q+1) + (2p+q+1)^2\|H\|^2 - \|\sigma\|^2. \tag{3.5}$$

On the other side, one has

$$2\tau = \sum_{1\le i<j\le n} R(e_i, e_j, e_j, e_i), \tag{3.6}$$

that gives

$$2\tau(p) = \{\left(\frac{k-3}{2}\right)(2p+q) + \left(\frac{k+1}{2}\right)(-2\epsilon - \epsilon^2 cos^2\theta)\}(2p+q+1) + \left(\frac{k+1}{2}\right)(4 - 2p - q + 2\epsilon^2 - \epsilon - \epsilon cos^2\theta) + \sum_{\alpha=n+1}^{m}\sum_{1\le i<j\le n}\left[h_{ii}^{\alpha}h_{jj}^{\alpha} - (h_{ij}^{\alpha})^2\right]. \tag{3.7}$$

It is easy to verify that

$$n^2||\mathcal{H}||^2 = \sum_{\alpha=n+1}^{m}\left(\sum_{i=1}^{n} h_{ii}^{\alpha}\right)^2 = \frac{1}{n-1}\sum_{\alpha=n+1}^{m}\sum_{1\le i<j\le n}(h_{ii}^{\alpha} - h_{jj}^{\alpha})^2 + \frac{2n}{n-1}\sum_{\alpha=n+1}^{m}\sum_{1\le i<j\le n} h_{ii}^{\alpha}h_{jj}^{\alpha}. \tag{3.8}$$

One also knows that [21]

$$\sum_{\alpha=n+1}^{m}\sum_{1\le i<j\le n}(h_{ii}^{\alpha} - h_{jj}^{\alpha})^2 + 2n\sum_{\alpha=n+1}^{m}\sum_{1\le i<j\le n}(h_{ij}^{\alpha})^2 \ge 2n\left\{\sum_{n+1\le\alpha<\beta\le m-n}\sum_{1\le i<j\le n}[\sum_{k=1}^{n}(h_{jk}^{\alpha}h_{ik}^{\beta} - h_{ik}^{\alpha}h_{jk}^{\beta})]^2\right\}^{\frac{1}{2}}. \tag{3.9}$$

Using (3.8), (3.9) and (2.15), one obtains

$$n^2||\mathcal{H}||^2 - n^2\rho_N \geq \frac{2n}{n-1}\sum_{\alpha=n+1}^{m-n}\sum_{1\leq i<j\leq n}[h_{ii}^{\alpha}h_{jj}^{\alpha} - (h_{ij}^{\alpha})^2]. \tag{3.10}$$

Thus, in the view of (2.16), (3.7) and (3.10) one can write

$$\rho_N - ||\mathcal{H}||^2 \leq (k-3) + \frac{k+1}{2p+q}(-2\epsilon - \epsilon^2 cos^2\theta) + \left(\frac{k+1}{(2p+q)(2p+q+1)}\right)(4 - 2p - q + 2\epsilon^2 - \epsilon - \epsilon cos^2\theta),$$

and hence the desired result.

4 Some Applications and Remarks

Remark 4.1 One can construct generalized Wintgen inequalities for different classes of submanifolds (invariant, anti-invariant, proper-slant, semi-invariant, totally umbilical CPS etc.) in (ϵ)-PSSF $\widetilde{M}(k)$ using the same technique of in Theorem 3.1.

Remark 4.2 Submanifolds that satisfy the equality case of generalized Wintgen inequalities are known as Wintgen ideal submanifolds. It is very typical job to classify them completely. But, some classifications of such submanifolds are given in [10, 11].

5 Ricci Solitons on Submanifold of (ϵ)-PSSF

Assuming that $(\widetilde{M}, g)$ is an m-dimensional Riemannian manifold and $\varphi : M \to \widetilde{M}$ be an isometric immersion. Then, the Ricci tensor S can be expressed as

$$\tilde{S}(\ell_1, \ell_2) = \tilde{S}|_{T_pM}(\ell_1, \ell_2) + \tilde{S}|_{T_p^{\perp}M}(\ell_1, \ell_2) \tag{5.1}$$

$\forall \ell_1, \ell_2 \in T_pM$.

Theorem 5.1 *Consider a Ricci soliton $(\tilde{M}, g, V, \Lambda)$ and let M be a submanifold of an (ϵ)-PSSF $\widetilde{M}(k)$. Then, one obtains*

$$\tau = \frac{n^2}{2}\|H\|^2 - \|\sigma\|^2 - div(V) - n\Lambda = 0, \tag{5.2}$$

$V \in TM$ represents the potential vector field of Ricci soliton.

Proof 5.1 Due to the fact that $(\tilde{M}, g, V, \Lambda)$ is a Ricci soliton. Then (1.1) generates

$$\frac{1}{2}\sum_{i=1}^{n}\{g(\nabla_{e_i}V, e_i) + g(e_i, \nabla_{e_i}V\} + \sum_{i=1}^{n}\tilde{S}(e_i, e_i) + \sum_{i=1}^{n}\Lambda g(e_i, e_i) = 0. \quad (5.3)$$

Utilizing (2.17), (2.18) and (5.3), we have

$$div(V) + \tilde{S}|_{T_pM}(\ell_1, \ell_2) + \Lambda\sum_{i=1}^{n}g(e_i, e_i) = 0. \quad (5.4)$$

The use of Gauss formula produces

$$Div(V) + 2\tau - \sum_{i,j=1}^{n}\{g(h(e_i, e_i)h(e_j, e_j)) - g(h(e_i, e_j, h(e_i, e_j))\}\Lambda\sum_{i=1}^{n}g(e_i, e_i) = 0, \quad (5.5)$$

whereby proving the required result.

As a consequence of Eq. (5.2) we articulate the following.

Theorem 5.2 *Assume that* $(\tilde{M}, g, V, \Lambda)$ *be a Ricci soliton and* M *be a submanifold of an* (ϵ)*-PSSF* $\widetilde{M}(k)$ *adopting a Ricci soliton with a potential vector field* $V \in TM$ *of Ricci soliton. Thus the Ricci soliton is expanding, steady and shrinking according as*

1. $\frac{n}{2}\|H\|^2 < \frac{1}{n}(\|\sigma\|^2 + div(V) + \tau)$,
2. $\frac{n}{2}\|H\|^2 = \frac{1}{n}(\|\sigma\|^2 + div(V) + \tau)$, *and*
3. $\frac{n}{2}\|H\|^2 > \frac{1}{n}(\|\sigma\|^2 + div(V) + \tau)$, *respectively.*

Next, in light of Theorem 5.1, (1.2) and (5.2) we obtain the result for gradient Ricci soliton if $V \in TM$ of Ricci soliton is of type $V = \nabla\psi$, where ψ is some smooth function on submanifold M.

Theorem 5.3 *Let* $(\tilde{M}, g, V, \Lambda)$ *be a gradient Ricci soliton with* $V = \nabla(\psi) \in TM$ *of gradient type and* M *be a submanifold of an* (ϵ)*-PSSF* $\widetilde{M}(k)$*. Then one can obtain*

$$\tau = \frac{n^2}{2}\|H\|^2 - \|\sigma\|^2 - \Delta(\psi) - n\Lambda = 0. \quad (5.6)$$

Corollary 5.1 *Let* $(\tilde{M}, g, V, \Lambda)$ *be a gradient Ricci soliton with a potential vector field* $V = \nabla(\psi) \in TM$ *of gradient type and* M *be a submanifold of an* (ϵ)*-PSSF*

$\widetilde{M}(k)$. *Then the submanifold* M *of* (ϵ)*-PSSF* $\widetilde{M}(k)$ *admits gradient Ricci soliton if and only if*

$$\frac{n}{2}\|H\|^2 < \frac{1}{n}(\|\sigma\|^2 + \Delta(\psi) + \tau). \tag{5.7}$$

Our next result based on pinching conditions of the Weyl curvature W, defined by the next result (for more details see [7])

$$W(\ell_1, \ell_2)\ell_3 = R(\ell_1, \ell_2)\ell_3 - \frac{1}{2n-1}\{g(Q\ell_2, \ell_3)\ell_1 - g(Q\ell_1, \ell_3)\ell_2 + g(\ell_2, \ell_3)\ell_1\}$$

$$- g(\ell_1, \ell_3)Q\ell_2 + \frac{\tau}{(n-1)(n-2)}\{g(\ell_2, \ell_3)\ell_1 - g(\ell_1, \ell_3)\ell_2\} \tag{5.8}$$

$\forall \ell_1, \ell_2, \ell_3 \in TM$, where R and τ represent the Riemannian curvature and scalar curvature tensor, respectively. Moreover, $\overset{\circ}{S} = S - \frac{1}{n}\tau g$ is the traceless Ricci tensor.

In 2013, Catino (Theorem 1.1 [7]) proved that any complete n-dimensional gradient Ricci soliton with non-negative Ricci curvature holding

$$|W|\tau \leq \sqrt{\frac{2(n-1)}{n-2}}\Big(|\overset{\circ}{S}| - \frac{1}{\sqrt{n(n-1)}}\Big)^2 \tag{5.9}$$

is a finite quotient of $\mathbb{R}^n$ or $\mathbb{S}^n$.

In 3-dimension, the Weyl tensor vanishes, making the pinching (5.9) trivially satisfied. However, in $n \leq 4$, the results regarding the locally conformally flat gradient shrinking Ricci soliton with non-negative Ricci tensor and very general pinching condition on the Weyl tensor in force are generalized.

Thus in the light of the above facts and adopting Theorem 5.3 with (5.6) and (5.9) we turn up an interesting result.

Theorem 5.4 *Let* $(\tilde{M}, g, V = \nabla(\psi), \Lambda)$ *be a gradient shrinking Ricci soliton with* $V = \nabla(\psi) \in TM$ *of gradient type and* M *be a submanifold of an* (ϵ)*-PSSF* $\widetilde{M}(k)$*. Then we have*

$$|W| \leq \frac{2}{n\|H\|^2 - 2\|\sigma\|^2 - 2\Delta(\psi) - 2n\Lambda}\sqrt{\frac{2(n-1)}{n-2}}. \tag{5.10}$$

Now from (2.20) and (5.9) we gain the following theorems.

Theorem 5.5 *Let* $(\tilde{M}, g, V = \nabla(\psi), \Lambda)$ *be a gradient Ricci soliton with* $V = \nabla(\psi) \in TM$ *of gradient type and* M *be a CPS-submanifold of an* (ϵ)*-PSSF* $\widetilde{M}(k)$*. Then we have*

$$|W| \leq \{\left(\frac{k-3}{4}\right)(2p+q) + \left(\frac{k+1}{4}\right)(-2\epsilon - \epsilon^2 cos^2\theta)\}(2p+q+1) \quad (5.11)$$

$$+\left(\frac{k+1}{4}\right)(4-2p-q+2\epsilon^2-\epsilon-\epsilon cos^2\theta)+(2p+q+1)^2\|H\|^2-\|\sigma\|^2.$$

Theorem 5.6 *Let* $(\tilde{M}, g, V = \nabla(\psi), \Lambda)$ *be a gradient shrinking Ricci soliton with* $V = \nabla(\psi) \in TM$ *of gradient type and* M *be a totally umbilical CPS-submanifold of an* (ϵ)*-PSSF* $\widetilde{M}(c)$*. Then we have*

$$|W| \leq \{\left(\frac{k-3}{4}\right)(2p+q) + \left(\frac{k+1}{4}\right)(-2\epsilon - \epsilon^2 cos^2\theta)\}(2p+q+1) \quad (5.12)$$

$$+\left(\frac{k+1}{4}\right)(4-2p-q+2\epsilon^2-\epsilon-\epsilon cos^2\theta)+(2p+q+1)^2\|H\|^2-\|\sigma\|^2.$$

References

1. J.K. Beem, P.E. Ehrlich, *Global Lorentzian Geometry*. Pure and Applied Mathematics, vol. 67 (Marcel Dekker, New York, 1981)
2. C.L. Bejan, M. Crasmareanu, Second order parallel tensors and Ricci solitons in 3-dimensional normal paracontact geometry. Anal. Glob. Anal. Geom. **46**, 117–128 (2014)
3. B.Y. Chen, S. Deshmukh, Ricci solitons and concurrent vector Field. Balkan J. Geom. Its Appl. **20**, 14–25 (2015)
4. B.Y. Chen, S. Deshmukh, Classification of Ricci solitons on Euclidean hypersurfaces. Int. J. Math. **25**(22), 1 (2014)
5. B.Y. Chen, Classification of Wintgen ideal surfaces in Euclidean 4-space with equal Gauss and normal curvatures. Ann. Global Anal. Geom. **38**, 145–160 (2010)
6. C. Calin, M. Crasmareanu, From the Eisenhart problem to Ricci solitons in f -Kenmotsu manifolds. Bull. Malays. Math. Sci. Soc. **33**, 361–368 (2010)
7. G. Catino, Complete gradient shrinking Ricci solitons with pinched curvature. Math. Ann. **355**(2), 629–635 (2013)
8. M.A. Choudhary, O. Bahadir, H. Alsulami, Generalized Wintgen inequality for some submanifolds in Golden Riemannian space forms. Balkan J. Geom. Appl. **25**(2), 1–11 (2020)
9. M.A. Choudhary, A.M. Blaga, Generalized Wintgen inequality for slant submanifolds in metallic Riemannian space forms. J. Geom. **112**, 26 (2021). https://doi.org/10.1007/s00022-021-00590-7
10. S. Decu, S. Haesen, L. Verstraelen, Optimal inequalities involving Casorati curvatures. Bull. Transylv. Univ. Brasv Ser. B **14**(2007), 85–93 (2014)
11. S. Decu, S. Haesen, L. Verstraelen, Optimal inequalities characterising quasi-umbilical submanifolds. J. Inequal. Pure. Appl. Math. **9**, Article ID 79 (2008)
12. K. De, M.N.I. Khan, U.C. De, Almost co-Kähler manifolds and (m, ρ)-quasi-Einstein solitons. Chaos Solitons Fractals **167**, 113050 (2023)
13. U.C. De, M.N.I. Khan, A. Sardar, h-almost Ricci-Yamabe solitons in paracontact geometry. Mathematics **10**(18), 3388
14. S. Dirik, M. Atceken, U. Yildirim, Contact pseudo-slant submanifolds of a Kenmotsu manifold. J. Math. Comput. Sci. **16**(3), 386–394 (2016)
15. S. Dirik, M. Atceken, U. Yildirim, Contact pseudo-slant submanifolds of an (ϵ)-PSSF. J. Int. Math. Virtual Inst. **10**(1), 59–74 (2020)

16. J. Ge, Z.Z. Tang, A proof of the DDVV conjecture and its equality case. Pacific J. Math. **237**, 87–95 (2008)
17. I.V. Guadalupe, L. Rodriguez, Normal curvature of surfaces in space forms. Pacific J. Math. **106**, 95–103 (1983)
18. R.S. Hamilton, Three Manifold, with positive Ricci curvature. J. Differ. Geom. **17**(2), 255–306 (1982)
19. R.S. Hamilton, The Ricci flow on surfaces. Contemp. Math. **71**, 237–261 (1988)
20. V.A. Khan, M.A. Khan, Pseudo-slant submanifolds of a Sasakian manifold. Indian J. Prue Appl. Math. **38**(1), 31–42 (2007)
21. Z. Lu, Normal scalar curvature conjecture and its applications. J. Funct. Anal. **261**, 1284–1308 (2011)
22. G. Macsim, V. Ghisoiu, Generalized Wintgen inequality for Lagrangian submanifolds in quaternionic space forms. Math. Inequal. Appl. **22**, 803–813 (2019)
23. I. Mihai, On the generalized Wintgen inequality for Lagrangian submanifolds in complex space forms. Nonlinear Anal. **95**, 714–720 (2014)
24. I. Mihai, On the generalized Wintgen inequality for Legendrian submanifolds in Sasakian space forms. Tohoku Math. J. **69**, 43–53 (2017)
25. I. Mihai, F.R. Al-Solamy, M.H. Shahid, On Ricci curvature of a quaternion CR-submanifold in a quaternion space form. Rad. Mat. **12**(1), 91–98 (2003)
26. B. Rouxel, *Sur une Famille des A-Surfaces d'un Espace Euclidien* E^4 (Insbruck, Austria, Osterreischer Mathematiker Kongress, 1981)
27. P.J. De Smet, F. Dillen, L. Verstraelen, L. Vrancken, A pointwise inequality in submanifold theory. Arch. Math. (Brno) **35**, 115–128 (1999)
28. I. Sato, On a structure similar to the almost contact structure. Tensor N.S. **30**(3), 219–224 (1976)
29. M.M. Tripathi, E. Kilic, S.Y. Perktas, S. Keles, Indefinite almost para-contact metric manidolds. Int. J. Math. Math. Sci. **2010**, Article ID 846195, 19 (2010)
30. P. Wintgen, Sur l'inegalite de Chen-Willmore, C. R. Acad. Sci. Paris Ser. A-B **288**, A993–A995 (1979)

Certain Optimal Inequalities for Casorati Curvatures in Quaternion Geometry

Mohd Danish Siddiqi, Aliya Naaz Siddiqui, and Kamran Ahmad

2010 Mathematics Subject Classification 34B10 · 34B15

1 Introduction

Ishihara [7] proposed a quaternion manifold, sometimes known as a quaternion Kaehler manifold, as a Riemannian manifold whose holonomy group is a subgroup of $Sp(l).Sp(m) = Sp(l) \times Sp(m)/\{\pm 1\}$. It is generally known that a 3-dimensional vector bundle V of tensor of type $(1, 1)$ exists on a quaternion manifold with local cross-section of essentially Hermitian structures following certain axioms [3]. If the quaternion sectional curvature c of a quaternion manifold is constant, it is a quaternion space form. A submanifold $\mathcal{M}$ in a quaternion manifold $\tilde{\mathcal{M}}$ is carried into itself by each section of V.

The establishment of a straightforward relationship between the essential intrinsic invariants and the important extrinsic invariants of the submanifolds is a central problem in the general theory of Riemannian submanifolds. Certain forms of inequalities can provide such a correlation.

M. D. Siddiqi
Department of Mathematics, College of Science, Jazan University,
Jazan, Saudi Arabia
e-mail: msiddiqi@jazanu.edu.sa

A. N. Siddiqui (✉) · K. Ahmad
Division of Mathematics, School of Basic Sciences, Galgotias University,
Greater Noida 203201, UP, India
e-mail: aliyanaazsiddiqui9@gmail.com

K. Ahmad
e-mail: qskamrankhan@gmail.com

B.-Y. Chen et al. (eds.), *Geometry of Submanifolds and Applications*, Infosys Science Foundation Series, https://doi.org/10.1007/978-981-99-9750-3_8

It is known that the Casorati curvature of a submanifold in a Riemannian manifold is an extrinsic invariant defined as the normalized square of the length of the second fundamental form. Moreover, this notion extends the concept of the principal direction of a hypersurface of a Riemannian manifold to submanifolds of a Riemannian manifold and it was preferred by Casorati over the traditional Gauss curvature because it corresponds better with the common intuition of curvature. Therefore, it is of great interest to obtain optimal inequalities for the Casorati curvatures of submanifolds in different ambient spaces. The Casorati curvature [5] has been widely investigated in the discipline of geometry as well as other fields such as human/computer vision. In the literature, several results in terms of isotropical Casorati curvature of production surfaces have been obtained. A geometric version of the Cauchy-Schwarz inequality is also examined, as well as the connection of this sort of curvature. Several geometers have established relationship between the normalized scalar curvature and (generalized) normalized δ-Casorati curvatures (for example, [15–22]). For this reason, the aim of this paper is to derive two inequalities for Casorati curvatures of quaternion bi-slant submanifolds in quaternion space forms. In addition, we determine the inequality for particular cases such as semi-slant, hemi-slant, CR-slant, invariant and totally real submanifolds in the same ambient space form.

2 Preliminaries

In this section, we will discuss some basic definitions and notions.

2.1 Quaternion Manifolds

Let $\tilde{\mathcal{M}}$ be a $4n$-dimensional Riemannian manifold with metric tensor g. Then $\tilde{\mathcal{M}}$ is said to be a quaternion Kaehler manifold, if there exists a 3-dimensional vector bundle V consisting of tensor of type $(1, 1)$ with local basis of almost Hermitian structures $\mathcal{J}_1$, $\mathcal{J}_2$ and $\mathcal{J}_3$ such that

$$\textbf{(a)}\quad \mathcal{J}_1^2 = -I, \qquad \mathcal{J}_2^2 = -I, \qquad \mathcal{J}_3^2 = -I, \tag{2.1}$$

$$\mathcal{J}_1\mathcal{J}_2 = -\mathcal{J}_2\mathcal{J}_1 = \mathcal{J}_3, \quad \mathcal{J}_2\mathcal{J}_3 = -\mathcal{J}_3\mathcal{J}_2 = \mathcal{J}_1, \quad \mathcal{J}_3\mathcal{J}_1 = -\mathcal{J}_1\mathcal{J}_3 = \mathcal{J}_2, \tag{2.2}$$

$$g(\mathcal{J}_1 E, \mathcal{J}_1 F) = g(\mathcal{J}_2 E, \mathcal{J}_2 F) = g(\mathcal{J}_3 E, \mathcal{J}_3 F) = g(E, F), \tag{2.3}$$

where I indicates the identity tensor field of type $(1, 1)$ on $\tilde{\mathcal{M}}$.

(b) For any local cross-section $\mathcal{J}$ of V and any vector E tangent to $\tilde{\mathcal{M}}$, $\bar{\nabla}_E\mathcal{J}$ is also a local cross-section of V, where $\bar{\nabla}$ denotes the Riemannian connection on $\tilde{\mathcal{M}}$.

(c) There exist local 1-forms u v and w such that

$$\bar{\nabla}_E \mathcal{J}_1 = w(E)\mathcal{J}_2 - v(E)\mathcal{J}_3, \tag{2.4}$$

$$\bar{\nabla}_E \mathcal{J}_2 = -w(E)\mathcal{J}_1 + u(E)\mathcal{J}_3, \tag{2.5}$$

$$\bar{\nabla}_E \mathcal{J}_3 = v(E)\mathcal{J}_1 - u(E)\mathcal{J}_2. \tag{2.6}$$

Now, let E be an unit vector tangent to the quaternion manifold $\tilde{\mathcal{M}}$, then $\{E, \mathcal{J}_1 E, \mathcal{J}_2 E, \mathcal{J}_3 E\}$ form an orthonormal frame. We denote by $\mathcal{Q}(E)$ the 4-plane spanned by this frame and call $\mathcal{Q}(E)$ the quaternion section determined by E. For any orthonormal vectors E, F tangent to $\tilde{\mathcal{M}}$, the plane spanned by E, F is said to be totally real if $\mathcal{Q}(E)$ and $\mathcal{Q}(F)$ are orthogonal. Any plane in a quaternion section is called a quaternion plane. The sectional curvature of a quaternion plane is called a quaternion sectional curvature.

Let $\tilde{\mathcal{M}}(c)$ be a $4n$-dimensional quaternion space form of constant quaternion sectional curvature c. The curvature tensor of $\tilde{\mathcal{M}}(c)$ has the following form [7]:

$$\begin{aligned}\mathcal{R}(E, F)G = \frac{c}{4}[&g(F, G)E - g(E, G)F \\ &+ g(\mathcal{J}_1 F, G)\mathcal{J}_1 E - g(\mathcal{J}_1 E, G)\mathcal{J}_1 F + 2g(E, \mathcal{J}_1 F)\mathcal{J}_1 G \\ &+ g(\mathcal{J}_2 F, G)\mathcal{J}_2 E - g(\mathcal{J}_2 E, G)\mathcal{J}_2 F + 2g(E, \mathcal{J}_2 F)\mathcal{J}_2 G \\ &+ g(\mathcal{J}_3 F, G)\mathcal{J}_3 E - g(\mathcal{J}_3 E, G)\mathcal{J}_3 F + 2g(E, \mathcal{J}_3 F)\mathcal{J}_3 G],\end{aligned} \tag{2.7}$$

for any $E, F, G \in \Gamma(T\tilde{\mathcal{M}})$.

Equation (2.7) can be written as

$$\begin{aligned}\mathcal{R}(E, F)G = \frac{c}{4}[&g(F, G)E - g(E, G)F \\ &+ \sum_{\alpha=1}^{3} g(\mathcal{J}_\alpha F, G)\mathcal{J}_\alpha E - g(\mathcal{J}_\alpha E, G)\mathcal{J}_\alpha F + 2g(E, \mathcal{J}_1 F)\mathcal{J}_\alpha G].\end{aligned} \tag{2.8}$$

2.2 *Submanifolds of Quaternion Manifolds*

Let $\mathcal{M}$ be an m-dimensional submanifold of a $4n$-dimensional quaternion space form $\tilde{\mathcal{M}}(c)$, and the induced Levi-Civita connection on $\mathcal{M}$ is denoted by $\nabla^{\mathcal{M}}$ and we use same notation g for the induced Riemannian metric on $\mathcal{M}$. Suppose that $\mathcal{R}$ and $\mathcal{R}^{\mathcal{M}}$

are the curvature tensors of $\bar{\nabla}$ and $\nabla^{\mathcal{M}}$. The Gauss equation is given by

$$\tilde{\mathcal{R}}(E, F, G, H) = \mathcal{R}(E, F, G, H) - g(h(E, H), h(F, G)) + g(h(F, H), h(E, G)), \tag{2.9}$$

where h is the second fundamental form of $\mathcal{M}$ in $\tilde{\mathcal{M}}$.

The geometry of $\mathcal{M}$ of $\tilde{\mathcal{M}}$ is mainly based on the action of the local basis $\{\mathcal{J}_1, \mathcal{J}_2, \mathcal{J}_3\}$ on each tangent space to $\mathcal{M}$. Let $\mathcal{M}^m$ be a submanifold of a quaternion space form $\mathcal{M}$ of dimension $4n$. For any $E \in \Gamma(T\mathcal{M})$, we can write $\mathcal{J}_\alpha E = R_\alpha E + S_\alpha E$, where the $R_\alpha E$ (resp. $S_\alpha E$) is the tangential component (resp. normal component) of $\mathcal{J}_\alpha E$. If $\mathcal{J}_\alpha(T\mathcal{M}) \subset T^\perp\mathcal{M}$, then $\mathcal{M}$ is totally real ($R_\alpha = 0$), and if $\mathcal{J}_\alpha(T\mathcal{M}) \subset T\mathcal{M}$, then $\mathcal{M}$ is quaternion ($S_\alpha = 0$). An n-dimensional totally real submanifold of a $4m$-dimensional $\tilde{\mathcal{M}}$ is said to be a Lagrangian submanifold if $n = m$.

The squared norm of R_α at $p \in \mathcal{M}$ is given as

$$||R_\alpha||^2 = \sum_{\alpha=1}^{3} \sum_{i,j=1}^{m} g^2(\mathcal{J}_\alpha e_i, e_j), \tag{2.10}$$

where $\{e_1, \ldots, e_m\}$ is any orthonormal basis of the tangent space $T_p\mathcal{M}$.

The notion of bi-slant submanifolds was introduced by J. L. Carriazo et al. as a natural generalization of CR, semi-slant, slant and hemi-slant submanifolds (see [13]). Moreover, Papaghiuc [9] developed an other classification of submanifolds in an almost Hermitian manifold, called the semi-slant submanifolds, which include proper CR-submanifolds and proper slant submanifolds as particular cases. In fact, semi-slant submanifolds [13], hemi-slant submanifolds, CR-submanifolds and slant submanifolds [12, 14] can be obtained from bi-slant submanifolds in particular. In [2], M. Barros et al. discussed about quaternion CR-submanifolds of quaternion Kaehler manifolds (some results are also given in [8]). Later on, Sahin [10] came up with the study of slant submanifolds in quaternion Kaehler manifolds. Since then, many authors have studied such submanifolds in quaternion Kaehler manifolds.

A submanifold $\mathcal{M}$ of a quaternion manifold $\tilde{\mathcal{M}}$ is called a quaternion bi-slant submanifold, whenever we have

1. $T\mathcal{M} = \mathcal{D}_{\theta_1} \oplus \mathcal{D}_{\theta_2}$, for any $i = 1, 2$, the distribution $\mathcal{D}_i$ is slant distribution with slant angle θ_i.
2. $\mathcal{J}_\alpha \mathcal{D}_{\theta_1} \perp \mathcal{D}_{\theta_2}$ and $\mathcal{J}_\alpha \mathcal{D}_{\theta_2} \perp \mathcal{D}_{\theta_1}$, for $\alpha = 1, 2, 3$,

where $\mathcal{D}_{\theta_1}$ and $\mathcal{D}_{\theta_2}$ are two orthogonal distributions of $\mathcal{M}$ with slant angles θ_1 and θ_2, respectively.

Let $\mathcal{M}$ be a quaternion bi-slant submanifold of a quaternion space form $\tilde{\mathcal{M}}$. Also, let $\dim \mathcal{M} = m = (4m_1 + 4m_2)$ and $\{e_1, \ldots, e_m\}$ be an orthonormal basis of $T_p\mathcal{M}$ at p in $\mathcal{M}$ such as

$$g^2(\mathcal{J}_\alpha e_{i+1}, e_i) = \begin{cases} cos^2\theta_1, & for\ i = 1, 2, \ldots, 4m_1 - 1, \quad \alpha = 1, 2, 3 \\ cos^2\theta_2, & for\ i = 4m_1 + 1, \ldots, 4m_1 + 4m_2 - 1, \quad \alpha = 1, 2, 3. \end{cases}$$

Thus, we have

$$\sum_{\alpha=1}^{3} \sum_{i,j=1}^{m} g^2(\mathcal{J}_\alpha e_i, e_j) = 12\{m_1 cos^2\theta_1 + m_2 cos^2\theta_2\}.$$

S.N	$\mathcal{M}^m$	D_{θ_1}	D_{θ_2}	θ_1	θ_2
(1)	bi-slant	slant distribution	slant distribution	slant angle	slant angle
(2)	semi-slant	invariant distribution	slant distribution	0	slant angle
(3)	hemi-slant	slant distribution	totally real distribution	slant angle	$\frac{\pi}{2}$
(4)	quaternion CR	invariant distribution D under $\mathcal{J}_\alpha$, that is, $\mathcal{J}_\alpha D \subseteq D$	totally real distribution $D^\perp$, that is, $\mathcal{J}_\alpha D^\perp \subseteq T^\perp\mathcal{M}$	0	$\frac{\pi}{2}$
(5)	slant	either $D_{\theta_1} = 0$ or $D_{\theta_2} = 0$		either $\theta_1 = \theta_2 = \theta$ or $\theta_1 = \theta_2 \neq \theta$	

Remark The special cases of slant submanifold are quaternion and totally real if $\theta = 0$ and $\theta = \frac{\pi}{2}$, respectively. The slant submanifold is said to be proper slant and proper bi-slant submanifold, respectively, if $0 < \theta < \frac{\pi}{2}$ and θ_i lies between 0 and $\frac{\pi}{2}$.

Remark We note that the quaternion CR-submanifolds contain quaternion and totally real submanifold as subcases. On the other hand, it is easy to observe that the quaternion CR-submanifolds do not include real hypersurfaces of quaternion Kaehler manifolds.

3 Main Results

In this section, we study the Casorati curvature of any submanifold $\mathcal{M}$ of dimension m in $4n$-dimensional quaternion space form $\tilde{\mathcal{M}}(c)$. We consider the local orthonormal tangent frame $\{e_1, \ldots, e_m\}$ of the tangent bundle $T\mathcal{M}$ of $\mathcal{M}$ and a local orthonormal normal frame $\{e_{m+l}, \ldots, e_{4n}\}$ of the normal bundle $T^\perp\mathcal{M}$ of $\mathcal{M}$ in $\tilde{\mathcal{M}}$. At any $p \in \mathcal{M}$, the scalar curvature τ at that point is given by

$$\tau = \sum_{1 \le i < j \le m} \mathcal{R}(e_i, e_j, e_j, e_i). \tag{3.1}$$

The mean curvature $\mathcal{H}$ of submanifold is given by $m\mathcal{H} = \sum_{i=1}^{m} h(e_i, e_i)$. Conveniently, let us put $h_{ij}^r = g(h(e_i, e_i), e_r)$, for any $i, j = \{1, \ldots, m\}$ and $r = \{m + 1, \ldots, 4n\}$.

We denote by κ the sectional curvature tensor on $\mathcal{M}$. Then the normalized scalar curvature ρ is given by

$$\rho = \frac{2\tau}{m(m-1)} = \frac{2}{m(m-1)} \sum_{1 \le i < j \le m} \kappa(e_i \wedge e_j). \tag{3.2}$$

The Casorati curvature is defined as

$$\mathcal{C} = \frac{1}{m}||h||^2 = \frac{1}{m} \sum_{r=m+1}^{4n} \sum_{i,j=1}^{4n} \left(h_{ij}^r\right)^2. \tag{3.3}$$

Let $\mathcal{L}$ be a s-dimensional subspace of $T\mathcal{M}$, $s \ge 2$ and $\{e_1, \ldots, e_s\}$ be an orthonormal basis of $\mathcal{L}$. Then the scalar curvature of the $s-$plane section $\mathcal{L}$ is given by

$$\tau(\mathcal{L}) = \sum_{1 \le i < j \le s} R(e_i, e_j, e_j, e_i)$$

and the Casorati curvature of the subspace $\mathcal{L}$ is defined by

$$\mathcal{C}(\mathcal{L}) = \frac{1}{s} \sum_{r=m+1}^{4n} \sum_{i,j=1}^{s} \left(h_{ij}^r\right)^2.$$

The normalized δ-Casorati curvatures $\delta_{\mathcal{C}}(m)$ and $\widehat{\delta_{\mathcal{C}}}(m)$ are defined as

$$[\delta_{\mathcal{C}}(m-1)]_p = \frac{1}{2}\mathcal{C}_p + \frac{m+1}{2m} \inf\{\mathcal{C}(\mathcal{L})|\mathcal{L} : \text{a hyperplane of } T_p\mathcal{M}\}$$

and

$$[\widehat{\delta_{\mathcal{C}}}(m-1)]_p = 2\mathcal{C}_p + \frac{2m-1}{2m} \sup\{\mathcal{C}(\mathcal{L})|\mathcal{L} : \text{a hyperplane of } T_p\mathcal{M}\}.$$

For a positive real number $t \neq m(m-1)$, put

$$b(t) = \frac{1}{mt}(m-1)(m+t)(m^2 - m - t), \tag{3.4}$$

then the generalized normalized δ-Casorati curvatures $\delta_{\mathcal{C}}(t; m-1)$ and $\tilde{\delta}_{\mathcal{C}}(t; m-1)$ are given as $[\delta_{\mathcal{C}}(t; m-1)]_p = t\mathcal{C}_p + b(t) \inf\{\mathcal{C}(\mathcal{L})|\mathcal{L} : \text{a hyperplane of } T_p\mathcal{M}\}$, if $0 < t < m^2 - m$, and

$[\hat{\delta}_{\mathcal{C}}(t; m-1)]_p = t\mathcal{C}_p + b(t) \sup\{\mathcal{C}(\mathcal{L}) | \mathcal{L} : \text{a hyperplane of } T_p\mathcal{M}\}$, if $t > m^2 - m$.

We note that the submanifold $\mathcal{M}$ is said to be *invariantly quasi-umbilical* if there exist $4n - m$ mutually orthogonal unit normal vectors $e_{m+1}, \ldots, e_{4n}$ such that the shape operator with respect to the direction e_s has an eigenvalue of multiplicity $m-1$ and that for each e_s the distinguished eigen-direction is the same (details in [4]).

Theorem 3.1 *Let $\mathcal{M}$ be an m-dimensional quaternion bi-slant submanifold in quaternion space form $\tilde{\mathcal{M}}(c)$ of dimension $4n$. Then for the generalized normalized δ-Casorati curvatures, we have the following:*

1. *For any real number t such that $0 < t < m(m-1)$,*

$$\rho \leq \frac{\delta_{\mathcal{C}}(t; m-1)}{m(m-1)} + \frac{c}{4}\left\{1 + \frac{36}{m(m-1)}(m_1 cos^2\theta_1 + m_2 cos^2\theta_2)\right\}. \quad (3.5)$$

The equality holds in (3.5) if and only if $\mathcal{M}$ is invariantly quasi-umbilical with trivial normal connection in $\mathcal{M}(c)$, such that with respect to suitable tangent orthonormal frame $\{e_1, \ldots, e_m\}$ and normal orthonormal frame $\{e_{m+1}, \ldots, e_{4n}\}$, the shape operator $\mathcal{S}_r \equiv \mathcal{S}_{e_r}$, $r \in \{m+1, \ldots, 4n\}$, take the following form:

$$\mathcal{S}_{m+1} = \begin{pmatrix} \beta & 0 & 0 & \ldots & 0 & 0 \\ 0 & \beta & 0 & \ldots & 0 & 0 \\ 0 & 0 & \beta & \ldots & 0 & 0 \\ \cdot & \cdot & \cdot & \cdot & \cdot & \cdot \\ \cdot & \cdot & \cdot & \cdot & \cdot & \cdot \\ \cdot & \cdot & \cdot & \cdot & \cdot & \cdot \\ 0 & 0 & 0 & \ldots & \beta & \cdot \\ 0 & 0 & 0 & \ldots & 0 & \frac{m(m-1)}{t}\beta \end{pmatrix} \qquad \mathcal{S}_{m+2} = \cdots = \mathcal{S}_{4n} = 0.$$

2. *For any real number $t > m(m-1)$,*

$$\rho \leq \frac{\hat{\delta}_{\mathcal{C}}(t; m-1)}{m(m-1)} + \frac{c}{4}\left\{1 + \frac{36}{m(m-1)}(m_1 cos^2\theta_1 + m_2 cos^2\theta_2)\right\}. \quad (3.6)$$

The equality holds in (3.6) if and only if $\mathcal{M}$ is invariantly quasi-umbilical with trivial normal connection in $\mathcal{M}$, such that with respect to suitable tangent orthonormal frame $\{e_1, \ldots, e_m\}$ and normal orthonormal frame $\{e_{m+1}, \ldots, e_{4n}\}$, the shape operator $\mathcal{S}_r$, $r \in \{m+1, \ldots, 4n\}$, take the following form:

$$\mathcal{S}_{m+1} = \begin{pmatrix} \beta\, 0\, 0 \ldots 0\;\; 0 \\ 0\, \beta\, 0 \ldots 0\;\; 0 \\ 0\, 0\, \beta \ldots 0\;\; 0 \\ . \;\; . \;\; . \;\; . \;\; . \quad . \;\; . \\ . \;\; . \;\; . \;\; . \;\; . \;\; . \;\; . \\ . \;\; . \;\; . \quad . \;\; . \;\; . \\ 0\, 0\, 0 \ldots \beta\;\; . \\ 0\, 0\, 0 \ldots 0\; 2\beta \end{pmatrix}, \quad \mathcal{S}_{m+2} = \cdots = \mathcal{S}_{4n} = 0. \tag{3.7}$$

Proof The scalar curvature is given by (3.1), so we write it for the desired submanifold as

$$\tau = \frac{c}{4}\left\{m(m-1) + 36(m_1 cos^2\theta_1 + m_2 cos^2\theta_2)\right\} + m^2||\mathcal{H}||^2 - m\mathcal{C}. \tag{3.8}$$

Now, we define the following function $\mathcal{F}$, which is a quadratic polynomial in the components of the second fundamental form:

$$\mathcal{F} = t\mathcal{C} + b(t)\mathcal{C}(\mathcal{L}) - 2\tau(p) + \frac{c}{4}\left\{m(m-1) + 36(m_1 cos^2\theta_1 + m_2 cos^2\theta_2)\right\}, \tag{3.9}$$

where $\mathcal{L}$ is the hyperplane of $T_p\mathcal{M}$. Without loss of generality, we suppose that $\mathcal{L}$ is spanned by $\{e_1, \ldots, e_{m-1}\}$, then (3.9) becomes

$$\mathcal{F} = \frac{(m+t)}{m}\sum_{r=m+1}^{4n}\sum_{i,j=1}^{m}(h_{ij}^r)^2 + \frac{b(t)}{(m-1)}\sum_{r=m+1}^{4n}\sum_{i,j=1}^{m-1}(h_{ij}^r)^2 - \sum_{r=m+1}^{4n}(\sum_{i=1}^{m}(h_{ii}^r)^2,$$

which can be easily written as

$$\begin{aligned}\mathcal{F} = &\sum_{r=m+1}^{4n}\sum_{i,j=1}^{m-1}\left[\left(\frac{m+t}{m} + \frac{b(t)}{m-1}\right)(h_{ii}^r)^2 + 2\left(\frac{m+t}{m}\right)(h_{im}^r)^2\right] \\ &+ \sum_{r=m+1}^{4n}\left[2\left(\frac{m+t}{m} + \frac{b(t)}{m-1} - 1\right)\sum_{i<j}^{m-1}(h_{ij}^r)^2 - 2\sum_{i<j}^{m} h_{ii}^r h_{jj}^r \right. \\ &\left. + \frac{t}{m}(h_{mm}^r)^2\right].\end{aligned} \tag{3.10}$$

From (3.10), we can see the critical points

$$h^c = \{h_{11}^{m+1}, h_{22}^{m+1}, \dots, h_{mm}^{m+1}, \dots, h_{11}^{4n}, h_{22}^{4n}, \dots, h_{mm}^{4n}\}$$

of $\mathcal{F}$ are the solutions of the following system of homogeneous equations:

$$\left.\begin{aligned} \tfrac{\partial \mathcal{F}}{\partial h_{ii}} &= 2\big(\tfrac{m+t}{m} + \tfrac{b(t)}{m-1} - 1\big)h_{ii}^r - 2\textstyle\sum_{q=1}^{m} h_{qq}^r = 0 \\ \tfrac{\partial \mathcal{F}}{\partial h_{mm}} &= \tfrac{2t}{m} h_{mm}^r - 2\textstyle\sum_{q=1}^{m-1} h_{qq}^r = 0 \\ \tfrac{\partial \mathcal{F}}{\partial h_{ij}} &= 4\big(\tfrac{m+t}{m} + \tfrac{b(t)}{m-1}\big)h_{ij}^r \\ \tfrac{\partial \mathcal{F}}{\partial h_{im}} &= 4\big(\tfrac{m+t}{m}\big)h_{im}^r. \end{aligned}\right\}, \tag{3.11}$$

where $i, j = \{1, 2, \dots, m-1\}, i \neq j$ and $r \in \{m+1, \dots, 4n\}$.

Hence, every solution h^c and $h_{ij}^r = 0$ for $i \neq j$ and the corresponding determinant to the first two equations of the above system is zero. Moreover, the Hessian matrix of $\mathcal{F}$ is of the following form:

$$\mathcal{H}ess(\mathcal{F}) = \begin{pmatrix} Hess_1 & O & O \\ O & Hess_2 & O \\ O & O & Hess_3 \end{pmatrix},$$

where

$$Hess_1 = \begin{pmatrix} 2\big(\tfrac{m+t}{m} + \tfrac{b(t)}{m-1}\big) - 2 & -2 & \dots & -2 & -2 \\ -2 & 2\big(\tfrac{m+t}{m} + \tfrac{b(t)}{m-1}\big) - 2 & \dots & -2 & -2 \\ . & . & . & . & . \\ . & . & . & . & . \\ . & . & . & . & . \\ -2 & -2 & \dots & 2\big(\tfrac{m+t}{m} + \tfrac{b(t)}{m-1}\big) - 2 & -2 \\ -2 & -2 & \dots & -2 & \tfrac{2t}{m} \end{pmatrix} \tag{3.12}$$

$Hess_2$ and $Hess_3$ are the diagonal matrices and O is the null matrix of the respective dimensions.

$$Hess_2 = diag\Big[4\Big(\frac{m+t}{m} + \frac{b(t)}{m-1} - 1\Big), 4\Big(\frac{m+t}{m} + \frac{b(t)}{m-1} - 1\Big), \dots, \\ 4\Big(\frac{m+t}{m} + \frac{b(t)}{m-1} - 1\Big)\Big]$$

and

$$Hess_3 = diag\left[4\frac{(m+t)}{m}, 4\frac{(m+t)}{m}, \dots, 4\frac{(m+t)}{m}\right].$$

Hence, we find that $\mathcal{H}ess(\mathcal{F})$ has the following eigenvalues:

$$\lambda_{11} = 0,\ \ \lambda_{22} = 2\left(\frac{2t}{m} + \frac{b(t)}{m-1}\right),\ \ \lambda_{33} = \cdots = \lambda_{mm} = 2\left(\frac{m+t}{m} + \frac{b(t)}{m-1}\right),$$

$$\lambda_{ij} = 4\left(\frac{m+t}{m} + \frac{b(t)}{m-1}\right),\ \ \lambda_{im} = 4\left(\frac{m+t}{m}\right)\ \ \forall\ i, j \in \{1, 2, \dots, m-1\},\ \ i \neq j.$$

Thus, $\mathcal{F}$ is parabolic and reaches at minimum $\mathcal{F}(h^c) = 0$ for some solution h^c of the system (3.11). Therefore, $\mathcal{F} \geq 0$ and hence

$$2\tau(p) \leq t\mathcal{C} + b(t)\mathcal{C}(\mathcal{L}) + \frac{c}{4}\Big\{m(m-1) + 36(m_1 cos^2\theta_1 + m_2 cos^2\theta_2)\Big\},$$

whereby, we obtain

$$\rho \leq \frac{t}{m(m-1)}\mathcal{C} + \frac{b(t)}{m(m-1)}\mathcal{C}(\mathcal{L}) + \frac{c}{4}\left\{1 + \frac{36}{m(m-1)}(m_1 cos^2\theta_1 + m_2 cos^2\theta_2)\right\}.$$

For each tangent hyperplane $\mathcal{L}$ of $\mathcal{M}$, if we take infimum over all tangent hyperplanes $\mathcal{L}$, the result trivially follows. Moreover, the equality sign holds if and only if

$$h^r_{ij} = 0,\quad \forall i, j \in \{1, \dots, m\},\quad i \neq j,\quad r \in \{m+1, \dots, 4n\}, \tag{3.13}$$

and

$$h^r_{mm} = 2h^r_{11} = \cdots = 2h^r_{m-1\,m-1},\quad \forall r \in \{m+1, \dots, 4n\}. \tag{3.14}$$

From (3.13) and (3.14), we conclude that the equality sign holds if and only if the submanifold M is invariantly quasi-umbilical submanifold with normal connection such that shape operator takes (3.6) with respect to the orthonormal tangent and orthonormal normal frames.

In the same manner, we can establish the inequality (3.6) as a second part of the theorem. □

Corollary 3.2 *Let $\mathcal{M}$ be an m-dimensional submanifold in quaternion space form $\tilde{\mathcal{M}}(c)$ of dimension $4n$, then we have the following Table 1:*

Table 1 Inequalities in different submanifolds

S.N.	$\mathcal{M}$	Inequality
(1)	semi-slant	• For any real number t such that $0 < t < m(m-1)$: $\rho \leq \frac{\delta_C(t;m-1)}{m(m-1)} + \frac{c}{4}\left\{1 + \frac{36}{m(m-1)}(m_1 + m_2 cos^2\theta)\right\}$ • For any real number $t > m(m-1)$: $\rho \leq \frac{\hat{\delta}_C(t;m-1)}{m(m-1)} + \frac{c}{4}\left\{1 + \frac{36}{m(m-1)}(m_1 + m_2 cos^2\theta)\right\}$
(2)	hemi-slant	• For any real number t such that $0 < t < m(m-1)$: $\rho \leq \frac{\delta_C(t;m-1)}{m(m-1)} + \frac{c}{4}\left\{1 + \frac{36}{m(m-1)}(m_1 cos^2\theta)\right\}$ • For any real number $t > m(m-1)$: $\rho \leq \frac{\hat{\delta}_C(t;m-1)}{m(m-1)} + \frac{c}{4}\left\{1 + \frac{36}{m(m-1)}(m_1 cos^2\theta)\right\}$
(3)	quaternion CR	• For any real number t such that $0 < t < m(m-1)$: $\rho \leq \frac{\delta_C(t;m-1)}{m(m-1)} + \frac{c}{4}\left(1 + \frac{36m_1}{m(m-1)}\right)$ • For any real number $t > m(m-1)$: $\rho \leq \frac{\hat{\delta}_C(t;m-1)}{m(m-1)} + \frac{c}{4}\left(1 + \frac{36m_1}{m(m-1)}\right)$
(4)	slant	• For any real number t such that $0 < t < m(m-1)$: $\rho \leq \frac{\delta_C(t;m-1)}{m(m-1)} + \frac{c}{4}\left(1 + \frac{9}{m-1}cos^2\theta\right)$ • For any real number $t > m(m-1)$: $\rho \leq \frac{\hat{\delta}_C(t;m-1)}{m(m-1)} + \frac{c}{4}\left(1 + \frac{9}{m-1}cos^2\theta\right)$
(5)	invariant	• For any real number t such that $0 < t < m(m-1)$: $\rho \leq \frac{\delta_C(t;m-1)}{m(m-1)} + \frac{c}{4}\left(1 + \frac{9}{m-1}\right)$ • For any real number $t > m(m-1)$: $\rho \leq \frac{\hat{\delta}_C(t;m-1)}{m(m-1)} + \frac{c}{4}\left(1 + \frac{9}{m-1}\right)$
(6)	totally real	• For any real number t such that $0 < t < m(m-1)$: $\rho \leq \frac{\delta_C(t;m-1)}{m(m-1)} + \frac{c}{4}$ • For any real number $t > m(m-1)$: $\rho \leq \frac{\hat{\delta}_C(t;m-1)}{m(m-1)} + \frac{c}{4}$

References

1. M. Aslam, A.N. Siddiqui, Bounds for generalized normalized δ-Casorati curvature for submanifold in real space forms endowed with a quarter-symmetric connection. Balkan J. Geom. App. **27**(2), 1–12 (2022)
2. M. Barros, B.Y. Chen, F. Urbano, Quaternion CR-submanifolds of quaternion manifolds. Kodai Math. J. **4**(3), 339–417 (1981)
3. M. Barros, F. Urbano, Totally real submanifolds of quaternion Kaehlerian manifolds. Soochow J. Math. **5**, 63–78 (1979)
4. D. Blair, Quasi-umbilical, minimal submanifolds of Euclidean space. Bull. Belgian Math. Soc.—Simon Stevin **51**, 3–22 (1977)
5. F. Casorati, Nuova definizione della curvatura delle superficie e suo confronto con quella di Gauss. (New definition of the curvature of the surface and its comparison with that of Gauss). Rend. Inst. Matem. Accad. Lomb. Ser. II **22**(8), 335–346 (1889)
6. P.J. De Smet, F. Dillen, L. Verstraelen, L. Vrancken, A pointwise inequality in submanifold theory. Arch. Math. (Brno) **35**(2), 115–128 (1999)
7. S. Ishihara, Quaternion Kahlerian manifolds. J. Differ. Geom. **9**, 483–500 (1974)
8. I. Mihai, F. Al-Solamy, M.H. Shahid, On Ricci curvature of a quaternion CR-submnaifolds in a quaternion space form. Rad. Math. **12**(1), 91–98 (2003)
9. N. Papaghiuc, Semi-slant submanifolds of a Kaehlerian manifold. An. Stiint. Univ. Al. I. Cuza Iasi Sect. I a Math. **40**(1), 55–61 (1994)
10. B. Sahin, Slant submanifolds of quaternion Kaehler manifolds. Commun. Korean Math. Soc. **22**(1), 123–135 (2007)
11. D.W. Yoon, A basic inequality of submanifolds in quaternion space forms. Balkan J. Geom. Appl. **9**(2), 92–103 (2004)
12. J.L. Cabrerizo, A. Carriazo, L.M. Fernandez, Slant submanifolds in Sasakian manifolds. Glasgow Math. J. **42**(1), 125–138 (2000)
13. J.L. Cabrerizo, A. Carriazo, L.M. Fernandez, M. Fernandez, Semi-slant submanifolds of a Sasakian manifold. Geometriae Dedicata **78**, 183–199 (1999)
14. B.Y. Chen, *Geometry of Slant Submanifolds* (Katholieke Universiteit Leuven, 1990)
15. C.W. Lee, J.W. Lee, G. . Vilcu, Optimal inequalities for the normalized δ-Casorati curvatures of submanifolds in Kenmotsu space forms. Adv. Geom. **17** (2017). https://doi.org/10.1515/advgeom-2017-0008
16. J.W. Lee, G.E. Vilcu, Inequalities for generalized normalized δ-Casorati curvatures of slant submanifolds in quaternion space forms. Taiwanese J. Math. **19**, 691–702 (2015)
17. M.H. Shahid, A.N. Siddiqui, Optimizations on totally real submanifolds of LCS-manifolds using Casorati curvatures. Commun. Korean Math. Soc. **34**(2), 603–614 (2019)
18. A.N. Siddiqui, M.D. Siddiqi, M.H. Shahid, Optimization on submanifolds ofδ-Lorentzian trans-Sasakian manifolds with Casorati curvatures. Tamkang J. Math. **53**(4) (2022)
19. A.N. Siddiqui, M.H. Shahid, Optimizations on statistical hypersurfaces with Casorati curvatures. Kragujevac J. Math. **45**(3), 449–463 (2021)
20. A.N. Siddiqui, Optimal Casorati inequalities on bi-slant submanifolds of generalized Sasakian space forms. Tamkang J. Math. **49**(3), 245–255 (2018)
21. A.N. Siddiqui, Upper bound inequalities for δ-Casorati curvatures of submanifolds in generalized Sasakian space forms admitting a semi-symmetric metric connection. Inter. Elec. J. Geom. **11**(1), 57–67 (2018)
22. A.N. Siddiqui, M.H. Shahid, A lower bound of normalized scalar curvature for bi-slant submanifolds in generalized Sasakian space forms using Casorati curvatures. Acta Mathe. Univ. Comen. **LXXXVII**(1), 127–140 (2018)
23. K. Yano, M. Kon, *Structures on Manifolds* (Worlds Scientific, Singapore, 1984)

Gravity and Dark Matter in the Framework of Metric-Affine Geometry

Ghodratallah Fasihi-Ramandi and Vahid Pirhadi

Mathematical Subject Classification: 53B21 · 53B20 · 53C44

1 Introduction

General relativity proposes that for a metric g, space-time represents the four-dimensional Lorentzian manifold (M, g) with the Levi–Civita connection of g [14]. While affine connection takes care of the manifold's straightness and twisting, the metric tensor is in charge of measuring distance. The responsibilities that these two geometric structures perform in the dynamics of space-time are entirely distinct and they are both fundamentally independent variables. This phenomenon inspires the "metric-affine gravity" [3, 16]. When the torsionless requirement and the metric-compatibility condition are satisfied, the metric-affine geometry is transformed into the Riemannian geometry. As a result, it is possible to think of metric-affine geometry as a generalization of (semi-) Riemannain geometry. The Riemannian geometry, which is often employed to develop gravitational theories, is a wholly metric theory of gravity because the LC-connection is managed only by the metric and its partial derivatives [5].

By separating the relationship between the connection and the metric tensor, the metric-affine theory of gravity [5, 11, 12] embodies a more comprehensive view of gravitation. This breakdown naturally discloses the new dynamic structures, including the curvature tensor, nonmetricity tensor, and torsion tensor [1, 16]. We will therefore have a greater amount of freedom to examine gravitational dynamics according to this new theory. Due to its potential for creating models of inflation and dark energy, this idea has gained a lot of study interest. The authors have demonstrated in [7] that

G. Fasihi-Ramandi (✉)
Department of Pure Mathematics, Imam Khomeini International University, Qazvin, Iran
e-mail: fasihi@sci.ikiu.ac.ir

V. Pirhadi
Department of Mathematics, University of Kashan, Kashan, Iran
e-mail: v.pirhadi@kashanu.ac.ir

B.-Y. Chen et al. (eds.), *Geometry of Submanifolds and Applications*, Infosys Science Foundation Series, https://doi.org/10.1007/978-981-99-9750-3_9

the gauge gravitation theory, in which a general linear connection serves as a gauge field, is the direct source of metric-affine gravitation theory. In [8], a theory of scalar, vector, and tensor fields is derived from a metric-affine gravity theory made up of the invariants derived from nonmetricity, torsion, and curvature tensors. [1] researched scalar-tensor theories in the metric-affine formalism of gravity. The results of this study demonstrate that scalar-metric-affine theories can have a ghost-free structure thanks to the projective symmetry, a local symmetry under a shift of the connection. Furthermore, in the context of metric-affine geometry, an organized comprehension for classes of inflationary models is examined in [10] from the perspective of the local conformal symmetry and the slightly broken global symmetry.

In the current article, we explore a metric-affine geometry where the connection and the metric tensor are fundamentally distinct variables. With the use of this framework, we are able to locate Einstein's field equation and demonstrate how the idea of dark matter is closely linked to the affine part of our geometry.

2 Metric-Affine Connections

The semi-Riemannian manifold (M^4, g) is used to represent a space-time manifold. Imagine ∇ represents the Levi–Civita relationship with regard to g. Any arbitrary affine connection $\overline{\nabla}$ on M can be expressed as given below (as the space of all affine connections on M is an affine space with $\otimes_2^1 TM$ as its underlying vector space)

$$\overline{\nabla}_X Y = \nabla_X Y + A(X, Y), \tag{1}$$

$X, Y \in \mathcal{X}M$ and A is tensor field of type $(1, 2)$. The aforementioned relation leads one to the conclusion that metric-affine gravity is rich enough to provide the numerous vector and scalar fields required by gravitational or cosmic processes. An established metric-affine gravity model might be developed by examination of various impacts. It is to note that the tachyon and ghost problem in cosmology [13] is related to the degree of freedom that $\overline{\nabla}$ can spread. Additionally, [9] reports that metric-affine theories of gravity contain a ghost- and tachyon-free sector. The overall connection's varied consequences are in-depth examined in [15]. In this paper, we examine the scenario where we have

$$A(X, Y) := \alpha(X)Y - \alpha(Y)X,$$

α is a one form on M. As $A(X, Y)$ is anti-symmetric in this situation, we have $T^{\overline{\nabla}}(X, Y) = 2\alpha(X)Y - 2\alpha(Y)X$, and the nonmetricity tensor N as follows

$$N(X, Y, Z) = -(\overline{\nabla} g)(X, Y, Z) = 2\alpha(X)g(Z, Y) - \alpha(Y)g(X, Z) - \alpha(Z)g(X, Y).$$

Proposition 2.1 *The curvature tensor of* $\overline{\nabla}$, *satisfies the following relation*

$$\widehat{R}(X, Y)(Z) = R(X, Y)(Z) + d\alpha(X, Y).Z + (\nabla_X \alpha)(Z).Y - (\nabla_Y \alpha)(Z).X \\ + \alpha(Z)\alpha(Y)X - \alpha(Z)\alpha(X)Y, \qquad \forall X, Y, Z \in \mathcal{X}(M),$$

R is the Riemannain curvature tensor of (M, g).

Proof Simple calculation shows that

$$\begin{aligned} \overline{R}(X, Y)(Z) &= \overline{\nabla}_X \overline{\nabla}_Y Z - \overline{\nabla}_Y \overline{\nabla}_X Z - \overline{\nabla}_{[X,Y]} Z \\ &= \overline{\nabla}_X \big(\nabla_Y Z + A(Y, Z)\big) - \overline{\nabla}_Y \big(\nabla_X Z + A(X, Z)\big) - \nabla_{[X,Y]} Z - A([X, Y], Z) \\ &= \nabla_X \nabla_Y Z + A(X, \nabla_Y Z) + \nabla_X A(Y, Z) + A(X, A(Y, Z)) \\ &- \nabla_Y \nabla_X Z - A(Y, \nabla_X Z) - \nabla_Y A(X, Z) - A(Y, A(X, Z)) \\ &- \nabla_{[X,Y]} Z - A(\nabla_X Y - \nabla_Y X, Z) \\ &= R(X, Y)(Z) + (\nabla_X A)(Y, Z) - (\nabla_Y A)(X, Z) + A(X, A(Y, Z)) - A(Y, A(X, Z)). \end{aligned}$$

Put $A(X, Y) = \alpha(X)Y - \alpha(Y)X$, therefore the outcome was quickly attained.

Next, define the Ricci curvature tensor of $\widehat{\nabla}$, by

$$\widehat{\mathrm{Ric}}(X, Y) := \mathrm{tr}\big(Z \mapsto \widehat{R}(X, Z)(Y)\big).$$

It is important to keep in mind that for every basis $\{E_i\}_{i=1}^k$ of an inner vector space V, there is a basis $\{E_i\}_{i=1}^k$ of V that is regarded as its reciprocal basis and has the condition $< E_i, E^j >= \delta_i^j$, δ_i^j is the Kronecker delta symbol. For a fixed vector field X and a 1-form α, write

$$F_X : \mathcal{X}M \to C^\infty(M), \quad Y \mapsto (\nabla_Y \alpha)(X).$$

It is obvious that α and X help define the 1-form F_X and that this 1-form is linked to the derivative of α. We shall employ this 1-form in subsequent computations.

Proposition 2.2 *The following association gives the Ricci curvature tensor*

$$\widehat{\mathrm{Ric}}(X, Y) = \mathrm{Ric}(X, Y) + F_X(Y) - 4F_Y(X) + 3\alpha(X)\alpha(Y),$$

in which Ric *means the Ricci curvature tensor of* ∇.

Proof After simple computations, we have

$$\begin{aligned}
\widehat{\mathrm{Ric}}(X,Y) &= \sum_{i=1}^{4}\langle \widehat{R}(E_i,X)(Y),E^i\rangle \\
&= \sum_{i=1}^{4}\langle R(E_i,X)(Y)+(\nabla_{E_i}\alpha)(X).Y-(\nabla_X\alpha)(E_i).Y \\
&+(\nabla_{E_i}\alpha)(Y).X-(\nabla_X\alpha)(Y).E_i \\
&+\alpha(Y)\alpha(X)E_i-\alpha(E_i)\alpha(Y)X,E^i\rangle \\
&= \sum_{i=1}^{4}\langle R(E_i,X)(Y),E^i\rangle+\sum_{i=1}^{4}\langle(\nabla_{E_i}\alpha)(X).Y,E^i\rangle-\sum_{i=1}^{4}\langle(\nabla_X\alpha)(E_i).Y,E^i\rangle \\
&+\sum_{i=1}^{4}\langle(\nabla_{E_i}\alpha)(Y).X,E^i\rangle-\sum_{i=1}^{4}\langle(\nabla_X\alpha)(Y).E_i,E^i\rangle \\
&+\sum_{i=1}^{4}\langle\alpha(Y)\alpha(X)E_i,E^i\rangle-\sum_{i=1}^{4}\langle\alpha(E_i)\alpha(Y)X,E^i\rangle
\end{aligned}$$

$$\begin{aligned}
&= \mathrm{Ric}(X,Y)+\langle Y,\sum_{i=1}^{4}(\nabla_{E_i}\alpha)(X).E^i\rangle-\langle Y,\sum_{i=1}^{4}(\nabla_X\alpha)(E_i).E^i\rangle \\
&+\langle X,\sum_{i=1}^{4}(\nabla_{E_i}\alpha)(Y).E^i\rangle-(\nabla_X\alpha)(Y)\sum_{i=1}^{4}\langle E_i,E^i\rangle \\
&+\alpha(Y)\alpha(X)\sum_{i=1}^{4}\langle E_i,E^i\rangle-\alpha(Y)\langle X,\sum_{i=1}^{4}\alpha(E_i)E^i\rangle \\
&= \mathrm{Ric}(X,Y)+\langle Y,\sum_{i=1}^{4}F_X(E_i).E^i\rangle-\langle Y,(\nabla_X\alpha)^\sharp\rangle \\
&+\langle X,\sum_{i=1}^{4}F_Y(E_i).E^i\rangle-4(\nabla_X\alpha)(Y)+n\alpha(Y)\alpha(X)-\alpha(Y)\langle X,\alpha^\sharp\rangle \\
&= \mathrm{Ric}(X,Y)+\langle Y,F_X^\sharp\rangle-(\nabla_X\alpha)(Y)+\langle X,F_Y^\sharp\rangle-4(\nabla_X\alpha)(Y) \\
&+4\alpha(Y)\alpha(X)-\alpha(Y)\alpha(X) \\
&= \mathrm{Ric}(X,Y)+F_X(Y)+F_Y(X)+3\alpha(X)\alpha(Y)-5(\nabla_X\alpha)(Y).
\end{aligned}$$

In order to define the Ricci endomorphism, we utilize g as follows

$$\overline{Q}(X) := \sharp\big(\overline{\mathrm{Ric}}(X,.)\big).$$

The scalar curvature of $\overline{\nabla}$ is characterized via the trace of the Ricci endomorphism

$$\overline{R} := \mathrm{tr}\big(X \mapsto \overline{Q}(X)\big).$$

Proposition 2.3 *The scalar curvature $\widehat{R}$ of $\widehat{\nabla}$, meets the next relationship*

$$\widehat{R} = R + 3|\alpha|^2 - 3\mathrm{div}(\alpha), \tag{2}$$

R represents the scalar curvature of M.

3 Application to Relativity

Here, space-time manifold (M, g) have been considered. An affine connection

$$\overline{\nabla}_X Y = \nabla_X Y + df(X)Y - df(Y)X$$

on M is determined by each smooth function f on M. Letting $\mathcal{M}$ representing set of all such metric-affine structures over manifold M. Then it can be thought of as the collection of certain pairings (g, f), in which g acts as a semi-Riemannian metric and f provides a smooth function on M.

Additionally, on the oriented manifold M, identify the canonical volume form of a meter g with Ω_g. This is generally acknowledged that the metrics that go away from the divergence-free Einstein tensor

$$\mathrm{Ric} - \frac{1}{2}Rg = 0$$

are the critical points of the Einstein–Hilbert functional

$$\mathcal{L}(g) = \int_M R_g \Omega_g.$$

These ideas served as the impetus for us to define a novel Hilbert–Einstein functional

$$\mathcal{L} : \mathcal{M} \to \mathbb{R}$$

for metric-affine manifolds,

$$\mathcal{L}(g, f) = \int_M \overline{R}\Omega_g,$$

involving the scalar curvature of (g, f), represented by $\overline{R} = R + 3|df|^2 - 3\Delta(f)$. As Laplacian's integral disappears, our Hilbert–Einstein action changes to

$$\mathcal{L}(g, f) = \int_M (R + 3|df|^2)\Omega_g.$$

The *new field equations* are those equations that this functional's critical points satisfy. For a smooth function h and a symmetric 2-covariant tensor s on M, let us fix

$$\tilde{g}(t) = g + ts,$$
$$\tilde{f}(t) = f + th.$$

The metric-affine structure on M, $(\tilde{g}(t), \tilde{f}(t))$, which is a variant of (g, f), exists for sufficiently small t. Observe that, (g, f) is a critical point for Hilbert action iff for (s, h):

$$\frac{d}{dt}|_{t=0}\mathcal{L}\big(\tilde{g}(t), \tilde{f}(t)\big) = \frac{d}{dt}|_{t=0}\int_M (R(t) + 3|df(t)|^2)\Omega_{g+ts} = 0. \qquad (1)$$

We must calculate the derivatives of $R_{\tilde{g}(t)}$, $|d\tilde{f}(t)|^2$ and Ω_{g+ts} for $t = 0$ in order to get the derivative in Eq. 1. It is demonstrated in [4], that

$$(R_{\tilde{g}(t)})'(0) = -\langle s, \mathrm{Ric}\rangle + \mathrm{div}(X), \qquad (2)$$
$$(\Omega_{g+ts})'(0) = \frac{1}{2}\langle g, s\rangle\Omega_g, \qquad (3)$$

$X = \nabla(\mathrm{tr}s) + \mathrm{div}(s)$.

Further, we write

$$\frac{d}{dt}|_{t=0}|d\tilde{f}(t)|^2 = \Big(\tilde{g}^{ij}(t)\frac{\partial\tilde{f}(t)}{\partial x^i}\frac{\partial\tilde{f}(t)}{\partial x^j}\Big)'(0) = -s^{ij}\frac{\partial f}{\partial x^i}\frac{\partial f}{\partial x^j} + 2g^{ij}\frac{\partial h}{\partial x^i}\frac{\partial f}{\partial x^j}$$
$$= -\langle s, df \otimes df\rangle + 2\langle\nabla h, \nabla f\rangle.$$

Recall that on M, any vector fields' integrals of divergence are zero, and that any smooth function's integral of Laplacian is also zero. We also have [2]

$$\int_M \langle\nabla h, \nabla f\rangle dV_g = -\int_M \Delta(f)h dV_g.$$

We are now ready to identify critical metric-affine structure for the Hilbert action

$$
\begin{aligned}
0 = \frac{d}{dt}|_{t=0}\mathcal{L}(\tilde{g}(t), \tilde{f}(t)) &= \frac{d}{dt}|_{t=0}\int_M (R(t) + 3|df(t)|^2)\Omega_{g+ts} \\
&= \int_M (R + 3|df|^2)\frac{1}{2}\langle s, g\rangle\Omega_g \\
&+ \int_M \big(-\langle s, \mathrm{Ric}\rangle + \mathrm{div}(X) - 3\langle s, df \otimes df\rangle + 6\langle \nabla h, \nabla f\rangle\big)\Omega_g \\
&= \int_M \langle \frac{1}{2}Rg - \mathrm{Ric} + \frac{3}{2}|df|^2 g - 3df \otimes df, s\rangle\Omega_g \\
&- 6\int_M h\Delta(f)\Omega_g.
\end{aligned}
$$

The aforementioned expressions disappear for all pair (s, h) iff

$$\mathrm{Ric} - \frac{1}{2}Rg = \frac{3}{2}|df|^2 g - 3df \otimes df, \tag{4}$$

$$\Delta(f) = 0. \tag{5}$$

Similar relations have been discovered using a different framework in [6]. The result of tracing both sides of (4) yields $R = -3|df|^2 = -3|\nabla f|^2$. The relationship between the scalar curvature R and the distribution of matter at particular locations in space-time is well established. As a result, f is connected to the distribution of matter in space-time. Dark matter could be geometrically understood as the curving and whirling of a space-time manifold because df plays a vital role in the definition of the affine connection that is responsible for curving space-time. If ∇f is time-like, then it can be understood as a current of some mass that complies with the conservation law, based on Eq. 5. The density associated with this mass is given by $\rho = \sqrt{-|\nabla f|^2}$. Additionally, the term on the right-hand side of Eq. 4 can be understood as the mass's stress-energy tensor. The conservation law of the stress-energy tensor is reminded by the fact that the divergence of

$$T^f := \frac{3}{2}|df|^2 g - 3df \otimes df,$$

is zero.

Theorem 3.1 *The symmetric tensor T^f is divergence-free if f is a smooth function on a Riemannian manifold (M, g) such that $\Delta(f) = 0$.*

Proof Assuming that $\{E_i\}_{i=1}^n$ is an orthonormal base on M, its reciprocal base is represented by $\{E^i\}_{i=1}^n$. We are able to write

$$
\begin{aligned}
\operatorname{div}(df \otimes df)(X) &= \sum_{i=1}^{n} (\nabla_{E_i} df \otimes df)(E^i, X) \\
&= \sum_{i=1}^{n} \big((\nabla_{E_i} df) \otimes df + df \otimes (\nabla_{E_i} df)\big)(E^i, X) \\
&= \sum_{i=1}^{n} \big((\nabla_{E_i} df)(E^i) df(X) + df(E^i)(\nabla_{E_i} df)(X) \\
&= \Delta(f) df(X) + \operatorname{Hess} f(\nabla f, X) = \operatorname{Hess} f(\nabla f, X).
\end{aligned}
$$

Also,

$$
\begin{aligned}
\operatorname{div}(|df|^2 g)(X) = \operatorname{div}(|\nabla f|^2 g)(X) = d(|\nabla f|^2)(X) = X\langle \nabla f, \nabla f\rangle = 2\langle \nabla_X(\nabla f), \nabla f\rangle \\
= 2(\nabla_X df)(\nabla f) = 2\operatorname{Hess} f(\nabla f, X).
\end{aligned}
$$

These calculations demonstrate that $\operatorname{div}(T^f) = 0$.

We have the following theorem, which makes our resulting equations simpler.

Theorem 3.2 *The subsequent equations are satisfied by the Hilbert action's critical metric-affine structure*

$$
\operatorname{Ric} = -3 df \otimes df, \tag{6}
$$

$$
\Delta(f) = 0. \tag{7}
$$

Proof It is the same as Eq. 5 in the second equation. Assuming that Eq. 4 is true, we can obtain by taking a trace on both sides of this equation

$$
[R - \frac{4}{2} R = (4\frac{3}{2}|df|^2 - 3|df|^2) \Rightarrow R = -3|df|^2.
$$

We obtain Eq. 6 by substituting the aforementioned formula in Eq. 4. Let Eq. 6 hold for a moment. We discover using the computation traces of the two sides of Eq. 6,

$$
R = -3|df|^2.
$$

Now we get Eq. 4 by adding an appropriate expression to each side of Eq. 6.

4 An Example of Metric-Affine Space-Time

In this section we give an example of a manifold which satisfies Eqs. 6 and 7. This example is an Einstein–de Sitter model in general relativity.

Set $M = N \times (0, \infty)$where $(N, \hat{g})$ is a Ricci-flat Riemannian manifold of dimension three. Any tangent vector to M at a point (p, t) is of the form (v, λ) in which

$v \in T_pN$ and $\lambda \in \mathbb{R}$ is a scalar. Let ∂_t denotes the vector field (0, 1) on M. Also, denote the second projection map $(p, t) \mapsto t$ on M by t. We can interpret t as time.

Define a metric g on M as follows.

$$g = \sqrt[3]{16t^2}\hat{g} - dt \otimes dt.$$

Consider $f : N \times (0, \infty) \to \mathbb{R}$ which depends only on t and denote it by $f(t)$ as follows

$$f(t) = \frac{2\sqrt{3}}{3} \ln t.$$

One can easily check that for a Ricci-flat 3-dimensional manifold $(N, \hat{g})$, the metric $g = (4t)^{2/3}\hat{g} - dt \otimes dt$ on M and the function $f(t) = \frac{2\sqrt{3}}{3} \ln t$ satisfy field Eqs. 6 and 7. As t approaches to zero in this model, universe become smaller and density of matter increases to infinity. Time $t = 0$ is not in our space-time M and this time is the instant of Big-Bang.

Conclusion

Metric-affine gravity extends the general theory of relativity by accommodating an affine connection that generalizes the Levi–Civita connection. From this work, one concludes that metric-affine gravity is rich enough to supply various vector and scalar fields needed in cosmological phenomena. Considering a scalar field, we established a gravity theory in which dark matter become a part of our geometry. In this paper, we have considered a special case for tensorial part of the affine connection, so one can construct more general case which consists of possible vector and scalar fields and establish more gravity theories. Also, in four dimensions by means of a symplectic structure, one can quantize our derived equations and create a quantum gravity theory.

Conflict of Interest: The authors declare no conflict of interest.

Data Availability: Not applicable.

References

1. K. Aoki, K. Shimada, Scalar-metric-affine theories: Can we get ghost-free theories from symmetry? Phys. Rev. D **100**, 044037 (2019)
2. D. Bleecker, *Guage Theory and Variational Principles* (Addison-wesely, 1981)
3. A. Kleyn, Metric-affine manifold, https://arxiv.org/abs/gr-qc/0405028
4. N. Boroojerdian, Geometrization of mass in general relativity. Int. J. Theor. Phys. **55**, 2432–2445 (2013)

5. S.M. Carroll (Addison-Wesley, San Francisco, USA, 2004), p. 513
6. Gh. Fasihi-Ramandi, The Bakery-Emery Ricci tensor: application to mass distribution in space-time. Gravit. Cosmol. **27**, 42–46 (2021)
7. F.W. Hehl, J.D. McCrea, E.W. Mielke, Y. Ne'eman, metric-affine gauge theory of gravity: field equations, Noether identities, world spinors, and breaking of dilation invariance. Phys. Rep. **258**, 1–171 (1995)
8. N.K. Canan, A. Altas, D.A. Demir, Scalars, vectors and tensors from metric-affine gravity. Gen. Relativ. Gravit. **45**, 319–343 (2013)
9. C. Marzo, Radiatively stable ghost and tachyon freedom in metric affine gravity, arXiv:2110.14788
10. Y. Mikura, Y. Tada, S. Yokoyoma, Conformal inflation in the metric-affine geometry. Europhys. Lett. **132**, 39001–6 (2020)
11. A. Palatini, Rend. Circ. Mat. Palermo **43** (1919)
12. P. Peldan, Actions for gravity, with generalizations: a review. Classical Quant. Grav. **11**, 1087–1132 (1994)
13. P. Percacci, E. Sezgin, New class of ghost- and tachyon-free metric affine gravities. Phys. Rev. D **101**, 084040 (2020)
14. R.K. Sachs, H. Wu, *General Relativity for Mathematicians* (Springer, 1977)
15. T.P. Sotiriou, S. Liberati, J. Phys. Conf. Ser. **68**, 01 (2022). (arXiv:0611040v1 [gr-qc], 2007)
16. V. Vitagliano, The role of nonmetricity in metric-affine theories of gravity. Classical Quant. Gravit. **31**, 045006 (2014)
17. V. Vitagliano, T.P. Sotiriou, S. Liberati, The dynamics of metric-affine gravity. Ann. Phys. **326**, 1259–1273 (2011)

Submersion on Statistical Metallic Structure

Mohit Saxena

AMS Subject Classification: 53C22 · 53C26

1 Introduction

Golden ratio is well defined and known to everyone. This ratio has wide application space in the field of design, music, architecture, painting, nature, optimization, etc. In fact, the golden ratio has inspired thinkers of all disciplines to correlate the concept accordingly in his field of mathematics.

Reference [1] Crasmareanu and Hretcanu researched the geometry and the properties of the golden structure which is well defined. Reference [2] Ozkan, [3] Ozkan, Taylan, Asmum studied the prolongations of golden structure. Khan [4, 5] defined some novel theorems endowed some metallic structure. Also, [6] Gezer et al., studied Golden Hessian manifolds, Khan et al. [7–9] also studied lift and properties of tangent bundles [10] Mehmet et al., studied Golden Riemannian Manifolds [11] Shun-Ichi discussed about different methods of Geometry.

The present chapter outline is as follows: In the preliminary of this chapter there is a brief study of metallic structure on Riemannian Manifold related to the generalized Fibonacci sequence to some elementary properties. In subsection we define metallic structure on manifold. In Sect. 3 we define Statistical Metallic Riemannian Manifold, Statistical Metallic Structure. [12, 13] define another connection known as conjugate connection. In the next subsection we define 2D Statistical Metallic Matrix. Sect. 4 explains certain properties of statistical submersion and deduces several theorems on that one.

Scopus id: 58551145100 orcid: 0000-0002-8107-6554 Wos: AAF-3511-2019.

M. Saxena (✉)
Department of Mathematics and Computer Sciences, The Papua New Guinea University of Technology, Lae, Papua New Guinea
e-mail: mohit.saxena@pnguot.ac.pg; mohitsaxenamohit@gmail.com

B.-Y. Chen et al. (eds.), *Geometry of Submanifolds and Applications*, Infosys Science Foundation Series, https://doi.org/10.1007/978-981-99-9750-3_10

2 Preliminary

In this section, we discuss a little about the metallic means family by Fibonacci sequence in the generalized secondary form.

2.1 Metallic Mean

Fibonacci sequence provides the following relation:

$$F(\alpha + 1) = F(\alpha) + F(\alpha - 1).$$

The above equation leads to the result which can be generalized as

$$F(\alpha + 1) = pF(\alpha) + qF(\alpha - 1), \alpha \geq 1$$

where p, q, F(0) = r, and F(1) = s are all real numbers.

$$r, s, ps + qr, p(ps + qr) + qs, \ldots$$

are the parts of the Generalized Secondary Fibonacci Sequence.

Let p and q be positive integers and [14] the roots of the equation

$$y^2 - py - q = 0$$

is known as a member of the family of metallic means. This root is denoted by

$$\rho_{(p,q)} = \frac{p + \surd(p^2 + 4q)}{2} \tag{1}$$

and $\rho_{(p,q)}$ is known as metallic ratio.

By Eq. (1), p and q have different values and we obtain the different ratios as follows:-

If p = q = 1, $\rho_{(1,1)} = \frac{1 + \surd 5}{2}$ for equation to be $y^2 - y - 1 = 0$, we obtain golden mean ratio of two consecutive classical Fibonacci number as golden mean [10, 15].

If p = 2, q = 1, $\rho_{(2,1)} = 1 + \surd 2$ for equation to be $y^2 - 2y - 1 = 0$, we obtain silver mean the same as the ratio of two consecutive Pell number [3].

If p = 3, q = 1, $\rho_{(3,1)} = \frac{3 + \surd 13}{2}$ for equation to be $y^2 - 3y - 1 = 0$, we get Bronze [16].

If p = 4, q = 1, $\rho_{(4,1)} = 2 + \surd 5$ for equation to be $y^2 - 4y - 1 = 0$, we get Subtle mean [17].

If p = 1, q = 2, $\rho_{(1,2)} = 2$ for equation to be $y^2 - y - 2 = 0$, we get Copper mean [17].

If p = 1, q = 3, $\rho_{(1,3)} = \dfrac{1+\sqrt{13}}{2}$ for equation to be $y^2 - y - 3 = 0$, we get Nickel mean [17].

In the present chapter, all manifolds are assumed of class C^∞.

2.2 Metallic Structure on Metallic Manifold

In this part of the chapter we have some definitions in order to base the main result on the metallic structure.

Definition 2.1 Let an n-dimensional manifold M_n. A second order structure on said manifold M_n is known as metallic structure if it has tensor field J which is (1, 1), satisfies the second order quadratic as follows

$$J^2 = pJ + qI \tag{2}$$

here I represents identity operator and q, p are positive integers on the $\chi(M_n)$ (Lie algebra) of the vector fields on M_n.

Hence the proposition or theorems as follows produces the central and important property of the said metallic structure.

Theorem 2.1 *i. Eigenvalues of J are nothing else but the metallic ratio $\rho_{p,q}$ and p - $\rho_{p,q}$.*

ii. $qJ^2 + pJI = 0$, where $J^{-1} = \hat{J}$ and J is invertible.

iii. For every $y \in M_n$ J a metallic structure which holds the property of isomorphism $T_x(M_n)$ which is a tangent space.

An almost product structure:- F be an almost complex structure:- C, almost tangent structure:- T, which always comes in duality or we can say, -F, -C, and -T as an almost product, an almost complex, and an almost tangent structures [2].

This definition reflects the property of duality in metallic structure:

Theorem 2.2 *For J be a metallic structure $\tilde{J} = pI - J$ would also be same that is metallic structure.*

We also know that any polynomial structure defined on the manifold M_n deduces generalized structure F which must be almost product.

Hence we have the theorem between almost product structure and metallic structure on M_n.

Theorem 2.3 *For F defined over M_n, be structure of type almost product then we have*

$$J = \frac{p}{2}I + \frac{\rho_{p,q} - p}{2}F \tag{3}$$

and it will be a metallic structure.

Conversely, if we consider J defined over M_n as structure of metallic sense, then

$$F = \frac{2}{\rho_{p,q} - p} J - \frac{a}{\rho_{p,q} - p} I \quad (4)$$

is a structure which is almost product and defined over M_n.

From the above two theorems (Theorems 2.2 and 2.3) we can straight forward say about Tangent metallic structure and hence the following definition:

Definition 2.2 For T be an almost tangent structure on M_n. then

$$J_t = \frac{p}{2} I + \frac{\rho_{p,q} - p}{2} .T$$

is defined on (M_n, T) as tangent metallic structure.

above structure verifies the equation:

$$J_t^2 - pJ_t + \frac{p^2}{4} .I = 0.$$

Considering the above equation in the real field R. that is $y^2 - py + \frac{p^2}{4} = 0$ we can find out tangent real metallic ratio as $\rho_t = \frac{p}{2}$

Similarly we have another definition for complex metallic structure:

Definition 2.3 For an almost complex structure C on M_n,

$$J_c = \frac{p}{2} I + \frac{2\rho_{p,q} - p}{2} .C$$

is known as complex metallic structure on the manifold M_n.

Hence the equation satisfied by J_c is

$$J_c^2 - pJ_c + \frac{p^2 + 2q}{4} I = 0$$

If the manifold M_n = R, then we have the following equation

$$y^2 - py + \frac{p^2 + 2q}{4} = 0$$

with solutions,

$$y_1 = \frac{p}{2} + \frac{\sqrt{p^2 + 4q}}{2} .i, \; y_2 = \frac{p}{2} - \frac{\sqrt{p^2 + 4q}}{2} .i$$

The positive root of the equation above

$$\rho_c = \frac{p}{2} + \frac{\sqrt{p^2 + 4q}}{2}.i$$

is called a complex metallic ratio.

3 Statistical Metallic Riemannian Manifold

3.1 Statistical Metallic Structure

Let us consider a Metallic Manifold M_n along with an affine connection ∇, which is torsion free and also there exists non-degenerate metric g. Now the Statistical Metallic Manifold (M, g, ∇) (SMM), where ∇ will always be symmetric. Reference [12] Noguchi M. studied statistical manifold. The dual connection is defined over the statistical manifold, such dual connections are also known as conjugate connections. This concept of duality is studied in the field of information geometry [11]. In the information geometry the statistical models have a Riemannian metric in terms of Fisher metric which admits an affine connection. Such Affine connections are constructed from the probability distribution y considering the mean. Considering said conditions affine connection is known as α-connection also its dual connected to the Fisher metric is $(-\alpha)$-connection, where $\alpha$is a real number.

In the present chapter we analyze statistical submersion. In Sect. 2 we had already defined metallic structure on manifold M^n with the property that g is the metric tensor which is also a non-degenerate symmetric tensor field defined on M_n where the index is constant.

Semi Metallic Riemannian Manifold M^n_ν is defined if specific value ν of index g over M, where ν is positive natural number less than or equal to n. Said manifold will be Metallic Riemannian Manifold which is the case of nullity, that is when $\nu = 0$.

3.2 Tangent Vector

Tangent vectors may be space like or time like depending on the conditions on E.
1. Tangent vector E defined on M will be space like, if $g(E, E) > 0, \forall p \in M$.
2. If E is not equal to zero, then $g(E, E) < 0$ then tangent vector will be time like.
For real vector space R^n_ν which is n-dimensional along with their inner product of index $(\nu, n - \nu)$ is

$$\langle y, y \rangle = \Sigma_{i=1}^{\nu} + \Sigma_{i=\nu+1}^{n} y_i^2$$

for $y = \langle y_1, y_2, \ldots, y_n \rangle$ is n-dimensional semi Euclidean space as it is the perfectly natural coordinate of R^n_ν.

3.3 Statistical Metallic Manifold

Let us define Torsion free affine connection ∇ and M a Semi Riemannian Metallic Manifold then Statistical Metallic Manifold (M, g, ∇) such that ∇ satisfies the property of symmetries. Affine connection ∇^* defined over the Statistical Metallic Manifold (M, ∇, g) is define as

$$F_g(G, H) = g(G, \nabla_F H) + g(\nabla_F G, H) \tag{1}$$

where F, G, and H are vector fields on M.

∇^*, which is also an affine connection is known as dual or conjugate to ∇ in context to g. ∇^* is also torsion free. ∇^*g satisfies $(\nabla^*)^* = \nabla$ and so ∇^* is symmetric.

Tensor field J holds the following equation is defined on a metallic structure on manifold M_n

$$J^2 = pJ + qI_n \tag{2}$$

where I_n stands for identity transformation.

J defined in (2) preserves the metric g and Semi Riemannian Metallic Manifold defined on the Metallic Manifold M, i.e.,

$$g(JF, JG) = g(F, G) \tag{3}$$

for (1, 1) vector field G and F defined over Metallic Manifold (M).

Now let (M_n, g) be a Semi Riemannian Metallic Manifold along with metallic structure J. Metallic structure has another tensor field J^* which is also of type (1, 1), furthermore

$$g(JF, G) + g(F, J^*G) = 0 \tag{4}$$

where F and G are vector field. If ∇ and J are parallel, so the defined (M, g, ∇, J) known as Kahler Embedded Statistical Metallic Manifold (KESSM). Furthermore as JR(F, G) = R(F, G)J and (2), we get

$$g(\nabla_{HJ} F, G) + g(F, (\nabla^*_H J^*)G) = 0 \tag{5}$$

where F, G, H are vector field on M_n. Hence (M_n, g, ∇, J) is a Kahler Embedded Statistical Metallic Manifold iff (M_n, g, ∇^*, J^*).

Kahler Embedded Statistical Metallic Manifold and F, G, and H which are vector fields defined on it also the curvature tensor R in reference to ∇ defined as

$$R(F,G)H = \frac{c}{4}[g(G,H)F - g(F,JH)JG - g(G,JH)JF - g(F,H)G + g(F,JG) - g(JF,G)JH] \quad (6)$$

where R is curvature tensor and c is the constant change J to J^*.

3.4 2D Statistical Metallic Matrix

For J $\in R_n^n$ satisfying (2) then the Statistical Metallic Matrix is defined where I is the Identity matrix of R_n^n. If we get family of the statistical metallic structure with two parameters then by solving (2) we have the following cases

Case 1 For m, o $\in$ R, and n $\in$ R - {0},

$$J_{m,n} = \begin{pmatrix} m & -\frac{1}{s}(m^2 - pm - q) \\ n & p - m \end{pmatrix}$$

or

$$J_{n,o} = \begin{pmatrix} a - o & -\frac{1}{n}(o^2 - po - q) \\ n & p - o \end{pmatrix} \quad (7)$$

Case 2 For m = $\rho_{(p,q)}$, o $\in$ R

$$J\rho_{(p,q)}, n = \begin{pmatrix} \rho_{(p,q)} & o \\ n & p - \rho_{(p,q)} \end{pmatrix}$$

or

$$J_{p-\rho_{(p,q)}}, n = \begin{pmatrix} p - \rho_{(p,q)} & o \\ n & \rho_{(p,q)} \end{pmatrix}$$

or

$$J_{\rho_{(p,q)}}, n = \begin{pmatrix} \rho_{(p,q)} & n \\ o & p - \rho_{(p,q)} \end{pmatrix}$$

or

$$J_{p-\rho_{(p,q)}}, n = \begin{pmatrix} p - \rho_{(p,q)} & n \\ o & \rho_{(p,q)} \end{pmatrix}$$

Case 3 For m = $\rho_{(p,q),n}$ = 0

$$J_{\rho_{(p,q)}}, o = \begin{pmatrix} \rho_{(p,q)} & o \\ o & p - \rho_{(p,q)} \end{pmatrix}$$

or

$$J_{p-\rho_{(p,q)}}, o = \begin{pmatrix} p - \rho_{(p,q)} & o \\ o & \rho_{(p,q)} \end{pmatrix}$$

From (7) we get boundary case of J_1 and J_2.

Within F an Euclidean space, the reflection in context of statistical hyperplane H which holds normalness.

v ∈ F{0} gives the result

$$m_v(y) = y - \frac{\langle y, v\rangle}{\langle v, v\rangle}$$

for all y ∈ F.
and

$$r_v^2 = I_F$$

where I_F is Identity and $I_F \in$ F.

Now reflection of Statistical Metallic in context to v:-

$$J_v = \frac{1}{2}(pI_E + \sqrt{p^2 + 4qm_v})$$

where v is an eigen vector of J_v with the corresponding eigen value p - $\rho_{(p,q)}$. In [13] with the conditions that y is an orthogonal group of F.

Now the following r transformation can be stated as

$$J_v(x) = \rho_{(p,q)}y - \frac{\sqrt{p^2 + 4q}\langle y, v\rangle}{\langle v, v\rangle}v.$$

4 Statistical Submersion on Statistical Metallic Manifold

If π: (M, g, ∇) → (B, g_B, $\widehat{\nabla}$) be an Embedded Statistical Submersion on Statistical Metallic Manifold. Now well-defined curvature tensor over (M, g, ∇, J, ξ, η) satisfy

$$\begin{aligned} R(F,G)H = \frac{1}{4}(3+c)[g(G,H)F - g(F,H)G] + \frac{1}{4}(c-1)[\eta(F)\eta(H)G \\ -\eta(G)\eta(H)F + g(F,H)\eta(G)\xi - g(G,H)\eta(F)\xi \\ -g(G,JH)JF + g(F,JH)JG + g(F,JG)JH - g(JF,G)JH] \end{aligned}$$

along with constant c, for F, G, H, H' be a member of tangent manifold TM.

$$
\begin{aligned}
g(R(F,G)H,H') &= \frac{1}{4}(3+c)[g(G,H)g(F,H') - g(F,H)g(G,H')] \\
&+ \frac{1}{4}(1-c)[\eta(G)\eta(H)g(F,H') - \eta(F)\eta(H)g(G,H') \\
&+ g(G,H)\eta(F)\eta(H') - g(F,H)\eta(G)\eta(H') \\
&- g(F,JH)g(JG,H') + g(G,JH)g(JF,H') \\
&+ g(JF,G)g(JH,H') - g(F,JG)g(JH,H')]
\end{aligned}
$$

then we have the following theorem.

Theorem 4.1 *If π: $(M, g, \nabla) \rightarrow (B, g_B, \widehat{\nabla})$ is Statistical Submersion on Statistical Metallic Manifold. Then α, β, γ, γ' are vectors, belonging to Horizontal component of Manifold M_n, Φ, Φ, Θ, Θ' are vectors and belong to the vertical component of the Manifold M, T and A are (1, 2) tensor field then following equations holds:*

$$
\begin{aligned}
g(\overline{R}(\Phi,\Psi)\Theta,\Theta') &+ g(T_\Phi\Theta, T^*_\Psi\Theta') - g(T_\Psi\Theta, T^*_\Phi\Theta') \\
&= \frac{1}{4}(3+c)[g(\Psi,\Theta)g(\Phi,\Theta') - g(\Phi,\Theta)g(\Psi,\Theta')] \\
&+ \frac{1}{4}(1-c)[\eta(\Psi)\eta(\Theta)g(\Phi,\Theta') - \eta(\Phi)\eta(\Theta)g(\Psi,\Theta') \\
&+ g(\Psi,\Theta)\eta(\Phi)\eta(\Theta') - g(\Phi,\Theta)\eta(\Psi)\eta(\Theta') \\
&- g(\Phi,\overline{J}\Theta)g(\overline{J}\Psi,\Theta) + g(\Psi,\overline{J}\Theta)g(\overline{J}\Phi,\Theta') \\
&+ g(\overline{J}\Phi,\Psi)g(\overline{J}\Theta,\Theta') - g(\Phi,\overline{J}\Psi)g(\overline{J}\Theta,\Theta') \qquad (1)
\end{aligned}
$$

$$
g((\nabla)_\Phi T)_\Psi\Theta,\alpha) - g((\nabla)_\Psi T)_\Phi\Theta,\alpha) = 0 \qquad (2)
$$

$$
g((\nabla)_\Phi T)_\Psi\alpha,\Theta) - g((\nabla)_\Psi T)_\Phi\alpha,\Theta) = 0 \qquad (3)
$$

$$
\begin{aligned}
g((aT)_\alpha\Psi,\beta) &- g((\nabla_\Psi A)_X\Phi,\beta) + g(T_\Phi\alpha, T^*_\Psi\beta) - g(T_\Psi\alpha, T^*_\Phi\beta) \\
&- g(A_\alpha\Phi, A^*_\beta\Psi) + g(A_\alpha\Psi, A^*_\beta\Phi) \\
&= \frac{1}{4}(1-c)[g(\overline{J}\Phi,\Psi) - g(\Phi,\overline{J}\Psi)]g(J\alpha,\beta) \qquad (4)
\end{aligned}
$$

$$
g((\Psi\nabla_\alpha,\overline{\nabla}_\Phi)\Psi,\Theta) - g(T_\Phi\Psi, A^*_\alpha\Theta) - g(\nabla_{[\alpha,\Phi]}\Psi,\Theta) + g(T^*_\Phi\Theta, A_\alpha\Psi) = 0 \qquad (5)
$$

$$
\begin{aligned}
g((\nabla_\alpha T)_\Phi\Psi,\beta) &- g((\nabla_\Phi A)_\alpha\Psi,\beta) + g(A_\alpha\Phi, A^*_\beta\Psi) - g(T_\Phi\alpha\Theta, T^*_\Psi\beta) \\
&= \frac{1}{4}(c+3)g(\Phi,\Psi)g(\alpha,\beta) - \frac{1}{4}(c-1)[\eta(\Phi)\eta(\Psi)g(\alpha,\beta)] \\
&+ g(\Phi,\overline{J}\Psi)g(J\alpha,\beta) \qquad (6)
\end{aligned}
$$

$$g((\nabla_\alpha T)_\Phi \beta, \Psi) - g((\nabla_\Phi A)_\alpha \beta, \Psi) + g(T_\Phi \alpha, T_\Psi \beta) - g(A_\alpha \Phi, A_\beta \Psi)$$
$$= -\frac{1}{4}(c+3)g(\alpha, \beta)g(\Phi, \Psi) + \frac{1}{4}(c-1)[g(\alpha, \beta)]\eta(\Phi)\eta(\Psi)$$
$$+ g(\Phi, J\Psi)g(\overline{J}\Phi, \Psi) \quad (7)$$
$$g((\nabla_\alpha A)_\beta \Phi, \gamma) - g(T_\Phi \alpha, A^*_\beta \gamma) - g(T_\Phi \beta, A^*_\alpha \gamma) + g(A_\alpha \beta, T^*_\Phi \gamma) = 0 \quad (8)$$

$$g((\nabla_\alpha T)_\alpha \beta, \Psi) - g((\nabla_\beta T)_\Phi \alpha, \Psi) + g((\nabla_\Phi \Theta)_\alpha \beta, \Psi) - g(T_\Phi \alpha, T_\Psi \beta)$$
$$- g(T_\Psi \alpha, T_\Phi \beta) - g(A_\alpha \Phi, A_\beta \Psi) + g(A_\beta \Phi, A_\alpha \Psi)$$
$$= \frac{1}{4}(c-1)g(\alpha, J\beta)g(\overline{J}\Phi, \Psi) - g(J\alpha, \beta)g(\overline{J}\Phi, \Psi) \quad (9)$$

$$g((\nabla_\alpha A)_\beta \Phi, \gamma) - g((\nabla_\beta A)_\alpha \Phi, \gamma) + g(T^*_\alpha \gamma, \Theta_\alpha \beta) = 0 \quad (10)$$

$$g((\nabla_\alpha A)_\beta \gamma, \Phi) - g((\nabla_\beta A)_\alpha \gamma, \Phi) + g(T_\Phi \gamma, \Theta_\alpha \beta) = 0 \quad (11)$$

$$g(\widehat{R}(\alpha, \beta)\gamma, \gamma') + g(A_\beta \gamma, A^*_\alpha \gamma') - g(A_\alpha \gamma, A^*_\beta \gamma') + g(\Theta_\alpha \beta, A^*_\gamma \gamma')$$
$$= \frac{1}{4}(c+3)[g(\beta, \gamma)g(\alpha, \gamma') - g(\alpha, \gamma)g(\beta, \gamma')]$$
$$+ \frac{1}{4}(c-1)[-g(\beta, J\gamma) - g(J\alpha, \gamma') + g(\alpha, J\beta)g(J\gamma, \gamma')$$
$$- g(J\alpha, \beta)g(J\gamma, \gamma')] \quad (12)$$

for $\Phi, \Psi, \Theta, \Theta' \in$ ***V****(M) and* $\alpha, \beta, \gamma, \gamma' \in$ ***H****(M)*

Theorem 4.2 *If* $\pi: (M, g, \nabla) \to (B, g_B, \widehat{\nabla})$ *is Statistical Submersion on Statistical Metallic Manifold with the property that the curvature tensor of the total space satisfies (4.1).*

We assume that rank $(\overline{J} + \overline{J}^*) = dim\overline{M} - 1$ *and* $\boldsymbol{H}\nabla_\alpha N = 0$ *for all* α *belongs to Horizontal Manifold* ***H****(M). Then*

1. If 3 + c = 0, then this is the case when base space is supposed to be flat and each fiber of the space is a totally geodesic submanifold of M and also the curvature tensor satisfies the type (4.2) with c = –3.

2. If $tr\,\overline{J} = 0$ *for all values of* $s > 1$

a. $c + 3 \leq 0$ *then g is positive definite.*
b. $c + 3 < 0$ *and* α *is space like if and only if* $T^*\alpha$ *is space like.*
c. $c + 3 > 0$ *and* α *is time like if and only if* $T^*\alpha$ *is time like.*
d. the horizontal vector α *is null if and only if* $T^*\alpha$ *is null.*

From (4.7) we can have

$$g((\nabla^*_\alpha T^*)_\Phi \Psi, \beta) - g(T^*_\Phi \alpha, T_\Psi \beta) = \frac{1}{4}(3 + c)[[g(\Phi, \Psi) - \eta(\Phi)\eta(\Psi)] + g(\alpha, J\beta)g(\overline{J}\Phi, \Psi)].$$

Which implies

$$g(\nabla^*_\alpha N, \beta) - g(T\alpha, T\beta) = \frac{1}{4}(c + 3)[(s - 1)g(\alpha, \beta) - (tr\overline{J})g(\alpha, J\beta)] \quad (13)$$

Hence the result that is for $\boldsymbol{H}\nabla^*_\alpha \boldsymbol{N}^* = 0$, *we can get* $c + 3 = 0$ *which also states that* $tr\overline{J} = 0$.

Disclosure of potential conflicts of interest: No authors reported any conflicts of interest.

Funding Statement: No fund or grant has been used in this research work.

Data Availability Statement: No type of data has been required for this research work.

References

1. M. Crasmareanu, C.-E. Hreţcanu, Golden differential geometry. Chaos, Solitons Fractals **38**(5), 1229–1238 (2008)
2. M. Özkan, F. Yilmaz, Prolongations of golden structures to tangent bundles of order R. Commun. Fac. Sci. Univ. Ank. Ser. A1 Math. Stat. **65**(1), 35–48 (2016)
3. T.E.A. Ozkam, On lifts of silver structure. J. Sci. Arts (2017)
4. M.N.I. Khan, Novel theorems for the frame bundle endowed with metallic structures on an almost contact metric manifold. Chaos, Solitons Fractals **146**, 110872 (2021)
5. M.N.I. Khan, Novel theorems for metallic structures on the frame bundle of the second order. Filomat **36**(13), 4471–4482 (2022)
6. A. Gezer, C. Karaman, Golden-hessian structures. Proc. Natl. Acad. Sci., India Sect. A: Phys. Sci. **86**(1), 41–46 (2016)
7. M.N.I. Khan, Complete and horizontal lifts of metallic structures. Int. J. Math. Comput. Sci. **15**(4), 983–992 (2020)
8. M.N.I. Khan, M.A. Choudhary, S.K. Chaubey, Alternative equations for horizontal lifts of the metallic structures from manifold onto tangent bundle. J. Math. **2022** (2022)
9. M.N.I. Khan, U.C. De, Liftings of metallic structures to tangent bundles of order R. AIMS Math. **7**(5), 7888–7897 (2022)
10. B. Şahin, M.A. Akyol, Golden maps between golden Riemannian manifolds and constancy of certain maps. Math. Commun. **19**(2), 333–342 (2014)
11. S.-I. Amari, H. Nagaoka, *Methods of Information Geometry*, vol. 191 (American Mathematical Soc., 2000)
12. M. Noguchi, Geometry of statistical manifolds. Differ. Geom. Appl. **2**(3), 197–222 (1992)
13. M.N.I. Khan, U.C. De, Lifts of metallic structure on a cross-section. Filomat **36**(18), 6369–6373 (2022)

14. V.W. de Spinadel, The family of metallic means. Vis. Math. **1**(3), 1–16 (1999)
15. A.I.Ö. Mustafa, E. Taylan, Prolongations of golden structures to tangent bundles of order 2. Gazi Univ. J. Sci. **28**(2), 253–258 (2015)
16. O.M.F. Yilmaz, Bronze structure, in *Mathematics Symposium, Ni ⌣ gde, Turkey* (2015), pp. 109–111
17. C.-E. Hretcanu, M. Crasmareanu, Metallic structures on Riemannian manifolds. Rev. Un. Mat. Argentina **54**(2), 15–27 (2013)

Tangent Bundles Endowed with Quarter-Symmetric Non-metric ξ-Connection on 3-Dimensional Quasi-Sasakian Manifolds

Mohammad Nazrul Islam Khan and Ljubica Velimirović

MSC (2010): 53C05 · 53C25 · 58A30

1 Introduction

The geometry of tangent bundles provides a powerful method that allows us to prove many new problems in the study of modern differential geometry. The method of lifts such as complete, vertical and horizontal lifts of connections and geometric structures from the manifold and its tangent bundle proposed by Yano and Kobayashi [38]. The vertical and complete lifts of the quarter-symmetric nonmetric connection were studied in [24, 25, 32]. Altunbas et al. [1] studied the tangent bundle with the Kaluza-Klein metric over the Riemannian manifold and investigated certain curvature properties on it. The complete, vertical and horizotal lifts of metallic structure on LP- Sasakian manifold were studied by Altunbas and Sengul [2] and they investigated the conditions for parallelity and the fundamental 2-Form. The connections and geometric structures on the tangent bundle have been studied by Das and Khan [10], Khan [26, 28], Kazan and Karadag [19], Peyghan et al. [33], Tani [35] and many more.

On the other hand, Golab [15] introduced the notion of quarter-symmetric metric (briefly, *QSM*) connection and studied the various properties on it. Barman [5] analized a quarter-symmetric non-metric (briefly, *QSNM*) connection on a P-Sasakian manifold and established the first Bianchi identity of curvature tensors on such manifold with *QSNM* connection. Barman and De [6] studied *QSNM* ξ-connection on

M. N. I. Khan (✉)
Department of Computer Engineering, College of Computer, Qassim University, Buraydah 51452, Saudi Arabia
e-mail: m.nazrul@qu.edu.sa

L. Velimirović
Faculty of Sciences and Mathematics, Department of Mathematics, University of Niš, Niš, Serbia
e-mail: vljubica@pmf.ni.ac.rs

B.-Y. Chen et al. (eds.), *Geometry of Submanifolds and Applications*, Infosys Science Foundation Series, https://doi.org/10.1007/978-981-99-9750-3_11

3-dimensional quasi-Sasakian manifolds and investigated certain curvature properties of such connection.

Let M be an n-dimensional differentiable manifold of class C^∞ with the Levi-Civita connection ∇. Let χ_1 and χ_2 be vector fields on M then a linear connection $\tilde{\nabla}$ is known as symmetric connection on M if the torsion tensor T of $\tilde{\nabla}$ defined by

$$T(\chi_1, \chi_2) = \tilde{\nabla}_{\chi_1}\chi_2 - \tilde{\nabla}_{\chi_2}\chi_1 - [\chi_1, \chi_2] \tag{1.1}$$

is zero, or else it is non-symmetric. If the torsion tensor T satisfies

$$T(\chi_1, \chi_2) = \pi(\chi_2)\phi\chi_1 - \pi(\chi_1)\phi\chi_2 \tag{1.2}$$

where

$$\pi(\chi_1) = g(\xi, \chi_1), \forall\xi \in \Im_0^1(M), \tag{1.3}$$

π is 1-form, ϕ is a tensor field of type (1,1) and g is the Riemannian metric, then $\tilde{\nabla}$ is called a quarter-symmetric connection. Also, if the Riemannian metric g satisfies $\tilde{\nabla}g \neq 0$, then $\tilde{\nabla}$ is said to be a *QSNM* connection. A quarter-symmetric connection $\tilde{\nabla}$ is said to be a *QSM* connection if $\tilde{\nabla}g = 0$. If moreover, a quarter-symmetric connection $\tilde{\nabla}$ satisfies the condition $(\tilde{\nabla}\chi_1 g)(\chi_2, \chi_3) \neq 0; \forall\chi_1, \chi_2, \xi \in \Im_0^1(M)$, then $\tilde{\nabla}$ is said to be a *QSNM* connection. Various types of *QSM* connection and *QSNM* connection have been recently discussed in [4, 11–14, 16–18, 27, 34, 36, 39].

In this paper, we have executed complete lifts on *QSNM* ξ-connection from 3-dimensinal quasi-Sasakian manifold to its tangent bundle.

These are the paper's primary contributions, in brief:

- We have executed complete lifts on quasi-Sasakian manifold to its tangent bundle.
- A theorem on complete lifts of *QSNM* ξ-connection on 3-dimensional quasi-Sasakian manifolds to its tangent bundle is proved.
- We have calculated curvature tensor and Ricci tensor of 3-dimensional quasi-Sasakian manifolds admitting *QSNM* ξ-connection on *TM*.
- Some calculations on the complete lifts of projective curvature tensor on 3-dimensional non-cosymplectic quasi-Sasakian manifolds with respect to *QSNM* ξ-connection $\bar{\nabla}$ to its tangent bundle have been given.

2 General Theory

Consider *TM* be the tangent bundle of a manifold M. Tensor field of type (r, s) are symbolized as $\Im_s^r(M)$ and $\Im_s^r(TM)$ on M and *TM*, respectively. Let the function, 1-form, vector field and (1,1) type tensor field be symbolized as f, η, χ_1 and ϕ, respectively. The complete and vertical lifts of f, η, χ_1, ϕ are symbolized as f^C, η^C, χ_1^C, ϕ^C

and $f^V, \eta^V, \chi_1^V, \phi^V$, respectively. The following operations on f, η, χ_1 and ϕ are defined by [10, 20, 37, 38]

$$(f\chi_1)^V = f^V \chi_1^V, (f\chi_1)^C = f^C \chi_1^V + f^V \chi_1^C, \tag{2.1}$$

$$\chi_1^V f^V = 0, \chi_1^V f^C = \chi_1^C f^V = (\chi_{1f})^V, \chi_1^C f^C = (\chi_{1f})^C, \tag{2.2}$$

$$\eta^V(\chi_1^V) = 0, \eta^V(\chi_1^C) = \eta^C(\chi_1^V) = (\eta\chi_1)^V, \eta^C(\chi_1^C) = (\eta\chi_1)^C, \tag{2.3}$$

$$\phi^V \chi_1^C = (\phi\chi_1)^V, \phi^C \chi_1^C = (\phi\chi_1)^C, \tag{2.4}$$

$$[\chi_1, \chi_2]^V = [\chi_1^C, \chi_2^V] = [\chi_1^V, \chi_2^C], [\chi_1, \chi_2]^C = [\chi_1^C, \chi_2^C]. \tag{2.5}$$

$$\nabla^C_{\chi_1^C} \chi_2^C = (\nabla_{\chi_1}\chi_2)^C, \quad \nabla^C_{\chi_1^C} \chi_2^V = (\nabla_{\chi_1}\chi_2)^V, \tag{2.6}$$

where ∇ being the Levi-Civita connection.

2.1 *Quasi-Sasakian Manifolds*

Let M be an almost contact metric manifold (dim$= 2n+1$) with (ϕ, ξ, η, g)-structure with an amicable Riemannian metric g fulfuling [7, 8, 29]

$$\phi^2 = -I + \eta \otimes \xi, \tag{2.7}$$

$$\eta(\xi) = 1, \quad \phi\xi = 0, \quad g(\chi_1, \xi) = \eta(\chi_1), \quad \eta(\phi\chi_1) = 0, \tag{2.8}$$

$$g(\chi_1, \chi_2) = g(\phi\chi_1, \phi\chi_2) + \eta(\chi_1)\eta(\chi_2), \tag{2.9}$$

$$\Phi(\chi_1, \chi_2) = g(\chi_1, \phi\chi_2) = -g(\phi\chi_1, \chi_2), \tag{2.10}$$

$\forall \phi \in \Im_1^1(M), \chi_1, \chi_2, \xi \in \Im_0^1(M), \eta \in \Im_1^0(M)$ and Φ denote the fundamental 2-form. M is said to be quasi-Sasakian if the almost contact structure (ϕ, ξ, η, g) is normal and the fundamental 2-form Φ is closed i.e. $d\Phi = 0$ [9].

An almost contact metric manifold M is a 3-dimensional quasi-Sasakian manifold if and only if [21, 22, 30]

$$\nabla_{\chi_1}\xi = -\beta\phi\chi_1, \forall \chi_1 \in \Im_0^1(M), \tag{2.11}$$

for a certain function β on M, such that $\xi\beta = 0$, ∇ is the Levi-Civita connection of M. Such a quasi-Sasakian manifold is cosymplectic iff $\beta = 0$.

In the view of (2.11), we infer

$$(\nabla_{\chi_1}\phi)(Y) = \beta g(\chi_1, \chi_2)\xi - \beta\eta(\chi_2)\chi_1, \tag{2.12}$$

$$(\nabla_{\chi_1}\eta)\chi_2 = g(\nabla_{\chi_1}\xi, \chi_2) = -\beta g(\phi\chi_1, \chi_2) = \beta g(\chi_1, \phi\chi_2). \tag{2.13}$$

Let S be the Ricci tensor of M and given by [31]

$$S(\chi_2, \chi_3) = \left(\frac{r}{2} - \beta^2\right) g(\chi_2, \chi_3) + \left(3\beta^2 - \frac{r}{2}\right) \eta(\chi_2)\eta(\chi_3) \\ - \eta(\chi_2) d\beta(\phi\chi_3) - \eta(\chi_3) d\beta(\phi\chi_2), \tag{2.14}$$

where r is the scalar curvature of M. The curvature tensor R of M is given by [6]

$$R(\chi_1, \chi_2)\chi_3 = g(\chi_2, \chi_3)Q\chi_1 - g(\chi_1, \chi_3)Q\chi_2 + S(\chi_2, \chi_3)Q\chi_1 \\ - S(\chi_1, \chi_3)Q\chi_2 - \frac{r}{2}(g(\chi_2, \chi_3)\chi_1 - g(\chi_1, \chi_3)\chi_2), \tag{2.15}$$

where Q is the Ricci operator, i.e., $g(Q\chi_1, \chi_2) = S(\chi_1, \chi_2)$.

From (2.14), we infer

$$QY = \left(\frac{r}{2} - \beta^2\right) \chi_2 + \left(3\beta^2 - \frac{r}{2}\right) \eta(\chi_2)\xi + \eta(\chi_2)(\phi grad\beta) - d\beta(\phi\chi_2)\xi, \tag{2.16}$$

where df is the exterior derivative such that $df(\chi_1) = g(gradf, \chi_1)$.

In the view of (2.14), we infer

$$S(\phi\chi_2, \phi\chi_3) = S(\chi_2, \chi_3) - 2\beta^2\eta(\chi_2)\eta(\chi_3). \tag{2.17}$$

3 Complete Lifts of Quasi-Sasakian Manifolds to Its Tangent Bundle

Consider TM be the tangent bundle of quasi-Sasakian manifold M. Applying the complete lift on (2.7)–(2.11), the following equations are obtained

$$(\phi^C)^2 = -I + \eta^V \otimes \xi^C + \eta^C \otimes \xi^V, \tag{3.1}$$
$$\eta^C\xi^C = \eta^V\xi^V = 0, \eta^C\xi^V = \eta^V\xi^C = 1, \tag{3.2}$$
$$\phi^C\xi^V = \phi^V\xi^C = \phi^V\xi^V = \phi^C\xi^C = 0, \tag{3.3}$$
$$\eta^V(\phi\chi_1)^C = \eta^C(\phi\chi_1)^V = \eta^V(\phi\chi_1)^V = \eta^C(\phi\chi_1)^C = 0, \tag{3.4}$$
$$g^C(\chi_1^C, \xi^C) = \eta^C(\chi_1^C), \tag{3.5}$$
$$g^C(\chi_1^C, \chi_2^C) = g^C((\phi\chi_1)^C, (\phi\chi_2)^C) + \eta^C(\chi_1^C)\eta^V(\chi_2^C) \\ + \eta^V(\chi_1^C)\eta^C(\chi_2^C), \tag{3.6}$$
$$\Phi^C(\chi_1^C, \chi_2^C) = g^C(\chi_1^C, \phi^C\chi_2^C) = -g^C(\phi^C\chi_1^C, \chi_2^C), \tag{3.7}$$
$$\nabla^C_{\chi_1^C}\xi^C = -\beta(\phi\chi_1)^C, \tag{3.8}$$

where $\forall\chi_1^C, \chi_2^C \in \Im_0^1(TM)$, ϕ^C and Φ^C are the complete lift of ϕ and Φ, respectively.

Let ∇^C be the complete lift of the Levi-Civita connection ∇ on TM. Then from (2.12) and (2.13), we obtain the relations [3]

$$\begin{aligned}(\nabla^C_{\chi_1^C}\phi^C)(Y^C) &= \beta(g^C(\chi_1^C,\chi_2^C)\xi^V + g^C(\chi_1^V,\chi_2^C)\xi^C) \\ &\quad - \beta(\eta^C(\chi_2^C)\chi_1^V + \eta^V(\chi_2^C)\chi_1^C) \qquad (3.9)\\ (\nabla^C_{\chi_1^C}\eta^C)\chi_2^C &= g(\nabla^C_{\chi_1^C}\xi^C,\chi_2^C) = -\beta g((\phi\chi_1)^C,\chi_2^C) \\ &= \beta g^C(\chi_1^C,(\phi\chi_2)^C). \qquad (3.10)\end{aligned}$$

Applying complete lifts on (2.14), we infer

$$\begin{aligned}S^C(\chi_2^C,\chi_3^C) &= \left(\frac{r^C}{2}-\beta^2\right)g^C(\chi_2^C,\chi_3^C) + (3\beta^2 - \frac{r^C}{2})\{\eta^C(\chi_2^C)\eta^V(\chi_3^C) \\ &\quad + \eta^V(\chi_2^C)\eta^C(\chi_3^C)\} \\ &\quad - \eta^C(\chi_2^C)d\beta(\phi\chi_3)^V - \eta^V(\chi_2^C)d\beta(\phi\chi_3)^C \\ &\quad - \eta^C(\chi_3^C)d\beta(\phi\chi_2)^V - \eta^V(\chi_3^C)d\beta(\phi\chi_2)^C, \qquad (3.11)\end{aligned}$$

where r^C is the complete lift of r.

Applying the complete lift on (2.15) and (2.16), we infer

$$\begin{aligned}R^C(\chi_1^C,\chi_2^C)\chi_3^C &= g^C(\chi_2^C,\chi_3^C)(Q\chi_1)^V + g^C(\chi_2^V,\chi_3^C)(Q\chi_1)^C \\ &\quad - g^C(\chi_1^C,\chi_3^C)(Q\chi_2)^V - g^C(\chi_1^V,\chi_3^C)(Q\chi_2)^C \\ &\quad + S^C(\chi_2^C,\chi_3^C)\chi_1^V + S^C(\chi_2^V,\chi_3^C)\chi_1^C \\ &\quad - S^C(\chi_1^C,\chi_3^C)\chi_2^V - S^C(\chi_1^V,\chi_3^C)\chi_2^C \\ &\quad - \frac{r^C}{2}\{g^C(\chi_2^C,\chi_3^C)\chi_1^V + g^C(\chi_2^V,\chi_3^C)\chi_1^C \\ &\quad - g^C(\chi_1^C,\chi_3^C)\chi_2^V - g^C(\chi_1^V,\chi_3^C)\chi_2^C\}, \qquad (3.12)\end{aligned}$$

where Q is the Ricci operator, i.e., $g^C((Q\chi_1)^C,\chi_2^C) = S^C(\chi_1^C,\chi_2^C)$.

In the view of (3.11), we infer

$$\begin{aligned}(QY)^C &= \left(\frac{r^C}{2}-\beta^2\right)\chi_2^C + \left(3\beta^2-\frac{r}{2}\right)(\eta^C(\chi_2^C)\xi^V + \eta^V(\chi_2^C)\xi^C) \\ &\quad + \eta^C(\chi_2^C)(\phi grad\beta)^V + \eta^V(\chi_2^C)(\phi grad\beta)^C \\ &\quad - d\beta(\phi\chi_2)^C\xi^V - d\beta(\phi\chi_2)^V\xi^C. \qquad (3.13)\end{aligned}$$

In the view of (3.11), we infer

$$\begin{aligned}S^C((\phi\chi_2)^C,(\phi\chi_3)^C) &= S^C(\chi_2^C,\chi_3^C) - 2\beta^2\{\eta^C(\chi_2^C)\eta^V(\chi_3^C) \\ &\quad + \eta^V(\chi_2^C)\eta^C(\chi_3^C)\}, \qquad (3.14)\end{aligned}$$

where S^C, R^C and Q^C denote the complete lifts of the Ricci tensors S, R and Q of M on TM, respectively.

4 Proposed Theorem on the Complete Lifts of $QSNM\xi$-connection on 3-dimensional Quasi-Sasakian Manifolds to its Tangent Bundle

Let (M, g) be a quasi-Sasakian manifold and TM its tangent bundle. A linear connection $\bar{\nabla}$ on M given by [6]

$$\bar{\nabla}_{\chi_1}\chi_2 = \nabla_{\chi_1}\chi_2 + (\beta - 1)\eta(\chi_1)\phi\chi_2 + \beta\eta(\chi_2)\phi\chi_1, \tag{4.1}$$

where ∇ is Levi-Civita connection and β is a certain function on M.

Applying the complete lifts on (1.1) and (4.1), we infer

$$T^C(\chi_1^C, \chi_2^C) = \bar{\nabla}^C_{\chi_1^C}\chi_2^C - \bar{\nabla}^C_{\chi_2^C}\chi_1^C - [\chi_1^C, \chi_2^C], \tag{4.2}$$

and

$$\begin{aligned}\bar{\nabla}^C_{\chi_1^C}\chi_2^C &= \nabla^C_{\chi_1^C}\chi_2^C + (\beta - 1)\{\eta^C(\chi_1^C)(\phi\chi_2)^V + \eta^V(\chi_1^C)(\phi\chi_2)^C\}\\ &\quad + \beta\{\eta^C(\chi_2^C)(\phi\chi_1)^V + \eta^V(\chi_2^C)(\phi\chi_1)^C\},\end{aligned} \tag{4.3}$$

$\forall \chi_1, \chi_2, \xi \in \Im_0^1(M)$.

Using (4.2) and (4.3), we infer

$$\begin{aligned}T^C(\chi_1^C, \chi_2^C) &= \eta^C(\chi_2^C)(\phi\chi_1)^V + \eta^V(\chi_2^C)(\phi\chi_1)^C\\ &\quad - \eta^C(\chi_1^C)(\phi\chi_2)^V - \eta^V(\chi_1^C)(\phi\chi_2)^C.\end{aligned} \tag{4.4}$$

Therefore, the linear connection $\bar{\nabla}^C$ is a quarter-symmetric connection on TM [23].

It is well-known that

$$(\bar{\nabla}_{\chi_1}g)(\chi_2, \chi_3) = \bar{\nabla}_{\chi_1}g(\chi_2, \chi_3) - g(\bar{\nabla}_{\chi_1}\chi_2, \chi_3) - g(\chi_2, \bar{\nabla}_{\chi_1}\chi_3). \tag{4.5}$$

In view of (4.3) and (2.10), then (4.5) becomes

$$\begin{aligned}(\bar{\nabla}^C_{\chi_1^C}g^C)(\chi_2^C, \chi_3^C) &= -\beta\{\eta^C(\chi_2^C)g^C((\phi\chi_1)^V, \chi_3^C)\\ &\quad + \eta^V(\chi_2^C)g^C((\phi\chi_1)^C, \chi_3^C)\} - \beta\{\eta^C(\chi_3^C)g^C((\phi\chi_2)^V, \chi_1^C)\\ &\quad + \eta^V(\chi_3^C)g^C((\phi\chi_2)^C, \chi_1^C)\} \neq 0.\end{aligned} \tag{4.6}$$

The linear connection $\bar{\nabla}^C$ satisfying (4.4) and (4.6) is called a *QSNM* connection on a 3-dimensional non-cosymplectic quasi-Sasakian manifold on *TM*.

In view of (3.2)–(3.6), (3.8) and (4.3), we infer

$$\begin{aligned}\bar{\nabla}^C_{\chi_1^C}\xi^C &= \nabla^C_{\chi_1^C}\xi^C + (\beta-1)(\eta^C(\chi_1^C)(\phi\xi)^V - \eta^V(\chi_1^C)(\phi\xi)^C)\\ &+ \beta(\eta^C(\xi^C)(\phi\chi_1)^V - \eta^V(\xi^C)(\phi\chi_1)^C).\end{aligned} \tag{4.7}$$

The linear connection $\bar{\nabla}^C$ given by (4.3) satisfying (4.4), (4.6) and (4.7) is the *QSNM* ξ-connection on 3-dimensional quasi-Sasakian manifold on *TM*.

Conversely, it is enough to show that a linear connection $\bar{\nabla}^C$ defined on *TM* satisfying (4.4), (4.6) and (4.7) is given by (4.3).

Let H be a tensor field of type (1, 2) on M and given by

$$\bar{\nabla}_{\chi_1}\chi_2 = \nabla_{\chi_1}\chi_2 + H(\chi_1, \chi_2). \tag{4.8}$$

Applying the complete lifts on (4.8), we infer

$$\bar{\nabla}^C_{\chi_1^C}\chi_2^C = \nabla^C_{\chi_1^C}\chi_2^C + H^C(\chi_1^C, \chi_2^C). \tag{4.9}$$

Using (4.4) and (4.9), we infer

$$T^C(\chi_1^C, \chi_2^C) = H^C(\chi_1^C, \chi_2^C) - H^C(\chi_2^C, \chi_1^C). \tag{4.10}$$

In the view of (4.9), we infer

$$\begin{aligned}(\bar{\nabla}^C_{\chi_1^C}g^C)(\chi_2^C, \chi_3^C) &= \bar{\nabla}^C_{\chi_1^C}g^C(\chi_2^C, \chi_3^C) - g^C(\bar{\nabla}^C_{\chi_1^C}\chi_2^C, \chi_3^C) - g^C(\chi_2^C, \bar{\nabla}^C_{\chi_1^C}\chi_3^C)\\ &= -g^C(H^C(\chi_1^C, \chi_2^C), \chi_3^C) - g^C(\chi_2^C, H^C(\chi_1^C, \chi_3^C)).\end{aligned} \tag{4.11}$$

Making use of (4.6) and (4.7), we infer

$$\begin{aligned}g^C(H^C(\chi_1^C, \chi_2^C), \chi_3^C) + g^C(\chi_2^C, H^C(\chi_1^C, \chi_3^C)) &= \beta\{\eta^C(\chi_2^C)g^C((\phi\chi_1)^V, \chi_3^C)\\ &+ \eta^V(\chi_2^C)g^C((\phi\chi_1)^C, \chi_3^C) + \eta^C(\chi_3^C)g^C((\phi\chi_2)^V, \chi_1^C)\\ &+ \eta^V(\chi_3^C)g^C((\phi\chi_2)^C, \chi_1^C)\}.\end{aligned} \tag{4.12}$$

Further, from (4.10) and (4.12), we infer

$$\begin{aligned}g^C(T^C(\chi_1^C, \chi_2^C), \chi_3^C) &+ g^C(T^C(\chi_3^C, \chi_1^C), \chi_2^C) + g^C(T^C(\chi_3^C, \chi_2^C), \chi_1^C)\\ &= 2g^C(H^C(\chi_1^C, \chi_2^C), \chi_3^C) + 2\beta\{\eta^C(\chi_1^C)g^C(\chi_2^V, \phi\chi_3^C)\\ &+ \eta^V(\chi_1^C)g^C(\chi_2^C, \phi\chi_3^C) + \eta^C(\chi_2^C)g^C(\chi_1^V, \phi\chi_3^C)\\ &+ \eta^V(\chi_2^C)g^C(\chi_1^C, \phi\chi_3^C)\}.\end{aligned} \tag{4.13}$$

The above equation can be expressed as

$$g^C(H^C(\chi_1^C, \chi_2^C), \chi_3^C) = \frac{1}{2}\{g^C(T^C(\chi_1^C, \chi_2^C), \chi_3^C) + g^C(T^C(\chi_3^C, \chi_1^C), \chi_2^C)$$
$$+ g^C(T^C(\chi_3^C, \chi_2^C), \chi_1^C)\}$$
$$- \beta\{\eta^C(\chi_1^C)g^C(\chi_2^V, \phi\chi_3^C) + \eta^V(\chi_1^C)g^C(\chi_2^C, \phi\chi_3^C)$$
$$+ \eta^C(\chi_2^C)g^C(\chi_1^V, \phi\chi_3^C)$$
$$+ \eta^V(\chi_2^C)g^C(\chi_1^C, \phi\chi_3^C)\}. \quad (4.14)$$

Let T'^C be the complete lift of a tensor field T' of type (1, 2) given by

$$g^C(T'^C(\chi_1^C, \chi_2^C), \chi_3^C) = g^C(T^C(\chi_3^C, \chi_1^C), \chi_2^C). \quad (4.15)$$

Adding (3.7), (4.4) and (4.15), we infer

$$T'^C(\chi_1^C, \chi_2^C) = -\eta^C(\chi_1^C)(\phi\chi_2)^V - \eta^V(\chi_1^C)(\phi\chi_2)^C$$
$$- g^C((\phi\chi_1)^C, \chi_2^C)\xi^V - g^C((\phi\chi_1)^V, \chi_2^C)\xi^C. \quad (4.16)$$

In view of (4.15) and (4.16), from (4.14) we infer

$$g^C(H^C(\chi_1^C, \chi_2^C), \chi_3^C) = \frac{1}{2}\{g^C(T^C(\chi_1^C, \chi_2^C), \chi_3^C) + g^C(T'^C(\chi_1^C, \chi_2^C), \chi_3^C)$$
$$+ g^C(T'^C(\chi_2^C, \chi_1^C), \chi_3^C)\} - 2\beta\{\eta^C(\chi_1^C)g^C(\chi_2^V, (\phi\chi_3)^C)$$
$$+ \eta^V(\chi_1^C)g^C(\chi_2^C, (\phi\chi_3)^C)$$
$$+ \eta^C(\chi_2^C)g^C(\chi_1^V, \phi\chi_3^C) + \eta^V(\chi_2^C)g^C(\chi_1^C, \phi\chi_3^C)\}$$
$$= (\beta - 1)\{\eta^C(\chi_1^C)g^C((\phi\chi_2)^V, \chi_3^C)$$
$$+ +\eta^V(\chi_1^C)g^C((\phi\chi_2)^C, \chi_3^C)\} + \beta\{\eta^C(\chi_2^C)g^C((\phi\chi_1)^V, \chi_3^C)$$
$$+ \eta^V(\chi_2^C 1)g^C((\phi\chi_1)^C, \chi_3^C)\}. \quad (4.17)$$

Now, contracting χ_3 in (4.17) and using (3.7), we infer

$$H^C(\chi_1^C, \chi_2^C) = (\beta - 1)\{\eta^C(\chi_1^C)(\phi\chi_2)^V + \eta^V(\chi_1^C)(\phi\chi_2)^C\}$$
$$+ \beta\{\eta^C(\chi_2^C)(\phi\chi_1)^V + \eta^V(\chi_2^C)(\phi\chi_1)^C\}. \quad (4.18)$$

Combining (4.9) and (4.18), it follows that

$$\bar{\nabla}^C_{\chi_1^C}\chi_2^C = \nabla^C_{\chi_1^C}\chi_2^C + (\beta - 1)\{\eta^C(\chi_1^C)(\phi\chi_2)^V + \eta^V(\chi_1^C)(\phi\chi_2)^C\}$$
$$+ \beta\{\eta^C(\chi_2^C)(\phi\chi_1)^V + \eta^V(\chi_2^C)(\phi\chi_1)^C\}. \quad (4.19)$$

Therefore, we have the following theorem:

Theorem 4.1 *Let $\bar{\nabla}^C$ be the complete lift of ∇^C on a 3-dimensional non-cosymplectic quasi-Sasakian manifold. Then a unique linear connection $\bar{\nabla}^C$, satisfies (4.4), (4.6) and (4.7).*

If $\beta = 0$ in (4.3), then we infer

$$\bar{\nabla}^C_{\chi_1^C}\chi_2^C = \nabla^C_{\chi_1^C}\chi_2^C + \eta^C(\chi_1^C)(\phi\chi_2)^V + \eta^V(\chi_1^C)(\phi\chi_2)^C. \tag{4.20}$$

Hence, $\bar{\nabla}^C$ becomes a *QSM* connection on *TM*.

5 The Complete Lift of Curvature Tensor and Ricci Tensor of 3-Dimensional Quasi-Sasakian Manifolds Admitting *QSNM*ξ-connection to *TM*

The curvature tensor $\bar{R}$ of M with respect to the *QSNM*ξ-connection $\bar{\nabla}$ is given by [6]

$$\bar{R}(\chi_1, \chi_2)\chi_3 = \bar{\nabla}_{\chi_1}\bar{\nabla}_{\chi_2}\chi_3 - \bar{\nabla}_{\chi_2}\bar{\nabla}_{\chi_1}\chi_3 - \bar{\nabla}_{[\chi_1,\chi_2]}\chi_3, \tag{5.1}$$

where $\forall\chi_1, \chi_2, \chi_3 \in \Im_0^1(M)$.

Using (3.2)–(3.6) and (4.3) in (5.1), we infer

$$\begin{aligned}
\bar{R}^C(\chi_1^C, \chi_2^C)\chi_3^C &= R^C(\chi_1^C, \chi_2^C)\chi_3^C - (\beta - 1)\{\eta^C(\chi_1^C)((\nabla_{\chi_2}\phi)(\chi_3))^V \\
&+ \eta^V(\chi_1^C)((\nabla_{\chi_2}\phi)(\chi_3))^C - \eta^C(\chi_2^C)((\nabla_{\chi_1}\phi)(\chi_3)^V \\
&- \eta^V(\chi_2^C)((\nabla_{\chi_1}\phi)(\chi_3)^C\} - \beta\{((\nabla_{\chi_2}\eta)(\chi_3))^V(\phi\chi_1)^C \\
&+ ((\nabla_{\chi_2}\eta)(\chi_3))^C(\phi\chi_1)^V - ((\nabla_{\chi_1}\eta)(\chi_3))^V(\phi\chi_2)^C \\
&- ((\nabla_{\chi_1}\eta)(\chi_3))^C(\phi\chi_2)^V\} - (\beta - 1)\{((\nabla_{\chi_2}\eta)(\chi_1))^V(\phi\chi_3)^C \\
&+ ((\nabla_{\chi_2}\eta)(\chi_1))^C(\phi\chi_3)^V - ((\nabla_{\chi_1}\eta)(\chi_2))^V(\phi\chi_3)^C \\
&- ((\nabla_{\chi_1}\eta)(\chi_2))^C(\phi\chi_3)^V\} - \beta\{\eta^C(\chi_3^C)((\nabla_{\chi_2}\phi)(\chi_1))^V \\
&+ \eta^V(\chi_3^C)((\nabla_{\chi_2}\phi)(\chi_1))^C - \eta^C(\chi_3^C)((\nabla_{\chi_1}\phi)(\chi_2)^V \\
&- \eta^V(\chi_3^C)((\nabla_{\chi_1}\phi)(\chi_2)^C\} - \beta(\beta - 1)\{\eta^C(\chi_1^C)\eta^C(\chi_3^C)\chi_2^V \\
&+ \eta^C(\chi_1^C)\eta^V(\chi_3^C)\chi_2^C + \eta^V(\chi_1^C)\eta^C(\chi_3^C)\chi_2^C\} \\
&+ \beta(\beta - 1)\{\eta^C(\chi_2^C)\eta^C(\chi_3^C)\chi_1^V \\
&+ \eta^C(\chi_2^C)\eta^V(\chi_3^C)\chi_1^C + \eta^V(\chi_2^C)\eta^C(\chi_3^C)\chi_1^C\} \\
&+ (\chi_1^C(\beta - 1))\eta^C(\chi_2^C)(\phi\chi_3)^V + (\chi_1^C(\beta - 1))\eta^V(\chi_2^C)(\phi\chi_3)^C \\
&+ (\chi_1^V(\beta - 1))\eta^C(\chi_2^C)(\phi\chi_3)^C + (\chi_1^C\beta)\eta^C(\chi_3^C)(\phi\chi_2)^V \\
&+ (\chi_1^C\beta)\eta^C(\chi_3^V)(\phi\chi_2)^C + (\chi_1^V\beta)\eta^C(\chi_3^C)(\phi\chi_2)^C \\
&- (\chi_2^C(\beta - 1))\eta^C(\chi_1^C)(\phi\chi_3)^V - (\chi_2^C(\beta - 1))\eta^V(\chi_1^C)(\phi\chi_3)^C \\
&- (\chi_2^V(\beta - 1))\eta^C(\chi_1^C)(\phi\chi_3)^C - (\chi_2^C\beta)\eta^C(\chi_3^C)(\phi\chi_1)^V \\
&- (\chi_2^C\beta)\eta^V(\chi_3^C)(\phi\chi_1)^C - (\chi_2^V\beta)\eta^C(\chi_3^C)(\phi\chi_1)^C
\end{aligned} \tag{5.2}$$

Using (3.2)–(3.6), (3.9) and (3.10) in (5.2), we infer

$$\begin{aligned}
\bar{R}^C(\chi_1^C,\chi_2^C)\chi_3^C &= R^C(\chi_1^C,\chi_2^C)\chi_3^C + \beta^2\{g^C((\phi\chi_2)^C,\chi_3^C)(\phi\chi_1)^V)\\
&+ g^C((\phi\chi_2)^V,\chi_3^C)(\phi\chi_1)^C)\} - \beta^2\{g^C((\phi\chi_1)^C,\chi_3^C)(\phi\chi_2)^V)\\
&+ g^C((\phi\chi_1)^V,\chi_3^C)(\phi\chi_2)^C)\} + 2\beta(\beta-1)\{g^C((\phi\chi_2)^C,\chi_1^C)(\phi\chi_3)^V)\\
&+ g^C((\phi\chi_2)^V,\chi_1^C)(\phi\chi_3)^C)\} + \beta(\beta-1)\{\eta^C(\chi_2^C)g^C(\chi_1^C,\chi_3^C)\xi^V\\
&+ \eta^C(\chi_2^C)g^C(\chi_1^V,\chi_3^C)\xi^C\\
&+ \eta^V(\chi_2^C)g^C(\chi_1^C,\chi_3^C)\xi^C\} - \beta(\beta-1)\{\eta^C(\chi_1^C)g^C(\chi_2^C,\chi_3^C)\xi^V\\
&+ \eta^C(\chi_1^C)g^C(\chi_2^V,\chi_3^C)\xi^C + \eta^V(\chi_1^C)g^C(\chi_2^C,\chi_3^C)\xi^C\}\\
&- \beta^2\{\eta^C(\chi_2^C)\eta^C(\chi_3^C)\chi_1^V + \eta^C(\chi_2^C)\eta^v(\chi_3^C)\chi_1^C + \eta^(\chi_2^C)\eta^C(\chi_3^C)\chi_1^C\}\\
&+ \beta^2\{\eta^C(\chi_1^C)\eta^C(\chi_3^C)\chi_2^V + \eta^C(\chi_1^C)\eta^v(\chi_3^C)\chi_2^C + \eta^(\chi_1^C)\eta^C(\chi_3^C)\chi_2^C\}\\
&+ (\chi_1^C(\beta-1))\eta^C(\chi_2^C)(\phi\chi_3)^V + (\chi_1^C(\beta-1))\eta^V(\chi_2^C)(\phi\chi_3)^C\\
&+ (\chi_1^V(\beta-1))\eta^C(\chi_2^C)(\phi\chi_3)^C + (\chi_1^C\beta)\eta^C(\chi_3^C)(\phi\chi_2)^V\\
&+ (\chi_1^C\beta)\eta^C(\chi_3^V)(\phi\chi_2)^C + (\chi_1^V\beta)\eta^C(\chi_3^C)(\phi\chi_2)^C\\
&- (\chi_2^C(\beta-1))\eta^C(\chi_1^C)(\phi\chi_3)^V - (\chi_2^C(\beta-1))\eta^V(\chi_1^C)(\phi\chi_3)^C\\
&- (\chi_2^V(\beta-1))\eta^C(\chi_1^C)(\phi\chi_3)^C - (\chi_2^C\beta)\eta^C(\chi_3^C)(\phi\chi_1)^V\\
&- (\chi_2^C\beta)\eta^V(\chi_3^C)(\phi\chi_1)^C - (\chi_2^V\beta)\eta^C(\chi_3^C)(\phi\chi_1)^C
\end{aligned} \tag{5.3}$$

If β is constant, then from (5.3), we infer

$$\begin{aligned}
\bar{R}^C(\chi_1^C,\chi_2^C)\chi_3^C &= R^C(\chi_1^C,\chi_2^C)\chi_3^C + \beta^2\{g^C((\phi\chi_2)^C,\chi_3^C)(\phi\chi_1)^V)\\
&+ g^C((\phi\chi_2)^V,\chi_3^C)(\phi\chi_1)^C)\} - \beta^2\{g^C((\phi\chi_1)^C,\chi_3^C)(\phi\chi_2)^V)\\
&+ g^C((\phi\chi_1)^V,\chi_3^C)(\phi\chi_2)^C)\} + 2\beta(\beta-1)\{g^C((\phi\chi_2)^C,\chi_1^C)(\phi\chi_3)^V)\\
&+ g^C((\phi\chi_2)^V,\chi_1^C)(\phi\chi_3)^C)\} + \beta(\beta-1)\{\eta^C(\chi_2^C)g^C(\chi_1^C,\chi_3^C)\xi^V\\
&+ \eta^C(\chi_2^C)g^C(\chi_1^V,\chi_3^C)\xi^C + \eta^V(\chi_2^C)g^C(\chi_1^C,\chi_3^C)\xi^C\}\\
&- \beta(\beta-1)\{\eta^C(\chi_1^C)g^C(\chi_2^C,\chi_3^C)\xi^V + \eta^C(\chi_1^C)g^C(\chi_2^V,\chi_3^C)\xi^C\\
&+ \eta^V(\chi_1^C)g^C(\chi_2^C,\chi_3^C)\xi^C\} - \beta^2\{\eta^C(\chi_2^C)\eta^C(\chi_2^C)\chi_1^V\\
&+ \eta^C(\chi_2^C)\eta^v(\chi_2^C)\chi_1^C + \eta^(\chi_2^C)\eta^C(\chi_2^C)\chi_1^C\}\\
&+ \beta^2\{\eta^C(\chi_1^C)\eta^C(\chi_3^C)\chi_2^V + \eta^C(\chi_1^C)\eta^v(\chi_3^C)\chi_2^C\\
&+ \eta(\chi_1^C)\eta^C(\chi_3^C)\chi_2^C\}.
\end{aligned} \tag{5.4}$$

It is easily proved that, from (5.2)

$$\bar{R}^C(\chi_1^C,\chi_2^C)\chi_3^C = -\bar{R}^C(\chi_2^C,\chi_1^C)\chi_3^C.$$

Also, from (5.2), we deduce that

$$\bar{R}^C(\chi_1^C,\chi_2^C)\chi_3^C + \bar{R}^C(\chi_2^C,\chi_3^C)\chi_1^C + \bar{R}^C(\chi_3^C,\chi_1^C)\chi_2^C = 0.$$

Let $\{e_i^C, i = 1, 2, 3\}$ be a local orthonormal basis of TM and $\{e_i, i = 1, 2, 3\}$ of M. Then we set $\chi_1 = e_i$ in (5.2) and applying summation over i and in the view of (3.2)–(3.6), we infer

$$\begin{aligned}\bar{S}^C(\chi_2^C, \chi_3^C) &= S^C(\chi_2^C, \chi_3^C) + \beta(2\beta - 1)g^C(\chi_2^C, \chi_3^C) \\ &- \beta(4\beta - 1)\{\eta^C(\chi_2^C)\eta^V(\chi_3^C) + \eta^V(\chi_2^C)\eta^C(\chi_3^C)\} \\ &+ (\phi\chi_3)^C(\beta)\eta^V(\chi_2^C) + (\phi\chi_3)^V(\beta)\eta^C(\chi_2^C) \\ &+ (\phi\chi_2)^C(\beta)\eta^V(\chi_3^C) + (\phi\chi_2)^V(\beta)\eta^C(\chi_3^C),\end{aligned} \tag{5.5}$$

where $\bar{S}^C$ and S^C denote the complete lifts of the Ricci tensors $\bar{S}$ and S of M with respect to $\bar{\nabla}^C$ and ∇^C on TM, respectively. If β is constant, then from (5.5), we infer

$$\begin{aligned}\bar{S}^C(\chi_2^C, \chi_3^C) &= S^C(\chi_2^C, \chi_3^C) + \beta(2\beta - 1)g^C(\chi_2^C, \chi_3^C) \\ &- \beta(4\beta - 1)\{\eta^C(\chi_2^C)\eta^V(\chi_2^C) + \eta^V(\chi_2^C)\eta^C(\chi_2^C)\},\end{aligned} \tag{5.6}$$

Again contracting χ_2 and χ_3 in (5.5), we infer

$$\bar{r}^C = r^C - 2\beta(1 - \beta). \tag{5.7}$$

Therefore, we have the following theorem:

Theorem 5.1 *In a 3-dimensional quasi-Sasakian manifold M admitting the QSNM ξ-connection $\bar{\nabla}^C$:*

(i) The curvature tensor $\bar{R}^C$ is given by (5.2).
(ii) The Ricci tensor $\bar{S}^C$ is given by (5.5).
(iii) $\bar{R}^C(\chi_1^C, \chi_2^C)\chi_3^C = -\bar{R}^C(\chi_2^C, \chi_1^C)\chi_3^C$,
(iv) $\bar{R}^C(\chi_1^C, \chi_2^C)\chi_3^C + \bar{R}^C(\chi_2^C, \chi_3^C)\chi_1^C + \bar{R}^C(\chi_3^C, \chi_1^C)\chi_2^C = 0.$,
(v) $\bar{r}^C = r^C - 2\beta(1 - \beta)$.

Moreover, if β is constant, we infer

Corollary 5.1 *In a 3-dimensional non-cosymplectic quasi-Sasakian manifold M with constant structure function admitting the QSNM ξ-connection $\bar{\nabla}^C$:*

(i) The curvature tensor $\bar{R}^C$ is given by (5.4).
(ii) The Ricci tensor $\bar{S}^C$ is given by (5.6).
(iii) $\bar{R}^C(\chi_1^C, \chi_2^C)\chi_3^C = -\bar{R}^C(\chi_2^C, \chi_1^C)\chi_3^C$, *(iv)* $\bar{R}^C(\chi_1^C, \chi_2^C)\chi_3^C + \bar{R}^C(\chi_2^C, \chi_3^C)\chi_1^C + \bar{R}^C(\chi_3^C, \chi_1^C)\chi_2^C = 0.$,
(v) $\bar{r}^C = r^C - 2\beta(1 - \beta)$.

6 Some Calculations on the Complete Lift of Projective Curvature Tensor on 3-Dimensional Non-cosymplectic Quasi-Sasakian Manifolds with Respect to $QSNM\xi$-connection $\bar{\nabla}$ to TM

Barman and De [6] introduced the notion of the projective curvature tensor $\bar{P}$ on 3-dimension non-cosymplectic quasi-Sasakian manifold with respect to a special type of $QSNM\,\xi$-connection $\bar{\nabla}$, by

$$\bar{P}(\chi_1, \chi_2)\chi_3 = \bar{R}(\chi_1, \chi_2)\chi_3 - \frac{1}{2n}\{\bar{S}(\chi_2, \chi_3)\chi_1 - \bar{S}(\chi_1, \chi_3)\chi_2\}. \tag{6.1}$$

where $\bar{S}$ is the Ricci tensor.

Definition 6.1 ([6]) "A 3-dimensional quasi-Sasakian manifold M with respect to the $QSNM\,\xi$-connection $\bar{\nabla}$ is said to be ξ-projectively flat if the condition $\bar{P}(\chi_1, \chi_2)\xi = 0$ holds".

Applying the complete lift on (6.1), we infer

$$\begin{aligned}\bar{P}^C(\chi_1^C, \chi_2^C)\chi_3^C = {}& \bar{R}^C(\chi_1^C, \chi_2^C)\chi_3^C - \frac{1}{2}\{\bar{S}^C(\chi_2^C, \chi_3^C)\chi_1^V + \bar{S}^C(\chi_2^V, \chi_3^C)\chi_1^C \\ & - \bar{S}^C(\chi_1^C, \chi_3^C)\chi_2^V - \bar{S}^C(\chi_1^V, \chi_3^C)\chi_2^C\}.\end{aligned} \tag{6.2}$$

So using (5.3), (5.5), (3.2)–(3.6) in (6.2), we infer

$$\begin{aligned}\bar{P}^C(\chi_1^C, \chi_2^C)\chi_3^C = {}& P^C(\chi_1^C, \chi_2^C)\chi_3^C - \beta(\beta-1)\beta^2\{g^C(\phi\chi_2)^C, \chi_3^C)(\phi\chi_1)^V) \\ & + g^C(\phi\chi_2)^V, \chi_3^C)(\phi\chi_1)^C)\} - \beta^2\{g^C(\phi\chi_1)^C, \chi_3^C)(\phi\chi_2)^V) \\ & + g^C(\phi\chi_1)^V, \chi_3^C)(\phi\chi_2)^C)\} + 2\beta(\beta-1)\{g^C(\phi\chi_2)^C, \chi_1^C)(\phi\chi_3)^V) \\ & + g^C(\phi\chi_2)^V, \chi_1^C)(\phi\chi_3)^C)\} + \beta(\beta-1)\{\eta^C(\chi_2^C)g^C(\chi_1^C, \chi_3^C)\xi^V \\ & + \eta^C(\chi_2^C)g^C(\chi_1^V, \chi_3^C)\xi^C \\ & + \eta^V(\chi_2^C)g^C(\chi_1^C, \chi_3^C)\xi^C\} - \beta(\beta-1)\{\eta^C(\chi_1^C)g^C(\chi_2^C, \chi_3^C)\xi^V \\ & + \eta^C(\chi_1^C)g^C(\chi_2^V, \chi_3^C)\xi^C + \eta^V(\chi_1^C)g^C(\chi_2^C, \chi_3^C)\xi^C\} \\ & - \beta^2\{\eta^C(\chi_2^C)\eta^C(\chi_3^C)\chi_1^V + \eta^C(\chi_2^C)\eta^v(\chi_3^C)\chi_1^C + \eta^{(}\chi_2^C)\eta^C(\chi_3^C)\chi_1^C\} \\ & + \beta^2\{\eta^C(\chi_1^C)\eta^C(\chi_3^C)\chi_2^V + \eta^C(\chi_1^C)\eta^v(\chi_3^C)\chi_2^C + \eta^{(}\chi_1^C)\eta^C(\chi_3^C)\chi_2^C\} \\ & - \frac{1}{2}[\beta(2\beta-1)\{g^C(\chi_2^C, \chi_3^C)\chi_1^V + g^C(\chi_2^V, \chi_3^C)\chi_1^C\} \\ & - \beta(2\beta-1)\{g^C(\chi_1^C, \chi_3^C)\chi_2^V + g^C(\chi_1^V, \chi_3^C)\chi_2^C\} \\ & + \beta(4\beta-1)\{\eta^C(\chi_1^C)\eta^C(\chi_3^C)\chi_2^V + \eta^C(\chi_1^C)\eta^V(\chi_3^C)\chi_2^C \\ & + \eta^V(\chi_1^C)\eta^C(\chi_3^C)\chi_2^C\} - \beta(4\beta-1)\{\eta^C(\chi_2^C)\eta^C(\chi_3^C)\chi_1^V \\ & + \eta^C(\chi_2^C)\eta^V(\chi_3^C)\chi_1^C + \eta^V(\chi_2^C)\eta^C(\chi_3^C)\chi_1^C\}],\end{aligned} \tag{6.3}$$

where $P^C(\chi_1^C, \chi_2^C)\chi_3^C = R^C(\chi_1^C, \chi_2^C)\chi_3^C - \frac{1}{2n}\{S^C(\chi_2^C, \chi_3^C)\chi_1^C - S^C(\chi_1^C, \chi_3^C)\chi_2^C\}$ is the projective curvature tensor with respect to ∇^C on TM.

Putting $\chi_3 = \xi$ in (6.3) and using (3.2)–(3.6), we infer

$$\bar{P}^C(\chi_1^C, \chi_2^C)\xi^C = P^C(\chi_1^C, \chi_2^C)\xi^C.$$

Therefore, we have the following theorem:

Theorem 6.1 *Let TM be the tangent bundle of 3-dimensional non-cosymplectic quasi-Sasakian manifold M with constant structure function admitting the QSNM ξ^C-connection $\bar{\nabla}^C$. Then the tangent bundle is ξ^C-projectively flat iff ∇^C is so.*

Definition 6.2 “A 3-dimensional quasi-Sasakian manifold M with respect to the *QSNM* ξ-connection $\bar{\nabla}$ is said to be ϕ-projectively flat if it satisfies the condition

$$g(\bar{P}(\phi\chi_1, \phi\chi_2)\phi\chi_3, \phi\chi_4) = 0." \tag{6.4}$$

From (6.2), it follows that

$$\begin{aligned}\tilde{\bar{P}}^C(\chi_1^C, \chi_2^C, \chi_3^C, \chi_4) = \tilde{\bar{R}}^C(\chi_1^C, \chi_2^C, \chi_3^C, \chi_4) - \frac{1}{2}\{\tilde{\bar{S}}^C(\chi_2^C, \chi_3^C)g^C(\chi_1^V, \chi_4^C)\\ + \tilde{\bar{S}}^C(\chi_2^V, \chi_3^C)g^C(\chi_1^C, \chi_4^C) - \tilde{\bar{S}}^C(\chi_1^C, \chi_3^C)g^C(\chi_2^V, \chi_4^C)\\ - \tilde{\bar{S}}^C(\chi_1^V, \chi_3^C)g^C(\chi_2^C, \chi_4^C)\},\end{aligned} \tag{6.5}$$

where $\tilde{\bar{P}}^C(\chi_1^C, \chi_2^C, \chi_3^C, \chi_4) = g^C(\bar{P}^C(\chi_1^C, \chi_2^C)\chi_3^C, \chi_4), \forall \chi_1^C, \chi_2^C, \chi_3^C, \chi_4^C \in \Im_0^1(TM)$.

Setting $\chi_1 = \phi\chi_1, \chi_2 = \phi\chi_2, \chi_3 = \phi\chi_3$ and $\chi_4 = \phi\chi_4$ in (6.5), we infer

$$\begin{aligned}\tilde{\bar{P}}^C(\phi\chi_1^C, \phi\chi_2^C, \phi\chi_3^C, \phi\chi_4) = \tilde{\bar{R}}^C(\phi\chi_1^C, \phi\chi_2^C, \phi\chi_3^C, \phi\chi_4)\\ - \frac{1}{2}\{\tilde{\bar{S}}^C(\phi\chi_2^C, \phi\chi_3^C)g^C(\phi\chi_1^V, \phi\chi_4^C)\\ + \tilde{\bar{S}}^C(\phi\chi_2^V, \phi\chi_3^C)g^C(\phi\chi_1^C, \phi\chi_4^C)\\ - \tilde{\bar{S}}^C(\phi\chi_1^C, \phi\chi_3^C)g^C(\phi\chi_2^V, \phi\chi_4^C)\\ - \tilde{\bar{S}}^C(\phi\chi_1^V, \phi\chi_3^C)g^C(\phi\chi_2^C, \phi\chi_4^C)\},\end{aligned} \tag{6.6}$$

Let $\{e_1^C, e_2^C, \xi^C\}$ be a local orthonormal basis of vector fields in TM. Then $\{\phi e_1^C, \phi e_2^C, \phi\xi^C\}$ is also a local orthonormal basis. Setting $\chi_1 = \chi_4 = e_i$ in (6.6), applying summation over i and in the view of (3.2)–(3.6), we infer

$$\begin{aligned}\widetilde{\bar{P}}^C(\phi e_i^C, \phi\chi_2^C, \phi\chi_3^C, \phi e_i) &= \widetilde{R}^C(\phi e_i^C, \phi\chi_2^C, \phi\chi_3^C, \phi e_i)\\ &\quad - \frac{1}{2}\{\widetilde{\bar{S}}^C(\phi\chi_2^C, \phi\chi_3^C) g^C(\phi e_i^V, \phi e_i^C)\\ &\quad + \widetilde{\bar{S}}^C(\phi\chi_2^V, \phi\chi_3^C) g^C(\phi e_i^C, \phi e_i^C)\\ &\quad - \widetilde{\bar{S}}^C(\phi e_i^C, \phi\chi_3^C) g^C(\phi\chi_2^V, \phi e_i^C)\\ &\quad - \widetilde{\bar{S}}^C(\phi e_i^V, \phi\chi_3^C) g^C(\phi\chi_2^C, \phi e_i^C)\},\end{aligned} \tag{6.7}$$

In the view of (6.4) and (6.7) and using (3.2)–(3.6), we infer

$$\bar{S}^C((\phi\chi_2)^C, (\phi\chi_3)^C) = 0. \tag{6.8}$$

Combining (3.14), (3.14) and (6.8) and using (3.2)–(3.6), we infer

$$\begin{aligned}S^C(\chi_2^C, \chi_3^C) &- \beta^2\{\eta^C(\chi_2^C)\eta^V(\chi_3^C) + \eta^V(\chi_2^C)\eta^C(\chi_3^C)\}\\ &+ \beta(2\beta - 1) g^C((\phi\chi_2)^C, (\phi\chi_3)^C) = 0.\end{aligned} \tag{6.9}$$

By making use of (3.1)–(3.6) in (6.9), we can express

$$\begin{aligned}S^C(\chi_2^C, \chi_3^C) &- \beta(4\beta - 1)\{\eta^C(\chi_2^C)\eta^V(\chi_3^C) + \eta^V(\chi_2^C)\eta^C(\chi_3^C)\}\\ &+ \beta(2\beta - 1) g^C((\phi\chi_2)^C, (\phi\chi_3)^C) = 0.\end{aligned} \tag{6.10}$$

Interchanging χ_2 and $\chi_3$in (6.10) and using (3.10), it follows that

$$\begin{aligned}S^C(\chi_2^C, \chi_3^C) &- \beta(4\beta - 1)\{\eta^C(\chi_2^C)\eta^V(\chi_3^C) + \eta^V(\chi_2^C)\eta^C(\chi_3^C)\}\\ &+ \beta(2\beta - 1) g^C((\phi\chi_2)^C, (\phi\chi_3)^C) = 0.\end{aligned} \tag{6.11}$$

Adding (6.10) and (6.11), we conclude that

$$\begin{aligned}S^C(\chi_2^C, \chi_3^C) &= \beta(4\beta - 1)\{\eta^C(\chi_2^C)\eta^V(\chi_3^C) + \eta^V(\chi_2^C)\eta^C(\chi_3^C)\}\\ &- \beta(2\beta - 1) g^C((\phi\chi_2)^C, (\phi\chi_3)^C) = 0.\end{aligned} \tag{6.12}$$

Therefore, we have the following theorem.

Theorem 6.2 *A ϕ^C-projectively flat 3-dimensional non-cosymplectic quasi-Sasakian manifold M on TM with constant structure function admitting the QSNM ξ^C-connection $\bar{\nabla}^C$ is an η^C-Einstein manifold with respect to ∇^C on TM.*

References

1. M. Altunbas, L. Bilen, A. Gezer, Remarks about the Kaluza-Klein metric on tangent bundle. Int. J. Geo. Met. Mod. Phys. **16**(3), 1950040 (2019)
2. M. Altunbaş, Ç. Şengül, Metallic structures on tangent bundles of Lorentzian para-Sasakian manifolds. J. Mahani Math. Res. 137–149 (2022). https://doi.org/10.22103/jmmr.2022.19411.1247
3. R.C. Akpinar, Weyl connection to tangent bundle of hypersurface. Int. J. Maps Math. **4**(1), 2–13 (2021)
4. C.S. Bagewadi, D.G. Prakasha, A. Venkatesha, A. Study of Ricci quarter-symmetric metric connection on a Riemannian manifold. Indian J. Math. **50**(3), 607–615 (2008)
5. A. Barman, A special type of quarter-symmetric non-metric connection on P-Sasakian manifolds, in *Bulletin of the Transilvania University of Brasov Series III: Mathematical Information Physics*, vol. 11(60), pp. 11–22 (2018)
6. A. Barman, U.C. De, On a type of quarter-symmetric non-metric ξ-connection on 3-dimensional quasi-Sasakian manifolds. Diff. Geom.-Dyn. Syst. **22**, 26–41 (2020)
7. D.E. Blair, *Contact Manifolds in Riemannian Geometry*, Lecture notes in Mathametics, vol. 509 (Springer, Berlin-Heidelberg-New York, 1976)
8. D.E. Blair, *Riemannian Geometry of Contact and Sympletic Manifolds, Progress Mathematics*, vol. 203 (Birkhauser, Boston-Basel-Berlin, 2002)
9. D.E. Blair, The theory of quasi-Sasakian structure. J. Diff. Geom. **1**, 331–345 (1967)
10. L.S. Das, M.N.I. Khan, Almost r-contact structures on the Tangent bundle. Diff. Geom.-Dyn. Syst. **7**, 34–41 (2005)
11. U.C. De, A.K. Mondal, Quarter-symmetric metric connection on 3-dimensional quasi-Sasakian manifolds. SUT J. Mat. **46**, 35–52 (2010)
12. U.C. De, J. Sengupta, Quarter-symmetric metric connection on a Sasakian manifold. Commun. Fac. Sci. Univ. Ank. Ser. **A1**(49), 7–13 (2000)
13. U.C. De, A.K. Mondal, Quarter-symmetric metric connection on 3-dimensional quasi-Sasakian manifolds. SUT J. Math. **46**(1), 35–52 (2010)
14. A. Friedmann, J.A. Schouten, Uber die Geometrie der halbsymmetrischen Ubertragung. Math. Zeitschr **21**, 211–223 (1924)
15. S. Golab, On semi-symmetric and quarter-symmetric linear connections, Tensor N.S. **29**, 293–301 (1975)
16. Y. Han, T. Yun, P. Zhao, Some invariants of quarter-symmetric metric connections under the projective transformation. Filomat **27**(4), 679–691 (2013)
17. H.A. Hayden, Subspaces of a space with torsion. Proc. London Math. Soc. **34**, 27–50 (1932)
18. J.B. Jun, U.K. Kim, On 3-dimensional almost contact metric manifolds. Kyungpook Math. J. **34**, 213–301 (1994)
19. A. Kazan, H.B. Karadag, Locally decomposable golden tangent bundles with CheegerGromoll metric. Miskolc Math. Not. **17**(1), 399–411 (2016)
20. M.N.I. Khan, U.C. De, Liftings of metallic structures to tangent bundles of order r. AIMS Math. **7**(5), 7888–7897 (2022)
21. M.N.I. Khan, Novel theorems for metallic structures on the frame bundle of the second order. Filomat **36**(13), 4471–4482 (2022)
22. M.N.I. Khan, U.C. De, L.S. Velimirovic, Lifts of a quarter-symmetric metric connection from a Sasakian manifold to its tangent bundle. Mathematics **11**, 53 (2023)
23. M.N.I. Khan, Lifts of hypersurfaces with quarter-symmetric semi-metric connection to tangent bundles. Afrika Matematika **27**, 475–482 (2014)
24. M.N.I. Khan, Lifts of semi-symmetric non-metric connection on a Kähler manifold. Afrika Matematika **27**(3), 345–352 (2016)
25. M.N.I. Khan, Tangent bundle endowed with quarter-symmetric non-metric connection on an almost Hermitian manifold. Facta Universitis (NIS) Ser. Math. Inform. **35**(1), 167–178 (2020)
26. M.N.I. Khan, Complete and horizontal lifts of Metallic structures. Int. J. Math. Comput. Sci. **15**(4), 983–992 (2020)

27. M.N.I. Khan, Quarter-symmetric semi-metric connection on a Sasakian manifold, Tensor, N.S. **68**(2), 154–157 (2007)
28. M.N.I. Khan, Proposed theorems for lifts of the extended almost complex structures on the complex manifold. Asian-European J. Math. **15**(11), 2250200 (13 pages) (2022)
29. Z.F. Ozen, S.A. Uysal, S.A. Demirbag, On sectional curvature of a Riemannian manifold with semi-symmetric metric connection. Ann. Polon. Math. **101**, 131–138 (2011)
30. Z. Olszak, Normal almost contact metric manifolds of dimension three. Ann. Polon. Math. **47**, 41–50 (1986)
31. Z. Olszak, On three dimensional conformally flat quasi-Sasakian manifolds. Periodica Math. Hunger. **33**, 105–113 (1996)
32. P. Pandey, B.B. Chaturvedi, On a Kahler manifold equipped with lift of a quarter-symmetric non-metric connection. Facta Universitis (NIS) Ser. Math. Inform. **33**(4), 539–546 (2018)
33. E. Peyghan, F. Firuzi, U.C. De, Golden Riemannian structures on the tangent bundle with g-natural metrics. Filomat **33**(8), 2543–2554 (2019)
34. S. Sular, C. Özgür, U.C. De, Quarter-symmetric metric connection in a Kenmotsu manifold. SUT J. Math. **44**(2), 297–306 (2008)
35. M. Tani, Prolongations of hypersurfaces of tangent bundles. Kodai Math. Semp. Rep. **21**, 85–96 (1969)
36. K. Yano, T. Imai, Quarter-symmetric metric connections and their curvature tensors, Tensor, N.S. **38**, 13–18 (1982)
37. K. Yano, S. Ishihara, *Tangent and Cotangent Bundles* (Marcel Dekker Inc, New York, 1973)
38. K. Yano, S. Kayabashi, Prolongations of tensor fields and connections to tangent bundles, I, General Theory. J. Math. Soc. Jpn. **18**, 194–210 (1966)
39. G. Zhen, On conformal symmetric K-contact manifolds. Chinese Quart. J. Math. **7**, 5–10 (1992)
40. P.B. Zhao, H.Z. Song, X.P. Yang, Some invariant properties of the semisymmetric metric recurrent connection and curvature tensor expressions. Chinese Quart. J. Math. **19**, 355–361 (2004)

The Darboux Mate and the Higher Order Curvatures of Spherical Legendre Curves

Mircea Crasmareanu

2010 Mathematics Subject Classification 53A04 · 53A45 · 53A55

1 Introduction

The geometry of spherical curves is an important part of the classical theory of space curves. Also, a fresh research is being published, some of it with graphical applications, such as the class of spherical p-elastica curves. If the dimension of the involved sphere permits additional structures (Hermitian or almost contact) then the features of these curves become more interesting.

The present work concerns mainly with spherical Legendre curves related to the Euclidean unit sphere S^2 and we follow mainly the approach of Takahashi Masatomo from [10]. More precisely, we are interested in new curvatures which are naturally associated. There are two classes of such new curvatures: the first class (containing two elements) is provided by the Darboux mate of a given spherical Legendre curve; (2) the second class (containing three elements) is provided by the higher order Frenet equations. All these new curvatures are the contents of the first section below where for obtaining the higher order curvatures we follow the approach of [4]. We point out that a special attention is given to examples, some of them related to the usual theory of regular space curves.

The second section concerns an introduction to the Lorentz–Minkowski case provided by $\mathbb{R}^3$ with the metric of signature $(-, +, +)$. We remark that the computations become more difficult and hence we restrict to only two examples for which the geodesic curvature is computed.

---

http://www.math.uaic.ro/~mcrasm.

M. Crasmareanu (✉)
Faculty of Mathematics, University Al. I. Cuza, 700506 Iasi, Romania
e-mail: mcrasm@uaic.ro

B.-Y. Chen et al. (eds.), *Geometry of Submanifolds and Applications*, Infosys Science Foundation Series, https://doi.org/10.1007/978-981-99-9750-3_12

2 Euclidean Spherical Legendre Curves

The setting of this section is provided by the space $\mathbb{R}^3$ which is an Euclidean vector space with respect to the canonical inner product:

$$\langle u, v\rangle = u^1v^1 + u^2v^2 + u^3v^3, u = (u^1, u^2, u^3), v = (v^1, v^2, v^3) \in \mathbb{R}^3, 0 \leq \|u\|^2 = \langle u, u\rangle. \tag{2.1}$$

Let S^2 be the unit sphere of $\mathbb{E}^3 := (\mathbb{R}^3, \langle\cdot,\cdot\rangle)$ and its unit spherical bundle:

$$T_1S^2 := \{(u, v) \in S^2 \times S^2; \langle u, v\rangle = 0\}$$

which is a compact three-dimensional contact metric manifold; see, for example, [3]. There is a natural action of $O(3)$ on T_1S^2:

$$(A, (u, v)) \in O(3) \times T_1S^2 \to (Au, Av) \in T_1S^2, \quad \langle Au, Av\rangle = \langle u, v\rangle = 0. \tag{2.2}$$

The general notion of Legendre curves associated to a contact form is well-known (cf. [5]), but we will work directly in our framework using the approach of [10] (see also [8]).

Definition 2.1 The smooth map $LC := (\gamma, \nu) : I \subseteq \mathbb{R} \to T_1S^2$, $t \in I \to (\gamma(t), \nu(t))$ is *a spherical Legendre curve* if $\langle\gamma'(t), \nu(t)\rangle = 0$ for all t in the open interval I. The map γ is called *the frontal* and ν is *the dual of* γ.

Since $\mathbb{R}^3$ is endowed also with the cross product $\times$, we define $\mu = \gamma \times \nu$ and hence the triple $\mathcal{F} := \{\gamma, \nu, \mu\}^t$ is a positively oriented *moving frame* along the frontal γ; here t means the transposition, so $\mathcal{F}$ is a column matrix. Its moving equation is provided by the Proposition 2.2. of [10]:

$$\frac{d}{dt}\mathcal{F}(t) = \begin{pmatrix} 0 & 0 & k_1(t) \\ 0 & 0 & k_2(t) \\ -k_1(t) & -k_2(t) & 0 \end{pmatrix} \mathcal{F}(t), \quad \begin{pmatrix} 0 & 0 & k_1(t) \\ 0 & 0 & k_2(t) \\ -k_1(t) & -k_2(t) & 0 \end{pmatrix} \in o(3). \tag{2.3}$$

The pair of smooth functions (k_1, k_2) is called *the curvature* of the spherical Legendre curve $LC = (\gamma, \nu)$. Sometimes, it is more useful to denote a given LC with all its elements as $LC = (\gamma, \nu; \mu)$.

Remark 2.2 (1) For the given LC, there are other three spherical Legendre curves ([10]): (i) $LC^{-1} := (-\gamma, \nu)$ with the curvature $(k_1, -k_2)$; (ii) $LC_{-1} := (\gamma, -\nu)$ with the curvature $(-k_1, k_2)$; (iii) $LC^{-1}_{-1} := (\nu, \gamma)$ with the curvature $(-k_2, -k_1)$.
(2) A matrix $A \in SO(3)$ acts on the set of all curves LC following (2): $LC = (\gamma, \nu) \to LC(A) := (A\gamma, A\nu)$. Since A is special we have ([6, p. 134]) $(A\gamma) \times (A\nu) = A\mu$ and then the curvatures of LC and $LC(A)$ are equal. The reciprocal of this statement is the Uniqueness Theorem of [10].

(3) As in the theory of regular space curves, the matrix Eq. (2.3) can be expressed into a vectorial form:

$$\frac{d}{dt}\mathcal{F}(t) = \Omega(t) \times \mathcal{F}(t) \tag{2.4}$$

with *the Darboux vector field*:

$$\Omega = k_2\gamma - k_1\nu = \mu \times \mu'. \tag{2.5}$$

Its norm $\omega := \sqrt{k_1^2 + k_2^2}$ is strictly positive and then every LC has a *Darboux mate*: $DLC = \left(\frac{1}{\omega}\Omega, \mu; -\frac{1}{\omega}(k_1\gamma + k_2\nu)\right)$. We remark that the frontal $\frac{1}{\omega}\Omega$ of DLC is exactly *the* $+$*-evolute* $Ev(\gamma)$ of the frontal γ as is defined in the relation (4.1) of [10, p. 347] and the fact that DLC is a spherical Legendre curve appears in the proof of Proposition 4.8. of [10, p. 349]. The curvature of the Darboux mate is

$$((Dk)_1, (Dk)_2) = \left(\frac{k_1'k_2 - k_1k_2'}{\omega^2}, \omega\right) \tag{2.6}$$

and if $k_1 \neq 0$, then

$$(Dk)_1 = -\left(\frac{k_2}{k_1}\right)'\left(\frac{k_1}{\omega}\right)^2. \tag{2.7}$$

$\square$

This short note defines new curvature functions for LC. In the following, $n \in \mathbb{N}^*$ is a fixed positive integer.

Definition 2.3 The *n-order Frenet equation* of LC is

$$\frac{d^n}{dt^n}\mu(t) = K_1^n(t)\gamma(t) + K_2^n(t)\nu(t) + K_3^n(t)\mu(t). \tag{2.8}$$

and the triple (K_1^n, K_2^n, K_3^n) is called *the n-curvature of* LC. The *n-Darboux vector field* of LC is

$$\Omega(n) := -K_2^n\gamma + K_1^n\nu. \tag{2.9}$$

We note that the new functions $K_{1;2;3}^n$ are smooth, and from (2.2), it results the first ones are $(K_1^1, K_2^1, K_3^1) = (-k_1, -k_2, 0)$. For the general n, these curvatures are obtained through a recurrence process expressed by (Fig. 1).

Theorem 2.4 *The higher order curvatures are provided by the following recurrence relation:*

$$\begin{pmatrix} K_1^{n+1} \\ K_2^{n+1} \\ K_3^{n+1} \end{pmatrix} = \frac{d}{dt}\begin{pmatrix} K_1^n \\ K_2^n \\ K_3^n \end{pmatrix} - \begin{pmatrix} 0 & 0 & k_1(t) \\ 0 & 0 & k_2(t) \\ -k_1(t) & -k_2(t) & 0 \end{pmatrix}\begin{pmatrix} K_1^n \\ K_2^n \\ K_3^n \end{pmatrix}. \tag{2.10}$$

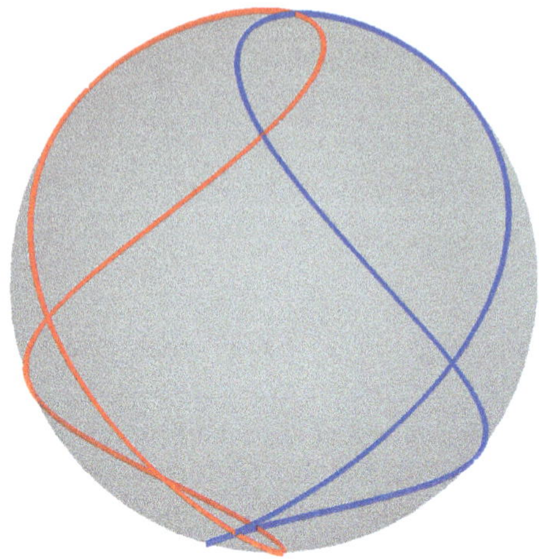

Fig. 1 The first two curves

Hence, the 2*- and* 3*-curvatures are*

$$\begin{cases} (K_1^2, K_2^2, K_3^2) = (-k_1', -k_2', -(k_1^2 + k_2^2)) = (-k_1', -k_2', -\omega^2), \\ (K_1^3, K_2^3, K_3^3) = (-k_1'' + k_1(k_1^2 + k_2^2), -k_2'' + k_2(k_1^2 + k_2^2), -3(k_1 k_1' + k_2 k_2')). \end{cases} \tag{2.11}$$

The derivative of the n-Darboux vector field is

$$\Omega'(n) = -(K_2^n)'\gamma + (K_1^n)'\nu + (k_2 K_1^n - k_1 K_2^n)\mu. \tag{2.12}$$

Example 2.5 Let γ be an usual Frenet curve in $\mathbb{E}^3$ having the curvature function k and the torsion function τ. After [1, p. 2], the curve γ is a Mannheim one if there exist $\lambda \in \mathbb{R}^*$ such that $k = \lambda(k^2 + \tau^2)$. Following this case, we call a spherical Legendre curve LC as being λ-*Mannheim* if

$$k_1 = \lambda(k_1^2 + k_2^2) \tag{2.13}$$

and then the sign of k_1 is that of λ. The curvatures (2.10) become

$$(K_1^2, K_2^2, K_3^2) = \left(-k_1', -k_2', -\frac{k_1}{\lambda}\right), (K_1^3, K_2^3, K_3^3) = \left(-k_1'' + \frac{k_1^2}{\lambda}, -k_2'' + \frac{k_1 k_2}{\lambda}, -3\frac{k_1'}{\lambda}\right). \tag{2.14}$$

Suppose that $\lambda > 0$. Then $K_1^3 \equiv 0$ if and only if $k_1(t) = \frac{6\lambda}{t^2} > 0$ for $t \in (0, \sqrt{6}\lambda]$ since $k_2(t) = \pm\frac{\sqrt{6(t^2 - 6\lambda^2)}}{t^2}$. Let us remark that if LC is λ-Mannheim then LC^{-1} is also λ-Mannheim and LC_{-1} is $(-\lambda)$-Mannheim. □

Example 2.6 Let us consider Example 2.8. of [10]. Starting with the natural numbers (k, m, n) satisfying $m = k + n$, the author defines the LC with

$$\gamma(t) = \frac{1}{\sqrt{1+t^{2n}+t^{2m}}}(1, t^n, t^m), \ \nu(t) = \frac{1}{\sqrt{n^2+m^2t^{2k}+k^2t^{2m}}}(kt^m, -mt^k, n), \ t \in \mathbb{R} \tag{2.15}$$

with the associated curvature:

$$k_1(t) = -\frac{t^{n-1}\sqrt{n^2+m^2t^{2k}+k^2t^{2m}}}{1+t^{2n}+t^{2m}}, \quad k_2(t) = \frac{kmnt^{k-1}\sqrt{1+t^{2n}+t^{2m}}}{n^2+m^2t^{2k}+k^2t^{2m}}. \tag{2.16}$$

All the new curvatures are extremely complicated and then we express only

$$\omega^2 = k_1^2(t) + k_2^2(t) = \frac{t^{2n-2}(n^2+m^2t^{2k}+k^2t^{2m})^3 + k^2m^2n^2t^{2k-2}(1+t^{2n}+t^{2m})^3}{(n^2+m^2t^{2k}+k^2t^{2m})^2(1+t^{2n}+t^{2m})^2}. \tag{2.17}$$

□

Example 2.7 Let us consider Example 2.7. of [10]. A spherical curve $\gamma : I \to S^2 \subset \mathbb{R}^3$ yields the spherical Legendre curve $LC(\gamma) = (\gamma, \mathbf{n}; -\mathbf{t})$ with $\mathbf{n} = \gamma \times \mathbf{t}$ for $\mathbf{t}(t) := \frac{\gamma'(t)}{\|\gamma'(t)\|}$. Then the curvature of $LC(\gamma)$ is related to the geodesic curvature k_g of γ through $k_2 = k_g|k_1|$ for

$$k_g(t) = \frac{\det(\gamma(t), \gamma'(t), \gamma''(t))}{\|\gamma'(t)\|^3} \tag{2.18}$$

while the usual curvature k of γ as space curve is: $k = \sqrt{k_g^2+1}$; if $k_g > 0$ then γ is *convex*. Then we get $k_1(t) = -\|\gamma'(t)\| < 0$ which yields

$$(K_1^2, K_2^2, K_3^2) = \left((\|\gamma'(t)\|)', (\|\gamma'(t)\|k_g)', -\|\gamma'(t)\|^2(1+k_g^2)\right), \quad (Dk)_1 = \frac{k_g'}{1+k_g^2} \tag{2.19}$$

Inspired by the case of regular curves, a spherical Legendre curve will be a λ-*helix* if $k_2 = \lambda k_1$ and then $(Dk)_1 \equiv 0$.

Three remarkable spherical curves are

$$r^1(t) = (-\sin^2 t, \sin t \cos t, \cos t), r^2(t) = (\sin t \cos t, -\cos^2 t, \sin t), r^3(t) = (\cos t, \sin t, 0), \tag{2.20}$$

since these three-dimensional unit vectors are orthogonal two by two: $\langle r_1, r_2\rangle = \langle r_2, r_3\rangle = \langle r_3, r_1\rangle = 0$; the last curve is the equator of S^2. So, these maps can be considered together as the columns of a symmetric matrix Γ from $SO(3)$. With the spinorial formalism of [11, p. 28], there exists two complex-valued curves $\psi_1 = \psi_1(t)$, $\psi_2 = \psi_2(t)$ such that

$$\begin{cases} r^1 + ir^2 = (\psi_1^2 - \psi_2^2, i(\psi_1^2 + \psi_2^2), -2\psi_1\psi_2), \\ r^3 = (\overline{\psi_1}\psi_2 + \psi_1\overline{\psi_2}, i(\psi_1\overline{\psi_2} - \overline{\psi_1}\psi_2), |\psi_1|^2 - |\psi_2|^2). \end{cases} \tag{2.21}$$

A straightforward computation gives the expression of these curves:

$$\psi_1 = constant = \frac{i}{\sqrt{2}}, \quad \psi_2(t) = \frac{i}{\sqrt{2}}e^{it} = \frac{1}{\sqrt{2}}e^{i\left(\frac{\pi}{2}+t\right)}. \tag{2.22}$$

We remark that we have a new spherical curve $t \to (\psi_1, \psi_2(t)) \in S^3$ and then we recall the famous Hopf bundle $H : S^3 \subset \mathbb{C}^2 \to S^2\left(\frac{1}{2}\right) \subset \mathbb{R} \times \mathbb{C}$:

$$H(z, w) = \left(\frac{1}{2}(|z|^2 - |w|^2), z\bar{w}\right). \tag{2.23}$$

It results in the curve:

$$t \to H(\psi_1, \psi_2(t)) = \left(0, \frac{1}{2}e^{-it}\right) \in S^2\left(\frac{1}{2}\right). \tag{2.24}$$

The geodesic and the usual curvatures for the columns of Γ are

$$\begin{cases} k_g^1(t) = \frac{\cos t(2+\sin^2 t)}{(1+\sin^2 t)^{\frac{3}{2}}}, \quad k_g^2(t) = \frac{\sin t(2+\cos^2 t)}{(1+\cos^2 t)^{\frac{3}{2}}}, \quad k_g^3 = 0, \\ k^1(t) = \frac{\sqrt{5+3\sin^2 t}}{(1+\sin^2 t)^{\frac{3}{2}}}, \quad k^2(t) = \frac{\sqrt{5+3\cos^2 t}}{(1+\cos^2 t)^{\frac{3}{2}}}, \quad k^3 = 1. \end{cases} \tag{2.25}$$

The lengths of these curves are

$$L^1 = \int_0^{2\pi} \sqrt{1+\sin^2 t}dt \simeq 7.6404, \quad L^2 = \int_0^{2\pi} \sqrt{1+\cos^2 t}dt \simeq 7.6404, \quad L^3 = 2\pi \simeq 6.28$$

while their total curvatures are

$$(Tk)^1 = \int_0^{2\pi} k^1(t)dt \simeq 9.4174, \quad (Tk)^2 = \int_0^{2\pi} k^2(t)dt \simeq 9.4174, \quad (Tk)^3 = L^3 = 2\pi.$$

A spherical curve with prescribed constant geodesic curvature $k_g = K$ and parametrized by the arc length s is

$$r_K(s) = \frac{1}{\sqrt{1+K^2}}\left(\cos(\sqrt{1+K^2}s), \sin(\sqrt{1+K^2}s), K\right) = \frac{1}{k}\left(\cos(ks), \sin(ks), K\right). \tag{2.26}$$

Following the approach of [7, p. 143] for a spherical curve parametrized by arc length, we define *the elastic curvature*

$$k_e := k_g'' + \frac{1}{2}(k_g^3 + k_g) \tag{2.27}$$

and a curve with a vanishing elastic curve will be called *elastic curve*. Then the elastic curvature of the curve (2.26) will be constant $k_e = \frac{K}{2}(K^2 + 1)$.

We recall also the characterization of a space curve γ to be a spherical one (not necessary on S^2) in terms of classical curvature and torsion:

$$\frac{\tau}{k}+\frac{d}{dt}\left[\frac{1}{\tau}\frac{d}{dt}\left(\frac{1}{k}\right)\right]=0 \tag{2.28}$$

as well as the usual parametrization of the unit sphere S^2:

$$S^2: r(u,v)=(\cos u\cos v, \sin u\cos v, \sin v), \quad u\in[0,2\pi], \quad v\in[-\pi,\pi]. \tag{2.29}$$

Then the above three curves are (Fig. 1)

$$r^1(t)=r\left(\frac{\pi}{2}+t,\frac{\pi}{2}-t\right), \quad r^2(t)=r\left(\frac{3\pi}{2}+t,t\right), \quad r^3(t)=r(t,0). \tag{2.30}$$

Also, the first curve is connected with the Viviani curve $\mathcal{V}$ which is the intersection of S^2 with the cylinder $x^2+y^2=x$. Since its parametrization is

$$\mathcal{V}(u)=(u,-\sqrt{u-u^2},-\sqrt{1-u})\subset S^2, \quad u\in[0,1] \tag{2.31}$$

it results $r^1(t)=-\mathcal{V}(\sin^2 t)$. The geodesic and the usual curvature of the Viviani curve are

$$k_g^{\mathcal{V}}(u)=-\frac{(2+u)\sqrt{1-u}}{(1+u)^{\frac{3}{2}}}<0, \quad k^{\mathcal{V}}(u)=\frac{\sqrt{5+3u}}{(1+u)^{\frac{3}{2}}} \tag{2.32}$$

and then $\mathcal{V}$ is not convex. The total geodesic and the total curvature of the Viviani curve are

$$\int_0^1 k_g^{\mathcal{V}}(u)du=-1, \quad \int_0^1 k^{\mathcal{V}}(u)du\simeq 1.4602.$$

□

Example 2.8 The simplest LC curve is that containing a constant dual curve:

$$\gamma(t)=(\cos t,\sin t,0)=(e^{it},0), \nu(t)=(0,0,1)=\Omega, \mu(t)=(\sin t,-\cos t,0)=-i\gamma \tag{2.33}$$

with i the complex unit of the plane $\mathbb{R}^2=\mathbb{C}$. It has the constant curvature $(-1,0)$ and then $K_2^n\equiv 0$ for all n. The Darboux mate $DLC=(\nu,\mu;\gamma)$ has the curvature $(0,1)$. Following the Darboux procedure, we have $DDLC=D^{(2)}LC=(\nu,\gamma;-\mu)=LC_{-1}^{-1}$ with the same curvature $(0,1)$. Another steps: $D^{(3)}LC=(\nu,-\mu;-\gamma)$ with the curvature $(0,1)$; $D^{(4)}LC=(\nu,-\gamma;\mu)$ with the curvature $(0,-1)$. It follows a Darboux periodicity of order 4: $D^{(5)}=D$.

The first two curves of the previous example can be joined to obtain a curve in $\mathbb{R}^6=\mathbb{C}^3$ which, in turn, can be put in correspondence with a special class of Legendre immersions. Namely, the Corollary 15 of [9, p. 301] expresses an arbitrary nonminimal biharmonic Legendre immersion $f_L: M^2\to S^5\subset\mathbb{C}^3$ as

$$f_L(x,y)=\frac{1}{\sqrt{2}}\left(e^{ix}, ie^{-ix}\sin(\sqrt{2}y), ie^{-ix}\cos(\sqrt{2}y)\right). \tag{2.34}$$

Then we have

$$(r^1,r^2)(t)=\sqrt{2}\begin{pmatrix} 0&0&1&0&0&0\\ 0&0&0&-1&0&0\\ 1&0&0&0&0&0\\ 0&0&0&0&1&0\\ 0&0&0&0&0&-1\\ 0&1&0&0&0&0 \end{pmatrix} f_L\left(t,-\frac{t}{\sqrt{2}}\right)\in S^5 \tag{2.35}$$

and the 6×6 matrix above belongs to $SO(6)$.

Moreover, all three curves of Example 2.7 can be joined through the Veronese map:

$$V:S^2\subset\mathbb{R}^3\to S^4\subset\mathbb{R}^5,\ V(u,v,w):=\left(\sqrt{3}vw,\sqrt{3}wu,\sqrt{3}uv,\frac{\sqrt{3}}{2}(u^2-v^2),w^2-\frac{u^2+v^2}{2}\right). \tag{2.36}$$

Then:

$$\begin{cases} V^1(t):=V\circ r^1(t)=\sqrt{3}\left(\sin t\cos^2 t,-\sin^2 t\cos t,-\sin^3 t\cos t,-\frac{1}{2}\sin^2 t\cos 2t,\frac{2\cos^2 t-\sin^2 t}{2\sqrt{3}}\right),\\ V^2(t):=V\circ r^2(t)=\sqrt{3}\left(-\sin t\cos^2 t,\sin^2 t\cos t,-\sin t\cos^3 t,-\frac{1}{2}\cos^2 t\cos 2t,\frac{2\sin^2 t-\cos^2 t}{2\sqrt{3}}\right),\\ V^3(t):=V\circ r^3(t)=\left(0,0,\frac{\sqrt{3}}{2}\sin 2t,\frac{\sqrt{3}}{2}\cos 2t,-\frac{1}{2}\right), \end{cases} \tag{2.37}$$

and hence: $V^1+V^2+V^3=0\in\mathbb{R}^5$. The angle between two distinct unit vectors V^i, $i=1,2,3$ is constant, namely $\frac{2\pi}{3}$.

Following [2] for a given spherical curve γ, we consider the spherical angular momentum with respect to the vertical axis:

$$\mathcal{K}(t):=-\det(\gamma(t),\mathbf{t}(t),\bar{k})=\frac{x'(t)y(t)-x(t)y'(t)}{\|\gamma'(t)\|}. \tag{2.38}$$

This momentum for the curves (2.20) is

$$\mathcal{K}^1(t)=-\frac{\sin^2 t}{\sqrt{1+\sin^2 t}}\le 0,\quad \mathcal{K}^2(t)=-\frac{\cos^2 t}{\sqrt{1+\cos^2 t}}\le 0,\quad \mathcal{K}^3=-1. \tag{2.39}$$

□

Example 2.9 Return to Example 2.7. Fix a regular curve $r:I\to\mathbb{R}^3$ parametrized by arc-length s and having the curvature function k and the torsion function τ; let $(T=r',N,B)$ its Frenet frame. Then its *tangent indicatrix* $\gamma:=r'$ will be a spherical curve with the geodesic curvature $k^T_g=\frac{\tau}{k}$; then γ is convex if and only if $\tau>0$. Following Example 2.7, we get the spherical Legendre curve $TLC:=(\gamma=r'=T,\mathbf{n}=B;-\mathbf{t}=-N)$ with the curvature:

$$k_1^T = -k < 0, \quad k_2^T = \tau. \tag{2.40}$$

The Darboux vector field (2.5) is exactly the Darboux vector field of r and if this initial curve r is λ-Mannheim (or λ-helix) then TLC is $(-\lambda)$-Mannheim (respectively, $(-\lambda)$-helix). □

3 A Lorentz–Minkowski Variant of Spherical Legendre Curves

The setting of this work is provided by the Lorentz–Minkoswki 3-space $E_1^3 := (\mathbb{R}^3, \langle u, v\rangle_L, \times_L)$ which is a pseudo-Euclidean vector space with

$$\langle u, v\rangle_L = -u^1v^1 + u^2v^2 + u^3v^3, \quad u \times_L v = \begin{vmatrix} -\bar{i} & \bar{j} & \bar{k} \\ u^1 & u^2 & u^3 \\ v^1 & v^2 & v^3 \end{vmatrix}. \tag{3.1}$$

It follows two types of unit spheres:

$$S_1^2 := \{u \in \mathbb{R}^3; \langle u, u\rangle_L = +1\}, \quad H_1^2 := \{u \in \mathbb{R}^3; \langle u, u\rangle_L = -1\} \tag{3.2}$$

and the norm is defined through: $0 \leq \|u\|_L^2 = |\langle u, u\rangle_L|$. Following the model of [10] and looking at Example 3.2 below, we denote $\Delta^+ := \{(u, v) \in S_1^2 \times H_1^2; \langle u, v\rangle_L = 0\}$ and then we introduce.

Definition 3.1 A *positive spherical Lorentz–Legendre curve* is a pair $LLC^+ := (\gamma, \nu) : I \subseteq \mathbb{R} \to \Delta^+$ of smooth positive spherical curves with $\langle \gamma'(t), \nu(t)\rangle_L = 0$ for all t in the open interval I.

Fix now a smooth positive spherical curve $\gamma : I \to S_1^2$ which we call *regular* if $\|\gamma'(t)\|_L > 0$ for all $t \in I$. As in the Euclidean case, its unit tangent vector field and normal vector field are

$$\mathbf{t}(t) := \frac{\gamma'(t)}{\|\gamma'(t)\|_L}, \quad \mathbf{n}(t) := \gamma(t) \times_L \mathbf{t}(t) \tag{3.3}$$

which yields the moving frame $\mathcal{F} := \{\gamma, \mathbf{t}, \mathbf{n}\}$ with the moving equations:

$$\frac{d}{dt}\mathcal{F}(t) = \|\gamma'(t)\|_L \begin{pmatrix} 0 & 1 & 0 \\ -1 & 0 & k_g^L(t) \\ 0 & k_g^L(t) & 0 \end{pmatrix} \mathcal{F}(t), \quad k_g^L(t) := \frac{\langle \mathbf{t}'(t), \mathbf{n}(t)\rangle_L}{\|\gamma'(t)\|_L \|\mathbf{n}(t)\|_L^2} \tag{3.4}$$

since $a \times_L (b \times_L c) = - < a, c >_L b + < a, b >_L c$ for any $a, b, c \in E_1^3$; we suppose also $\|\mathbf{n}(t)\|_L^2 \neq 0$. We call k_g^L as being *the Lorentz geodesic curve* of γ. It follows a positive spherical Lorentz–Legendre curve $LLC = (\gamma, \mathbf{n})$.

Example 3.2 Inspired by the curve r^1 of Example 2.7, let

$$\gamma(t) = (\sin t \sinh t, \sin t \cosh t, \cos t) \in S_1^2 \tag{3.5}$$

having then:

$$\begin{cases} \gamma'(t) = (\cos t \sinh t + \sin t \cosh t, \cos t \cosh t + \sin t \sinh t, -\sin t), \\ \langle \gamma'(t), \gamma'(t) \rangle_L = \cos^2 t \end{cases} \tag{3.6}$$

and then it is regular for $t \in \left(-\frac{\pi}{2}, \frac{\pi}{2}\right)$; both γ and $\mathbf{t}$ belong to S_1^2. Then $\|\gamma'(t)\|_L = \cos t$ and a straightforward computation gives that the normal is a timelike unit vector field, i.e., belongs to H_1^2:

$$\begin{cases} \mathbf{n}(t) = (\frac{\cosh t}{\cos t} + \sin t \sinh t, \frac{\sinh t}{\cos t} + \sin t \cosh t, -\sin t \tan t) \\ \langle \mathbf{n}(t), \mathbf{n}(t) \rangle_L = -1 \end{cases} \tag{3.7}$$

giving the geodesic curvature:

$$k_g^L(t) = \frac{2}{\cos t} \geq 0. \tag{3.8}$$

□

Example 3.3 A last example is

$$\gamma(t) = (\sin t, \cos t, \sqrt{2} \sin t) \in S_1^2, \quad t \in \mathbb{R}, \tag{3.9}$$

with

$$\begin{cases} \gamma'(t) = (\cos t, -\sin t, \sqrt{2} \cos t), \quad < \gamma'(t), \gamma'(t) >_L = 1, \\ \mathbf{n} = constant = (-\sqrt{2}, 0, -1), \quad < \mathbf{n}, \mathbf{n} >_L = -1, \quad k_g^L = 0. \end{cases} \tag{3.10}$$

□

References

1. C. Camci, B.-Y. Chen, K. Ilarslan, A. Uçum, Sequential natural mates of Frenet curves in Euclidean 3-space. J. Geom. **112**(3), 15 (2021). Paper No. 46. Zbl 07432886
2. I. Castro, I. Castro-Infantes, J. Castro-Infantes, Sherical curves whose curvature depends on distance to a great circle (2021). arXiv: 2111.00458v1
3. M. Crasmareanu, Adapted metrics and Webster curvature in Finslerian 2-dimensional geometry. Chin. Ann. Math. Ser. B **37**(3), 419–426 (2016). Zbl 1351.53088
4. M. Crasmareanu, Higher-order curvatures of plane and space parametrized curves. Algorithms **15**(11), (2022). Paper No. 436. https://doi.org/10.3390/a15110436

5. M. Crasmareanu, C. Frigioiu, Unitary vector fields are Fermi-Walker transported along Rytov-Legendre curves. Int. J. Geom. Methods Mod. Phys. **12**(10), 53025 (2015). Article ID 1550111. Zbl 1350
6. D.D. Holm, *Geometric Mechanics. Part II: Rotating, Translating and Rolling*, 2nd edn. (Imperial College Press, London, 2011). Zbl 1381.70001
7. G.R. Jensen, E. Musso, L. Nicolodi, *Surfaces in Classical Geometries. A Treatment by Moving Frames*. Universitext (Springer, Cham, 2016). Zbl 1347.53001
8. Y. Li, D. Pei, M. Takahashi, H. Yu, Envelopes of Legendre curves in the unit spherical bundle over the unit sphere. Q. J. Math. **69**(2), 631–653 (2018). Zbl 1401.53025
9. T. Sasahara, Legendre surface in Sasakian space forms whose mean curvature vectors are eigenvectors. Publ. Math. Debr. **67**(3–4), 285–303 (2005). Zbl 1082.53067
10. M. Takahashi, Legendre curves in the unit spherical bundle over the unit sphere and evolutes, in *Real and Complex Singularities*, vol. 675, ed. by A.C. Nabarro et al. (AMS Contemporary Mathematics, 2016), pp. 337–355. Zbl 1362.58022
11. G.F. Torres del Castillo, G. Sánchez Barrales, Spinor formulation of the differential geometry of curves. Rev. Colomb. Mat. **38**(1), 27–34 (2004). Zbl 1098.53001

Conformal η-Ricci-Yamabe Solitons in the Framework of Riemannian Manifolds

Sudhakar Kumar Chaubey and Abdul Haseeb

2010 MSC 53E20 · 53C25 · 53C21 · 53C20

1 Introduction

In response to his own work on Ricci flow [13], Hamilton introduced the notion of Yamabe flow [12] on Riemannian n-manifolds $\mathcal{M}^n$. In case of $\mathcal{M}^2$, there is an equivalence between Ricci and Yamabe flows. Although, for higher dimension case ($n > 2$), they do not coincide. The Ricci and Yamabe solitons are exclusive solutions of the Ricci and Yamabe flows, respectively. The geometrical and physical properties of these solitons have been studied by several authors. In 2022, some remarkable results of these solitons have been established by Suh ([26–28]). To generalize the notion of Ricci soliton, Cho and Kimura [8] introduced the notion of η-Ricci soliton (η is a 1-form), which has been further studied by several researchers (see, [4, 9, 17, 29]).

Recently in 2019, Güler and Crasmareanu [11] considered a scalar combination of Ricci and Yamabe flows, and introduced a new class of geometric flow called as Ricci-Yamabe flow of type (p, q) and it is defined by

$$\frac{\partial}{\partial t}g(t) + 2p\mathcal{S}(g(t)) + qr(t)g(t) = 0, \quad g(0) = g_0$$

S. K. Chaubey
Section of Mathematics, Department of Information Technology, University of Technology and Applied Sciences, Shinas, Oman
e-mail: sudhakar.chaubey@utas.edu.sa

A. Haseeb (✉)
College of Science, Department of Mathematics, Jazan University, P.O. Box. 114, Jazan 45142, Kingdom of Saudi Arabia
e-mail: malikhaseeb80@gmail.com; haseeb@jazanu.edu.sa

B.-Y. Chen et al. (eds.), *Geometry of Submanifolds and Applications*, Infosys Science Foundation Series, https://doi.org/10.1007/978-981-99-9750-3_13

for some scalars p and q; where $r(t)$ is the scalar curvature of the Riemannian metric $g(t)$, $\mathcal{S}$ is the Ricci tensor and t denotes the time.

A solution to the Ricci-Yamabe flow is called Ricci-Yamabe soliton (RYS) if it depends only on one parameter group of diffeomorphism and scaling. An $\mathcal{M}^n$ is said to have an RYS if

$$\pounds_F g + 2p\mathcal{S} + (2\sigma - qr)g = 0, \tag{1.1}$$

where $\pounds_F$ represents the Lie derivative operator along the smooth vector field F on $\mathcal{M}^n$ and $\sigma \in \mathbb{R}$ (the set of real numbers).

As a generalization of the classical Ricci flow, the concept of conformal Ricci flow was introduced by Fischer [10] in 2004. Later, in 2015, Basu and Bhattacharya [3] defined and studied the conformal Ricci soliton on Kenmotsu manifold. Bhattacharya and his collaborators have explored the conformal Ricci solitons on different spaces (see, [19–21]).

An $\mathcal{M}^n$ is said to have a conformal Ricci-Yamabe soliton (CRYS) of type (p, q) (briefly, (g, F, σ, p, q)) [32] if

$$\pounds_F g + 2p\mathcal{S} + \left\{2\sigma - qr - \left(\psi + \frac{2}{n}\right)\right\} g = 0, \tag{1.2}$$

where ψ defines a time dependent non-dynamical scalar field (also called the conformal pressure). To maintain the incompressibility of the fluid in fluid mechanics, real physical pressure plays a major role.

As a continuation of this study, we have tried to study the conformal η-Ricci-Yamabe solitons (CERYS) and gradient conformal η-Ricci-Yamabe solitons (gradient CERYS) in the frame-work of three-dimensional Riemannian geometry. An $\mathcal{M}^n$ is said to have a conformal η-Ricci-Yamabe soliton (CERYS) of type (p, q) (briefly, $(g, F, \sigma, \rho, p, q)$) if

$$\pounds_F g + 2p\mathcal{S} + \left\{2\sigma - qr - \left(\psi + \frac{2}{n}\right)\right\} g + 2\rho\eta \otimes \eta = 0, \tag{1.3}$$

where $\rho \in \mathbb{R}$.

If F is the gradient of a smooth function f on $\mathcal{M}^n$, then the above soliton equation named as the gradient conformal η-Ricci-Yamabe soliton (briefly, gradient CERYS), and Eq. (1.3) turns to

$$\nabla^2 f + p\mathcal{S} + \left\{\sigma - \frac{qr}{2} - \frac{1}{2}\left(\psi + \frac{2}{3}\right)\right\} g + \rho\eta \otimes \eta = 0, \tag{1.4}$$

where $\nabla^2 f$ is the Hessian of f and ∇ stands for the Levi-Civita connection. In case $\rho = 0$, the CERYS reduces to CRYS. A CRYS is said to be shrinking, steady or expanding if $\sigma < 0$, $\sigma = 0$ or $\sigma > 0$, respectively. A CRYS (gradient CRYS) is said to be

- conformal Ricci soliton (gradient conformal Ricci soliton) if $p = 1,\ q = 0$,
- conformal Yamabe soliton (gradient conformal Yamabe soliton) if $p = 0,\ q = 1$,
- conformal Einstein soliton (gradient conformal Einstein soliton) if $p = -q = 1$,
- conformal ϵ-Einstein soliton (gradient conformal ϵ-Einstein soliton) if $p = 1,\ q = -2\epsilon$ in Eqs. (1.3) and (1.4), respectively.

For more details about the related studies in the context of Riemannian and semi-Riemannian geometries we recommend ([1, 2, 14–16, 18, 22–25, 30]) and the references therein.

We characterize CERYS and gradient CERYS solitons on $\mathcal{M}^3$. To achieve the goals, we organize this chapter as follows: After preliminaries in Sect. 2, we study the properties of $\mathcal{M}^3$ with CERYS in Sect. 3. Section 4 deals with the study of gradient CERYS on $\mathcal{M}^3$. Finally, we construct a non-trivial example of $\mathcal{M}^3$ admitting a semi-symmetric metric ζ-connection and verify some of our results.

2 Preliminaries

In 1970, Yano [33] characterized $\mathcal{M}^n$ endowed with a semi-symmetric metric connection $\tilde{\nabla}$. He proved that the conformal curvature tensors corresponding to $\tilde{\nabla}$ and ∇ coincide. To address the invariant properties of the conformal curvature tensor, the projective curvature tensor, the quasi-conformal curvature tensor, the concircular curvature tensor and the M-projective curvature tensor, Chaubey et al. [5], in 2019, proposed the notion of semi-symmetric metric ζ-connection on $\mathcal{M}^n$. They have proved that an $\mathcal{M}^n$ endowed with a semi-symmetric metric ζ-connection (briefly, $(\mathcal{M}^n, \tilde{\nabla}, \zeta)$) satisfies

$$\nabla\zeta = -I + \eta \otimes \zeta, \tag{2.1}$$

where I is the identity transformation and η is the 1-form associated with the unit vector field ζ such that $g(\zeta, \zeta) = 1$ and $\eta(\cdot) = g(\cdot, \zeta)$. In view of (2.1), we directly get the following identities (see, [6, 7]).

$$R(K_1, K_2)\zeta = \eta(K_1)K_2 - \eta(K_2)K_1 \iff R(K_1, \zeta)K_2 = g(K_1, K_2)\zeta - \eta(K_2)K_1, \tag{2.2}$$

$$(\nabla_{K_1}\eta)K_2 = -g(K_1, K_2) + \eta(K_1)\eta(K_2), \text{ and} \tag{2.3}$$

$$(\pounds_{\zeta}g)(K_1, K_2) = -2\{g(K_1, K_2) - \eta(K_1)\eta(K_2)\} \tag{2.4}$$

for all vector fields K_1, K_2 on $\mathcal{M}^n$.

Since this chapter is about the study of three-dimensional Riemannian manifolds. Thus we take $n = 3$. The contraction of (2.2) over K_1 gives

$$\mathcal{S}(K_2, \zeta) = -2\eta(K_2) \iff \mathcal{Q}\zeta = -2\zeta, \tag{2.5}$$

where R, $\mathcal{S}$ and $\mathcal{Q}$ denote the curvature tensor, the Ricci tensor and the Ricci operator of g, respectively. Equation (2.5) indicates that -2 is the eigenvalue of $\mathcal{Q}$ corresponding to the unit vector field ζ. Thus we state.

Lemma 2.1 *-2 is the eigenvalue of the Ricci operator $\mathcal{Q}$ of $(\mathcal{M}^3, \tilde{\nabla}, \zeta)$ corresponding to the eigenvector ζ.*

Chaubey and De [6, 7] studied the properties of $(\mathcal{M}^3, \tilde{\nabla}, \zeta)$. They proved the following lemmas.

Lemma 2.2 *The non-vanishing Ricci tensor $\mathcal{S}$ of $(\mathcal{M}^3, \tilde{\nabla}, \zeta)$ satisfies*

$$\mathcal{S}(K_1, K_2) = \left(1 + \frac{r}{2}\right) g(K_1, K_2) - \left(3 + \frac{r}{2}\right) \eta(K_1)\eta(K_2) \tag{2.6}$$

for all K_1 and K_2 on $\mathcal{M}^3$ [6, 7].

Lemma 2.3 *In an $(\mathcal{M}^3, \tilde{\nabla}, \zeta)$, we have*

$$\zeta(r) = 2(6 + r), \tag{2.7}$$

where r is the scalar curvature of $\mathcal{M}^3$ [6, 7].

An $\mathcal{M}^n$ is said to be a quasi-Einstein manifold if its Ricci tensor $\mathcal{S}(\neq 0)$ satisfies

$$\mathcal{S} = l_1 g + l_2 \eta \otimes \eta \tag{2.8}$$

for some smooth functions l_1 and l_2 on $\mathcal{M}^n$. If we choose $l_1 =$ constant and $l_2 = 0$ in (2.8), then the quasi-Einstein manifold reduces to an Einstein manifold.

If r is constant, then from (2.6) and (2.7), it is clear that $r = -6$ and $\mathcal{S} = -2g$, respectively. Thus $(\mathcal{M}^3, \tilde{\nabla}, \zeta)$ is an Einstein manifold. Conversely, we suppose that $\mathcal{M}^3$ is Einstein, then in view of (2.8) we have $\mathcal{S} = l_1 g$. This equation together with (2.6) gives

$$l_1 g(K_1, K_2) = \left(1 + \frac{r}{2}\right) g(K_1, K_2) - \left(3 + \frac{r}{2}\right) \eta(K_1)\eta(K_2).$$

Contraction of this equation over K_1 and K_2 gives $r = 3l_1$. Again taking $K_2 = \zeta$ in the above equation and then using (2.5), we find $l_1 = -2$. Thus $r = -6 =$ constant. Now we state.

Lemma 2.4 *An $(\mathcal{M}^3, \tilde{\nabla}, \zeta)$ is an Einstein manifold if and only if its scalar curvature is constant.*

A smooth vector field V on an $\mathcal{M}^n$ is said to be a conformal vector field if

$$\pounds_V g = 2\lambda_1 g \tag{2.9}$$

for some smooth function λ_1 (called the potential function of the conformal vector field V) on $\mathcal{M}^n$. Particularly, V with $\lambda_1 = \text{constant}$ or $\lambda_1 = 0$ reduces to the homothetic or Killing vector fields, respectively. From (2.9), we state the following lemma.

Lemma 2.5 *Let an $\mathcal{M}^n$ admit a conformal vector field V. If $\mathcal{M}^n$ satisfies (2.1), then* $(\pounds_V g)(\zeta, \zeta) + \lambda_1 = 0$.

Proof The Lie derivative of $g(\zeta, \zeta) = \eta(\zeta) = 1$ along V gives $(\pounds_V g)(\zeta, \zeta) + 2g(\pounds_V \zeta, \zeta) = 0$ and $(\pounds_V \eta)(\zeta) + \eta(\pounds_V \zeta) = 0$, respectively. These equations together with Eq. (2.9) accomplish the proof. □

3 CERYS on $(\mathcal{M}^3, \tilde{\nabla}, \zeta)$

Let the metric of $(\mathcal{M}^3, \tilde{\nabla}, \zeta)$ be a CERYS. Then in view of (2.6), (1.3) takes the form

$$
\begin{aligned}
(\pounds_F g)(K_1, K_2) = &-2\left\{p\left(1+\frac{r}{2}\right) + \sigma - \frac{qr}{2} - \frac{1}{2}\left(\psi + \frac{2}{3}\right)\right\} g(K_1, K_2) \\
&+\{p(r+6) - 2\rho\}\eta(K_1)\eta(K_2)
\end{aligned} \tag{3.1}
$$

for any vector fields K_1 and K_2 on $\mathcal{M}^3$.

The covariant differentiation of (3.1) respecting to K_3 leads to

$$
\begin{aligned}
(\nabla_{K_3}\pounds_F g)(K_1, K_2) = &-p(K_3 r)(g(K_1, K_2) - \eta(K_1)\eta(K_2)) + q(K_3 r)g(K_1, K_2) \\
&+\{p(r+6) - 2\rho\}(2\eta(K_1)\eta(K_2)\eta(K_3) \\
&-g(K_1, K_3)\eta(K_2) - g(K_2, K_3)\eta(K_1)).
\end{aligned} \tag{3.2}
$$

As g is parallel with respect to ∇, therefore the following formula [31]

$$
\begin{aligned}
&(\pounds_F \nabla_{K_1} g - \nabla_{K_1}\pounds_F g - \nabla_{[F, K_1]} g)(K_2, K_3) \\
&= -g((\pounds_F \nabla)(K_1, K_2), K_3) - g((\pounds_F \nabla)(K_1, K_3), K_2)
\end{aligned}
$$

turns to

$$
(\nabla_{K_1}\pounds_F g)(K_2, K_3) = g((\pounds_F \nabla)(K_1, K_2), K_3) + g((\pounds_F \nabla)(K_1, K_3), K_2)
$$

for any K_1, K_2, K_3 on $\mathcal{M}^3$. Since the operator $\pounds_V \nabla$ is symmetric, therefore we have

$$
\begin{aligned}
2g((\pounds_F \nabla)(K_1, K_2), K_3) = &(\nabla_{K_1}\pounds_F g)(K_2, K_3) \\
&+(\nabla_{K_2}\pounds_F g)(K_1, K_3) - (\nabla_{K_3}\pounds_F g)(K_1, K_2),
\end{aligned}
$$

which in view of (3.2) turns to

$$2g((\pounds_F\nabla)(K_1,K_2),K_3) = -(K_1r)\{(p-q)g(K_2,K_3) - p\eta(K_2)\eta(K_3)\}$$
$$-(K_2r)\{(p-q)g(K_1,K_3) - p\eta(K_1)\eta(K_3)\}$$
$$+(K_3r)\{(p-q)g(K_1,K_2) - p\eta(K_1)\eta(K_2)\}$$
$$+2\{p(r+6) - 2\rho\}\{\eta(K_1)\eta(K_2) - g(K_1,K_2)\}\eta(K_3),$$

from which it follows that

$$2(\pounds_F\nabla)(K_1,K_2) = -(K_1r)\{p(K_2 - \eta(K_2)\zeta) - qK_2\}$$
$$-(K_2r)\{p(K_1 - \eta(K_1)\zeta) - qK_1\}$$
$$+(Dr)\{(p-q)g(K_1,K_2) - p\eta(K_1)\eta(K_2)\}$$
$$+2\{p(r+6) - 2\rho\}\{\eta(K_1)\eta(K_2)\zeta - g(K_1,K_2)\zeta\}, \tag{3.3}$$

where D represents the gradient operator of g.
Putting $K_1 = \zeta$ in (3.3) gives

$$2(\pounds_F\nabla)(K_2,\zeta) = qg(Dr,K_2)\zeta - q\eta(K_2)Dr$$
$$-2(r+6)\{(p-q)K_2 - p\eta(K_2)\zeta\}. \tag{3.4}$$

Let the scalar curvature r of $\mathcal{M}^3$ be constant, then from (2.7) we find $r = -6$. Hence (3.4) reduces to

$$(\pounds_F\nabla)(K_2,\zeta) = 0. \tag{3.5}$$

The covariant differentiation of (3.5) respecting to K_1 and making use of (2.1), (2.3), and (3.4) infers

$$(\nabla_{K_1}\pounds_F\nabla)(K_2,\zeta) = 0. \tag{3.6}$$

Again from Yano [31], we have

$$(\pounds_F R)(K_1,K_2)K_3 = (\nabla_{K_1}\pounds_F\nabla)(K_2,K_3) - (\nabla_{K_2}\pounds_F\nabla)(K_1,K_3),$$

which by putting $K_3 = \zeta$ and using (3.6) reduces to

$$(\pounds_F R)(K_1,K_2)\zeta = 0. \tag{3.7}$$

Next, taking the Lie derivative of $R(K_1,\zeta)\zeta = \eta(K_1)\zeta - K_1$ along F and using (3.7), we arrive at

$$(\pounds_F\eta)(K_1)\zeta = -2\eta(\pounds_F\zeta)K_1 + g(K_1,\pounds_F\zeta)\zeta. \tag{3.8}$$

By the Lie differentiation of $\eta(K_1) = g(K_1,\zeta)$, we have

$$(\pounds_F \eta)(K_1) = (\pounds_F g)(K_1, \zeta) + g(K_1, \pounds_F \zeta). \tag{3.9}$$

Now by replacing K_2 by ζ in (3.1), we find

$$(\pounds_F g)(K_1, \zeta) = 2\left\{2p - \sigma - \rho - 3q + \frac{1}{2}(\psi + \frac{2}{3})\right\} \eta(K_1). \tag{3.10}$$

The Lie derivative of $g(\zeta, \zeta) = 1$ leads to

$$(\pounds_F g)(\zeta, \zeta) = -2\eta(\pounds_F \zeta). \tag{3.11}$$

Now from the Eqs. (3.8)–(3.11), it follows that

$$\sigma + \rho = 2p - 3q + \frac{1}{2}\left(\psi + \frac{2}{3}\right) + \eta(\pounds_F \zeta). \tag{3.12}$$

This helps us to state.

Theorem 3.1 *The scalars σ and ρ on $(\mathcal{M}^3, \tilde{\nabla}, \zeta)$ admitting a CERYS are governed by* (3.12), *provided the scalar curvature of $\mathcal{M}^3$ is constant.*

Let the scalar curvature of $\mathcal{M}^3$ admitting a CERYS be constant. Then (3.3) assumes the form

$$(\pounds_F \nabla)(K_2, K_3) = 2\rho\{g(K_2, K_3) - \eta(K_2)\eta(K_3)\}\zeta,$$

which gives

$$\begin{aligned}(\nabla_{K_1} \pounds_F \nabla)(K_2, K_3) = &-2\rho\{(\nabla_{K_1}\eta)(K_2)\eta(K_3) - (\nabla_{K_1}\eta)(K_3)\eta(K_2)\}\zeta \\ &+2\rho\{g(K_2, K_3) - \eta(K_2)\eta(K_3)\}\nabla_{K_1}\zeta.\end{aligned}$$

This equation together with (2.1) and (2.3) takes the following form

$$\begin{aligned}(\nabla_{K_1} \pounds_F \nabla)(K_2, K_3) &= 2\rho\{g(K_1, K_3)\eta(K_2) + g(K_1, K_2)\eta(K_3) \\ &-2\eta(K_1)\eta(K_2)\eta(K_3)\}\zeta - 2\rho\{g(K_2, K_3) - \eta(K_2)\eta(K_3)\}(K_1 - \eta(K_1)\zeta).\end{aligned} \tag{3.13}$$

Using (3.13) in the curvature identity $(\pounds_F R)(K_1, K_2)K_3 = (\nabla_{K_1} \pounds_F \nabla)(K_2, K_3) - (\nabla_{K_2} \pounds_F \nabla)(K_1, K_3)$, we find

$$\begin{aligned}(\pounds_F R)(K_1, K_2)K_3 = 2\rho[&g(K_1, K_2)K_2 - g(K_2, K_3)K_1 \\ &+\eta(K_2)\eta(K_3)K_1 - \eta(K_1)\eta(K_3)K_2].\end{aligned} \tag{3.14}$$

On the other hand, it is clear that the conformal curvature tensor on $\mathcal{M}^3$ vanishes, that is, $C = 0$. Thus we have $R(K_1, K_2)K_3 = \mathcal{S}(K_2, K_3)K_1 - \mathcal{S}(K_1, K_3)K_2 + g(K_2, K_3)QK_1 - g(K_1, K_3)QK_2 - \frac{r}{2}\{g(K_2, K_3)K_1 - g(K_1, K_3)K_2\}$. This relation together with Theorem 3.1 gives

$$R(K_1, K_2)K_3 = g(K_1, K_3)K_2 - g(K_2, K_3)K_1,$$

which after Lie derivative along the soliton vector field F gives

$$(£_F R)(K_1, K_2)K_3 = (£_F g)(K_1, K_3)K_2 - (£_F g)(K_2, K_3)K_1. \quad (3.15)$$

In consequence of (3.14) and (3.15), we find

$$(£_F g)(K_1, K_3)K_2 - (£_F g)(K_2, K_3)K_1 = 2\rho[g(K_1, K_3)K_2 \\ -g(K_2, K_3)K_1 + \eta(K_2)\eta(K_3)K_1 - \eta(K_1)\eta(K_3)K_2].$$

Putting ζ in lieu of K_2 in the above equation and then taking the inner product with ζ, we find

$$(£_F g)(K_1, K_3) = 2\rho\{g(K_1, K_3) - \eta(K_1)\eta(K_3)\} + (£_F g)(\zeta, K_3)\eta(K_1).$$

In view of Eqs. (3.10), (3.12) and Theorem 3.1, the above equation assumes the form

$$£_F g = 2\rho g - 2(\rho + g(£_F\zeta, \zeta))\eta \otimes \eta. \quad (3.16)$$

If the soliton vector field F of CERYS $(g, F, \sigma, \rho, p, q)$ is a homothetic vector field with the potential function $\lambda_1 = \rho$, then from the above equation and Lemma 2.5 we have $\rho + g(£_F\zeta, \zeta) = 0$. Conversely, if $\mathcal{M}^3$ endowed with a CERYS $(g, F, \sigma, \rho, p, q)$ satisfies $\rho + g(£_F\zeta, \zeta) = 0$, then from the above equation it is obvious that F is a homothetic vector field with the potential function ρ. Thus, we conclude our results as.

Theorem 3.2 *Let the metric of $(\mathcal{M}^3, \tilde{\nabla}, \zeta)$ be a CERYS $(g, F, \sigma, \rho, p, q)$. If the scalar curvature of $\mathcal{M}^3$ is constant, then the soliton vector field F of $(g, F, \sigma, \rho, p, q)$ is a homothetic vector field with the potential function ρ if and only if $\rho + g(£_F\zeta, \zeta) = 0$.*

Let $(\mathcal{M}^3, \tilde{\nabla}, \zeta)$ admit a CERYS $(g, \zeta, \sigma, \rho, p, q)$. Then Eq. (3.16) together with $F = \zeta$ infers that

$$£_\zeta g = 2\rho(g - \eta \otimes \eta), \quad (3.17)$$

since $£_\zeta\zeta = [\zeta, \zeta] = 0$. Equation (2.4) together with (3.17) give $\rho = -1$. Thus we state the following corollary.

Corollary 3.3 *Let $(\mathcal{M}^3, \tilde{\nabla}, \zeta)$ possess the constant scalar curvature. If the metric of $(\mathcal{M}^3, \tilde{\nabla}, \zeta)$ is a CERYS $(g, \zeta, \sigma, \rho, p, q)$, then $\rho = -1$.*

Let the soliton vector field F of $(g, F, \sigma, \rho, p, q)$ be a scalar multiple of ζ, that is, $F = \mathfrak{a}\zeta$ for some scalar $\mathfrak{a}$ on $\mathcal{M}^3$. Then we have $\eta(£_F\zeta) = 0$, since $£_F\zeta = [F, \zeta] = [\mathfrak{a}\zeta, \zeta] = \mathfrak{a}[\zeta, \zeta] = 0$. Here $[\cdot\ ,\ \cdot]$ denotes the Lie bracket. Thus we state the following corollary as.

Corollary 3.4 *Let $(\mathcal{M}^3, \tilde{\nabla}, \zeta)$ admit a CERYS. If the soliton vector field of CERYS is a scalar multiple of ζ, then $\sigma + \rho = 2p - 3q + \frac{1}{2}(\psi + \frac{2}{3})$, provided the scalar curvature of $\mathcal{M}^3$ is constant.*

Particularly if we choose $\rho = 0$ in (1.3), then CERYS reduces to CRYS. Thus (3.12) assumes the form

$$\sigma = 2p - 3q + \frac{1}{2}\left(\psi + \frac{2}{3}\right) + \eta(\pounds_F\zeta). \quad (3.18)$$

Well-known that the CRYS (g, F, σ, p, q) on $\mathcal{M}^3$ is shrinking, expanding or steady if $\sigma < 0$, $\sigma > 0$ or $\sigma = 0$, respectively. Thus, we have

Corollary 3.5 *A CRYS (g, F, σ, p, q) on $(\mathcal{M}^3, \tilde{\nabla}, \zeta)$ is, respectively, expanding, shrinking or steady if $2p - 3q + \frac{1}{2}\left(\psi + \frac{2}{3}\right) + \eta(\pounds_F\zeta) > 0,\ < 0,$ or $= 0$, provided the scalar curvature of $\mathcal{M}^3$ is constant.*

Corollary 3.6 *Let $(\mathcal{M}^3, \tilde{\nabla}, \zeta)$ admit a CRYS (g, F, σ, p, q). If the scalar curvature of $\mathcal{M}^3$ is constant, then we have the following table.*

Values of p, q	Solitons	Soliton constant	Conditions for solitons to be expanding, shrinking or steady
$p = 1, q = 0$	Conformal Ricci soliton	$\sigma = \frac{1}{2}\psi + \frac{7}{3} + \eta(\pounds_F\zeta)$	(i) for $\frac{\psi}{2} + \frac{7}{3} > -\eta(\pounds_F\zeta)$, the conformal Ricci soliton is shrinking, (ii) for $\frac{\psi}{2} + \frac{7}{3} < -\eta(\pounds_F\zeta)$, the conformal Ricci soliton is shrinking, (iii) for $\frac{\psi}{2} + \frac{7}{3} = -\eta(\pounds_F\zeta)$, the conformal Ricci soliton is steady
$p = 0, q = 1$	Conformal Yamabe soliton	$\sigma = \frac{1}{2}\psi + \eta(\pounds_F\zeta) - \frac{8}{3}$	(i) for $\psi + 2\eta(\pounds_F\zeta) > \frac{16}{3}$, the conformal Yamabe soliton is expanding, (ii) for $\psi + 2\eta(\pounds_F\zeta) < \frac{16}{3}$, the conformal Yamabe soliton is shrinking, (iii) for $\psi + 2\eta(\pounds_F\zeta) = \frac{16}{3}$, the conformal Yamabe soliton is steady
$p = 1, q = -1$	Conformal Einstein soliton	$\sigma = \frac{1}{2}\psi + \eta(\pounds_F\zeta) + \frac{16}{3}$	(i) for $\psi + 2\eta(\pounds_F\zeta) > \frac{32}{3}$, conformal Einstein soliton is expanding, (ii) for $\psi + 2\eta(\pounds_F\zeta) < \frac{32}{3}$, conformal Einstein soliton is shrinking, (iii) for $\psi + 2\eta(\pounds_F\zeta) = \frac{32}{3}$, conformal Einstein soliton is steady
$p = 1, q = -2\epsilon$	Conformal ϵ-Einstein soliton	$\sigma = 6[\epsilon + \frac{1}{6}(\frac{1}{2}\psi + \eta(\pounds_F\zeta) + \frac{7}{3})]$	(i) for $\epsilon > -\frac{1}{6}(\frac{1}{2}\psi + \eta(\pounds_F\zeta) + \frac{7}{3})$, the conformal ϵ-Einstein soliton is expanding, (ii) $\epsilon < -\frac{1}{6}(\frac{1}{2}\psi + \eta(\pounds_F\zeta) + \frac{7}{3})$, the conformal ϵ-Einstein soliton is shrinking, (iii) $\epsilon = -\frac{1}{6}(\frac{1}{2}\psi + \eta(\pounds_F\zeta) + \frac{7}{3})$, the conformal ϵ-Einstein soliton is steady

4 Gradient CERYS on $(\mathcal{M}^3, \tilde{\nabla}, \zeta)$

Let the metric g be a gradient CERYS on $(\mathcal{M}^3, \tilde{\nabla}, \zeta)$. Then Eq. (1.2) can be written as

$$\nabla_{K_2} Df + p\mathcal{Q}K_2 + \left\{\sigma - \frac{qr}{2} - \frac{1}{2}\left(\psi + \frac{2}{3}\right)\right\} K_2 + \rho\eta(K_2)\zeta = 0 \tag{4.1}$$

for any K_2 on $\mathcal{M}^3$.
By the covariant differentiation of (4.1) with respect to K_1, we have

$$\begin{aligned}\nabla_{K_1}\nabla_{K_2} Df &= -p\{(\nabla_{K_1}\mathcal{Q})K_2 + \mathcal{Q}(\nabla_{K_1}K_2)\} - \left\{\sigma - \frac{qr}{2} - \frac{1}{2}\left(\sigma + \frac{2}{3}\right)\right\} \nabla_{K_1}K_2 \\ &+ q\frac{K_1(r)}{2}K_2 + \rho\{g(K_1, K_2)\zeta - 2\eta(K_1)\eta(K_2)\zeta - \eta(\nabla_{K_1}K_2)\zeta + \eta(K_2)K_1\}.\end{aligned} \tag{4.2}$$

Interchanging K_1 and K_2 in (4.2), we have

$$\begin{aligned}\nabla_{K_2}\nabla_{K_1} Df &= -p\{(\nabla_{K_2}\mathcal{Q})K_1 + \mathcal{Q}(\nabla_{K_2}K_1)\} - \left\{\sigma - \frac{qr}{2} - \frac{1}{2}\left(\sigma + \frac{2}{3}\right)\right\} \nabla_{K_2}K_1 \\ &+ q\frac{K_2(r)}{2}K_1 + \rho\{g(K_1, K_2)\zeta - 2\eta(K_2)\eta(K_1)\zeta - \eta(\nabla_{K_2}K_1)\zeta + \eta(K_1)K_2\}.\end{aligned} \tag{4.3}$$

By making use of (4.1)–(4.3), we find

$$\begin{aligned}R(K_1, K_2)Df &= p\{(\nabla_{K_2}\mathcal{Q})K_1 - (\nabla_{K_1}\mathcal{Q})K_2\} + \rho(\eta(K_2)K_1 - \eta(K_1)K_2) \\ &+ \frac{q}{2}(K_1(r)K_2 - K_2(r)K_1).\end{aligned} \tag{4.4}$$

Contracting (4.4) along K_1 implies

$$\mathcal{S}(K_2, Df) = \left(\frac{p}{2} - q\right) K_2(r) + 2\rho\eta(K_2). \tag{4.5}$$

From (2.6), we also have

$$\mathcal{S}(K_2, Df) = \left(1 + \frac{r}{2}\right) K_2(f) - \left(3 + \frac{r}{2}\right)\eta(K_2)\zeta(f). \tag{4.6}$$

By equating (4.5) and (4.6) then putting $K_2 = \zeta$, we deduce

$$\zeta(f) = -\frac{1}{2}(r + 6)(p - 2q) - \rho. \tag{4.7}$$

Again from (2.6) it follows that

$$\mathcal{Q}K_1 = \left(\frac{r}{2} + 1\right) K_1 - \left(\frac{r}{2} + 3\right)\eta(K_1)\zeta. \tag{4.8}$$

The covariant derivative of (4.8) with respect to K_2 gives

$$(\nabla_{K_2} Q)K_1 = \frac{K_2(r)}{2}\{K_1 - \eta(K_1)\zeta\} + \left(\frac{r}{2} + 3\right)\{g(K_1, K_2)\zeta + \eta(K_1)K_2 - 2\eta(K_1)\eta(K_2)\zeta\}. \tag{4.9}$$

By fixing K_1 by ζ in (4.9), we infer

$$(\nabla_{K_2} Q)\zeta = \left(\frac{r}{2} + 3\right)\{K_2 - \eta(K_2)\zeta\}. \tag{4.10}$$

Again setting $K_2 = \zeta$ in (4.9), we have

$$(\nabla_{\zeta} Q)K_1 = \frac{\zeta(r)}{2}\{K_1 - \eta(K_1)\zeta\}. \tag{4.11}$$

Now from the last two equations, we have

$$(\nabla_{K_2} Q)\zeta - (\nabla_{\zeta} Q)K_2 = -\left(\frac{r}{2} + 3\right)\{K_2 - \eta(K_2)\zeta\}. \tag{4.12}$$

Now putting $K_1 = \zeta$ in (4.4) then making use of (4.12) and (2.7), we have

$$R(\zeta, K_2)Df = \left\{p\left(\frac{r}{2} + 3\right) + \rho\right\}(\eta(K_2)\zeta - K_2) + q(r+6)K_2 - \frac{qK_2(r)\zeta}{2}. \tag{4.13}$$

The inner product of (4.13) with ζ gives

$$g(R(\zeta, K_2)Df, \zeta) = q(r+6)\eta(K_2) - \frac{qK_2(r)}{2}.$$

Also from (2.2), we have

$$g(R(\zeta, K_2)Df, \zeta) = -K_2(f) + \zeta(f)\eta(K_2).$$

By equating the last two equations, we find

$$K_2(f) = -\left\{\frac{p(r+6)}{2} + \rho\right\}\eta(K_2) + \frac{qK_2(r)}{2}.$$

Let r be constant. Then in view of (2.7), the foregoing equation implies

$$F = Df = -\rho\zeta.$$

This shows that the gradient of the smooth function f of a gradient CERYS is a constant multiple of ζ.

The covariant derivative of the above equation gives

$$\nabla Df = \rho(I - \eta \otimes \zeta), \tag{4.14}$$

since Eq. (2.1) is used and I stands for identity transformation. From Eqs. (4.1) and (4.14), we infer

$$pQ = -\left\{\sigma + \rho + 3q - \frac{1}{2}\left(\psi + \frac{2}{3}\right)\right\} I, \tag{4.15}$$

which after contraction leads to

$$\sigma = 2p - \rho - 3q + \frac{1}{2}\left(\psi + \frac{2}{3}\right). \tag{4.16}$$

The last two equations show that $\mathcal{S} = -2g$, and hence $(\mathcal{M}^3, \tilde{\nabla}, \zeta)$ is an Einstein manifold. Thus we state.

Theorem 4.1 *An $(\mathcal{M}^3, \tilde{\nabla}, \zeta)$ admitting a gradient CERYS is an Einstein manifold. Also, the gradient of the smooth function f of gradient CERYS is a constant multiple of ζ.*

From Theorem 4.1, we have.

Corollary 4.2 *Let $(\mathcal{M}^3, \tilde{\nabla}, \zeta)$ admit a gradient CERYS. If the scalar curvature of M^3 is constant, then $\sigma = 2p - \rho - 3q + \frac{1}{2}(\psi + \frac{2}{3})$.*

Particularly if we choose $\rho = 0$ in (4.1), then the gradient CERYS changes to the gradient conformal Ricci-Yamabe soliton. This fact with (4.14) shows that $Df = 0 \Longleftrightarrow f =$ constant, and the soliton is trivial. Thus, we have.

Corollary 4.3 *There does not exist a non-trivial gradient conformal Ricci-Yamabe soliton on $(\mathcal{M}^3, \tilde{\nabla}, \zeta)$, provided the scalar curvature of $\mathcal{M}^3$ is constant.*

Let $(\mathcal{M}^3, \tilde{\nabla}, \zeta)$ admit a gradient CERYS. If the scalar curvature of $\mathcal{M}^3$ is constant, then Theorem 4.1 and Eq. (4.14) infer that

$$2Hess(f) + \rho\mathcal{S} + 2\rho\eta \otimes \eta = 0 \text{ and } \triangle f = 2\rho.$$

Here $Hess$ and $\triangle$ denote the Hessian and Laplacian operators, respectively.

An $\mathcal{M}^n$ is said to possess the Poisson equation if a smooth function Ψ satisfies the equation $\triangle\Psi = \omega$, where ω is a smooth function (or source term) on $\mathcal{M}^n$. The Poisson equation reduces to the Laplace equation if $\omega = 0$.

Thus we state.

Corollary 4.4 *Let $(\mathcal{M}^3, \tilde{\nabla}, \zeta)$ admit a gradient CERYS. If the scalar curvature of $\mathcal{M}^3$ is constant, then the gradient function f of gradient CERYS satisfies the Poisson equation and $2Hess(f) + \rho\mathcal{S} + 2\rho\eta \otimes \eta = 0$.*

Corollary 4.5 *Let the scalar curvature of* $(\mathcal{M}^3, \tilde{\nabla}, \zeta)$ *be constant and the metric of* $(\mathcal{M}^3, \tilde{\nabla}, \zeta)$ *is a gradient CRYS, then the gradient smooth function* f *of CRYS satisfies the Laplace equation.*

5 Example

We consider a 3-dimensional differentiable manifold $\mathcal{M}^3 = \{(x_1, x_2, x_3) \in \mathbb{R}^3 : x_3 \neq 0\}$, where (x_1, x_2, x_3) represents a point of $\mathbb{R}^3$. Let υ_1, υ_2 and υ_3 be the vector fields on $\mathcal{M}^3$ given by

$$\upsilon_1 = e^{-x_3}\left(\frac{\partial}{\partial x_1} + \frac{\partial}{\partial x_2}\right), \quad \upsilon_2 = e^{-x_3}\left(-\frac{\partial}{\partial x_1} + \frac{\partial}{\partial x_2}\right), \quad \upsilon_3 = -\frac{\partial}{\partial x_3} = \zeta.$$

Obvious that the vector fields υ_1, υ_2 and υ_3 form a basis of $\mathcal{M}^3$.

Let g refer the Riemannian metric of $\mathcal{M}^3$ defined by

$$g(\upsilon_i, \upsilon_j) = \begin{cases} 1, & 1 \leq i = j \leq 3, \\ 0, & \text{otherwise.} \end{cases}$$

The 1-form η on $\mathcal{M}^3$ is defined by $\eta(K_1) = g(K_1, \upsilon_3) = g(K_1, \zeta)$, for any K_1 on $\mathcal{M}^3$. By direct computations, we easily obtain

$$[\upsilon_1, \upsilon_2] = 0, \quad [\upsilon_1, \upsilon_3] = -\upsilon_1, \quad [\upsilon_2, \upsilon_3] = -\upsilon_2.$$

From Koszul's formula, we easily calculate

$$\nabla_{\upsilon_1}\upsilon_1 = \upsilon_3, \quad \nabla_{\upsilon_1}\upsilon_2 = 0, \quad \nabla_{\upsilon_1}\upsilon_3 = -\upsilon_1,$$

$$\nabla_{\upsilon_2}\upsilon_1 = 0, \quad \nabla_{\upsilon_2}\upsilon_2 = \upsilon_3, \quad \nabla_{\upsilon_2}\upsilon_3 = -\upsilon_2,$$

$$\nabla_{\upsilon_3}\upsilon_1 = 0, \quad \nabla_{\upsilon_3}\upsilon_2 = 0, \quad \nabla_{\upsilon_3}\upsilon_3 = 0.$$

It follows that $\nabla_{K_1}\zeta = -K_1 + \eta(K_1)\zeta$. Hence $\mathcal{M}^3$ is a 3-dimensional Riemannian manifold satisfying Eq. (2.1). It can be easily shown that $\tilde{\nabla}$ defined on $\mathcal{M}^3$ satisfies the conditions

$$\tilde{T}(K_1, K_2) = -\eta(K_1)K_2 + \eta(K_2)K_1, \ \tilde{\nabla}g = 0, \ \text{and } \tilde{\nabla}\zeta = 0$$

for arbitrary vector fields K_1 and K_2 on $\mathcal{M}^3$, where $\tilde{T}$ denotes the torsion tensor of the semi-symmetric metric ζ-connection $\tilde{\nabla}$. Thus, we can say that $\tilde{\nabla}$ is a semi-symmetric metric ζ-connection on $\mathcal{M}^3$. Now, we can easily obtain the following non-vanishing components of curvature tensor R and Ricci tensor $\mathcal{S}$, that are,

$$R(\upsilon_1,\upsilon_2)\upsilon_1=\upsilon_2,\quad R(\upsilon_1,\upsilon_3)\upsilon_1=\upsilon_3,\quad R(\upsilon_2,\upsilon_3)\upsilon_2=\upsilon_3,$$

$$R(\upsilon_1,\upsilon_2)\upsilon_2=R(\upsilon_1,\upsilon_3)\upsilon_3=-\upsilon_1,\quad R(\upsilon_2,\upsilon_3)e_3=-\upsilon_2,$$

and

$$\mathcal{S}(\upsilon_1,\upsilon_1)=\mathcal{S}(\upsilon_2,\upsilon_2)=\mathcal{S}(\upsilon_3,\upsilon_3)=-2\implies r=-6.$$

These equations show that the Lemmas 2.1–2.4 are satisfied. Now, we are going to verify our main Theorem 3.1, that is, if $(\mathcal{M}^3,\tilde{\nabla},\zeta)$ satisfies Eq. (1.3), then $\sigma+\rho=2p-3q+\frac{1}{2}\left(\psi+\frac{2}{3}\right)+\eta(\pounds_F\zeta)$. Choose $F=Df=\rho\upsilon_3$. Then we have

$$\nabla_{\upsilon_1}Df=\rho\upsilon_1,$$

$$\nabla_{\upsilon_2}Df=\rho\upsilon_2,$$

and

$$\nabla_{\upsilon_3}Df=0.$$

Since the scalar curvature of $\mathcal{M}^3$ is -6, therefore from Lemma 2.2, we have $\mathcal{S}=-2g$. This facts together with (1.3) tell us that the smooth function f satisfies the following partial differential equation

$$\nabla_{K_1}Df-2pK_1+\{\sigma+3q-\frac{1}{2}\left(\psi+\frac{2}{3}\right)\}K_1+\rho\eta(K_1)\zeta=0$$

for all vector field K_1 of $\mathcal{M}^3$. Let $K_1=\upsilon_1$. Then the last two equations give

$$\rho-2p+\sigma+3q-\frac{1}{2}\left(\psi+\frac{2}{3}\right)=0,$$

since $\eta(\pounds_F\zeta)=\eta(\pounds_F\zeta,\zeta)=g(\pounds_{\rho\upsilon_3}\upsilon_3,\upsilon_3)=\rho g(\pounds_{\upsilon_3}\upsilon_3,\upsilon_3)=\rho g([\upsilon_3,\upsilon_3],\upsilon_3)=0$. That is, the Eq. (3.12) is satisfied for the vector field υ_1. In the similar way, we can easily prove that υ_2 and υ_3 will also satisfy (3.12). Hence Theorem 3.1 is verified.

References

1. G. Ayar, M. Yildirim, η-Ricci solitons on nearly Kenmotsu manifolds. Asian Eur. J. Math. **12**(6), (2019). Article ID 2040002
2. G. Ayar, M. Yildirim, Ricci solitons and gradient Ricci solitons on nearly Kenmotsu manifolds. Facta Univ. Ser. Math. Inform. **34**(3), 503–510 (2019)
3. N. Basu, A. Bhattacharyya, Conformal Ricci soliton in Kenmotsu manifold. Glob. J. Adv. Res. Class. Mod. Geom. **4**, 15–21 (2015)

4. A.M. Blaga, Some geometrical aspects of Einstein, Ricci and Yamabe solitons. J. Geom. Symmetry Phys. **52**, 17–26 (2019)
5. S.K. Chaubey, J.W. Lee, S. Yadav, Riemannian manifolds with a semisymmetric metric P-connection. J. Korean Math. Soc. **56**(4), 1113–1129 (2019)
6. S.K. Chaubey, U.C. De, Three dimensional Riemannian manifolds and Ricci solitons. Quaestiones Mathematicae 1–14, (2021)
7. S.K. Chaubey, U.C. De, Characterization of three-dimensional Riemannian manifolds with a type of semi-symmetric metric connection admitting Yamabe soliton. J. Geom. Phys. **157** (2020), 103846, 8 pp
8. J.T. Cho, M. Kimura, Ricci solitons and real hypersurfaces in a complex space form. Tohoku Math. J. **61**, 205–212 (2009)
9. S. Deshmukh, B.Y. Chen, A note on Yamabe solitons. Balk. J. Geom. Appl. **23**(1), 37–43 (2018)
10. A.E. Fischer, An introduction to conformal Ricci flow. Class. Quantum Grav. **21**, 171–218 (2004)
11. S. Güler, M. Crasmareanu, Ricci-Yamabe maps for Riemannian flows and their volume variation and volume entropy. Turk. J. Math. **43**, 2631–2641 (2019)
12. R.S. Hamilton, Lectures on geometric flows (Unpublished manuscript, 1989)
13. R.S. Hamilton, The Ricci flow on surfaces, Mathematics and general relativity (Santa Cruz, CA, 1986). Contemp. Math. A.M.S. **71**, 237–262 (1988)
14. A. Haseeb, S.K. Chaubey, Lorentzian para-Sasakian manifolds and $*$-Ricci solitons. Kragujev. J. Math. **48**(2), 167–179 (2024)
15. A. Haseeb, S.K. Chaubey, M.A. Khan, Riemannian 3-manifolds and Ricci-Yamabe solitons. Int. J. Geom. Methods Mod. Phys. **20**(1), 2350015 (2023)
16. A. Haseeb, M.A. Khan, Conformal η-Ricci-Yamabe solitons within the framework of ϵ-LP-Sasakian 3-manifolds. Adv. Math. Phys. **2022**, 8 pages. Article ID 3847889
17. A. Haseeb, S. Pandey, R. Prasad, Some results on η-Ricci solitons in quasi-Sasakian 3-manifolds. Commun. Korean Math. Soc. **36**(2), 377–387 (2021)
18. A. Haseeb, R. Prasad, F. Mofarreh, Sasakian manifolds admitting $*$-η-Ricci-Yamabe solitons. Adv. Math. Phys. **2022**, 7 pages. Article ID 5718736
19. Y. Li, D. Ganguly, S. Dey, A. Bhattacharyya, Conformal η-Ricci solitons within the framework of indefinite Kenmotsu manifolds. AIMS Math. **7**, 5408–5430 (2022)
20. Y. Li, S. Dey, S. Pahan, A. Ali, Geometry of conformal η-Ricci solitons and conformal η-Ricci almost solitons on paracontact geometry. Open Math. **20**(2022), 1–20. https://doi.org/10.1515/math-2022-0443
21. Y. Li, A. Abolarinwa, S. Azami, A. Ali, Yamabe constant evolution and monotonicity along the conformal Ricci flow. AIMS Math. **7**, 12077–12090 (2022)
22. A. Sardar, Ricci-Yamabe Solitons and Gradient Ricci-Yamabe solitons on Kenmotsu 3-manifolds. Kyungpook Math. J. **61**(4), 813–822 (2021)
23. M.D. Siddiqi, S.K. Chaubey, Almost η-conformal Ricci solitons in $(LCS)_3$-manifolds. Sarajevo J. Math. **16**(29), 245–260 (2020)
24. M.D. Siddiqi, S.K. Chaubey, Almost conformal η-Ricci solitons in three-dimensional Lorentzian concircular structures. Konuralp J. Math. **8**(1), 70–78 (2020)
25. M.D. Siddiqi, Conformal η-Ricci solitons in δ-Lorentzian trans Sasakian manifolds. Int. J. Maps Math. **1**(1), 15–34 (2018)
26. Y.J. Suh, Ricci-Bourguignon solitons on real hypersurfaces in the complex hyperbolic quadric. Rev. R. Acad. Cienc. Exactas Fís. Nat. Ser. A Mat. RACSAM **116**(3), (2022). Paper No. 110
27. Y.J. Suh, Yamabe and gradient Yamabe solitons in the complex hyperbolic two-plane Grassmannians. Rev. Math. Phys. (2022). https://doi.org/10.1142/S0129055X22500246
28. Y.J. Suh, Ricci soliton and pseudo-Einstein real hypersurfaces in the complex hyperbolic quadric. J. Geom. Phys. **162**, (2021). Paper No. 103888, 16 pp
29. M. Turan, C. Yetim, S.K. Chaubey, On quasi-Sasakian 3-manifolds admitting η-Ricci solitons. Filomat **33**(15), 4923–4930 (2019)

30. H.I. Yoldas, On Kenmotsu manifolds admitting η-Ricci-Yamabe solitons. Int. J. Geom. Methods Mod. Phys. **18**(12), 2150189 (2021)
31. K. Yano, *Integral Formulas in Riemannian Geometry*. Pure and Applied Mathematics, vol. I (Marcel Dekker, New York, 1970)
32. P. Zhang, Y. Li, S. Roy, S. Dey, A. Bhattacharyya, Geometrical structure in a perfect fluid spacetime with conformal Ricci-Yamabe soliton. Symmetry **14**, 594 (2022)
33. K. Yano, On semi-symmetric metric connection. Rev. Roum. de Math. Pures et Appl. **15**, 1579–1586 (1970)

www.ingramcontent.com/pod-product-compliance
Ingram Content Group UK Ltd.
Pitfield, Milton Keynes, MK11 3LW, UK
UKHW022211051026
15624UKWH00001B/113
9789819997497